Topics in Knot Theory

NATO ASI Series

Advanced Science Institutes Series

A Series presenting the results of activities sponsored by the NATO Science Committee, which aims at the dissemination of advanced scientific and technological knowledge, with a view to strengthening links between scientific communities.

The Series is published by an international board of publishers in conjunction with the NATO Scientific Affairs Division

A Life Sciences
B Physics

Plenum Publishing Corporation
London and New York

C Mathematical and Physical Sciences
D Behavioural and Social Sciences
E Applied Sciences

Kluwer Academic Publishers
Dordrecht, Boston and London

F Computer and Systems Sciences
G Ecological Sciences
H Cell Biology
I Global Environmental Change

Springer-Verlag
Berlin, Heidelberg, New York, London, Paris and Tokyo

NATO-PCO-DATA BASE

The electronic index to the NATO ASI Series provides full bibliographical references (with keywords and/or abstracts) to more than 30000 contributions from international scientists published in all sections of the NATO ASI Series.
Access to the NATO-PCO-DATA BASE is possible in two ways:

– via online FILE 128 (NATO-PCO-DATA BASE) hosted by ESRIN, Via Galileo Galilei, I-00044 Frascati, Italy.

– via CD-ROM "NATO-PCO-DATA BASE" with user-friendly retrieval software in English, French and German (© WTV GmbH and DATAWARE Technologies Inc. 1989).

The CD-ROM can be ordered through any member of the Board of Publishers or through NATO-PCO, Overijse, Belgium.

Series C: Mathematical and Physical Sciences - Vol. 399

Topics in Knot Theory

edited by

M.E. Bozhüyük

Atatürk Üniversitesi,
Fen-Edebiyat Fakültesi,
Matematik Bölümü,
Erzurum, Turkey

Kluwer Academic Publishers

Dordrecht / Boston / London

Published in cooperation with NATO Scientific Affairs Division

Proceedings of the NATO Advanced Study Institute on
Topics in Knot Theory
Erzurum, Turkey
September 1-12, 1992

Library of Congress Cataloging-in-Publication Data

NATO Advanced Study Institute on Topics in Knot Theory (1992 :
Erzurum, Turkey)
Topics in knot theory : proceedings of the NATO Advanced Study
Institute on Topics in Knot Theory, Erzurum, Turkey, September 1-12,
1992 / edited by M.E. Bozhüyük.
p. cm. -- (NATO ASI series. Series C, Mathematical and
physical sciences ; vol. 399)
"Published in cooperation with NATO Scientific Affairs Division."
Includes index.
ISBN 0-7923-2285-1 (acid-free)
1. Knot theory--Congresses. I. Bozhüyük, M. E. (Mehmet Emin),
1944- . II. Title. III. Series: NATO ASI series. Series C,
Mathematical and physical sciences ; no. 399.
QA612.2.N38 1992
514'.224--dc20 93-17084

ISBN 0-7923-2285-1

Published by Kluwer Academic Publishers,
P.O. Box 17, 3300 AA Dordrecht, The Netherlands.

Kluwer Academic Publishers incorporates the publishing programmes of
D. Reidel, Martinus Nijhoff, Dr W. Junk and MTP Press.

Sold and distributed in the U.S.A. and Canada
by Kluwer Academic Publishers,
101 Philip Drive, Norwell, MA 02061, U.S.A.

In all other countries, sold and distributed
by Kluwer Academic Publishers Group,
P.O. Box 322, 3300 AH Dordrecht, The Netherlands.

Printed on acid-free paper

TABLE OF CONTENTS

Preface and Acknowledgements

The Director and the Authors gratefully acknowledge the encouragement and assistance provided by the NATO Scientific Affairs Division, Brussels,Belgium. We are very grateful to Dr.L.V. da Cunha, Director of the NATO ASI Programme, for his advice and guidance during the preparations for the Advanced Study Institute.
Thanks are also due to the Rector of Erzurum Ataturk Uniersity, Professor Dr. Erol Oral, Vice Rector Professor Dr. Zeki Ertugay and the Dean of the Faculty of Sciences and Letters Professor Dr. Ahmet Cakir.
A special thanks is due to Nevim Kısmet Temel for all her help.
The invited and contributing lecturers spent many hours preparing their lectures and their written papers included in this book.
We are very grateful to them for their contributions and support.
A number of papers have been revised in the light of discussions at the institute. We also thank all the delegates who attended and contributed their papers and ideas to the meeting.

Prof.Dr. Mehmet Emin Bozhuyuk
Erzurum, Turkey
25 January 1993

About the Director and Editor

Prof. Dr. Mehmet Emin Bozhüyük

Dr. Mehmet Emin Bozhüyük is a Professor of Mathemetics at Atatürk University, Erzurum, Turkey. He holds the M.A. and Ph.D. degrees from Princeton University. He has been a Visiting Fellow at Dartmouth College, Hanover, New Hampshire, US and a Visiting Proffessor at J.W. Goethe Universität in Frankfurt am Main, Germany in 1974 and 1989 respectively. His main interests are Knot Theory and Computer Programming. He has worked with world fameous knot theorists Professors Fox, Crowell, Burde and now Murasugi, Zieschang, Moran, Fenn, Morton, Rolfsen. He has about 20 papers on knot theory.

NATO ASI TOPICS IN KNOT THEORY
DIRECTOR EDITOR
M.E.Bozhüyük, Atatürk University, Erzurum-Turkey
ORGANIZING COMMITTEE
M.E.Bozhüyük , Atatürk University, Erzurum-Turkey
G.Burde, J.W.Goethe Universität, Frankfurt-Germany
D.Rolfsen, University of British Columbia, Vancouver-Canada
H.Zieschang, Ruhr Universität, Bochum-Germany
NAME INSTITUTION-ADDRESS
LECTURERS and PARTIPANTS
CANADA
K.Murasugi, Department of Mathematics, University of Toronto
Toronto, Ontario, Canada M5S 1A1 e-mail: murasugi @ math. utoronto. ca
D.Rolfsen, University of British Columbia, Department of Mathematics
No. 121-1984 Mathematics Road, Vancouver, B.C. Canada V6T 1Y4
e-mail: math@mtsg.ubc.ca
D.Kim, Math. Dept. University of British Columbia
No. 121-1984 Mathematics Road, Vancouver, B.C. Canada V6T 1Z2
e-mail: djun @ unixg.ubc.ca or djun@raven.math.ubc.ca
FRANCE
M.Boileau, Universite Paul Sabatier, Laboratorie Topologie
et Geometrie Ufr Mig: 118 Route de Narbonne 31062 Toulouse-France
C.H.Legrand, Univesite Paul Sabatier, Laboratiore Topology
et Geometre Ufr Mig: 118 Route de Narbonne 31062 Toulouse-France
D.Chenoit, Universite de Nice, Rue Du Canal, Le Casset F 05220
Le Monetier-Les-Bains Nice Cedex-France
Y.Grandati, L.P.L.I.-Institut de Physique 1 Bd F. Arago ,Technopole
Metz 2000,57070 Metz-France
F.Jaeger, Isd BP 53X, Greenoble Cedex, France e-mail: jaeger @imag.fr
M.Jambu, Universite de Nantes, Department de Mathematique, Nantes
Cedex-France e-mal: Jambu@ FRCICB81. BITNET
A.Legrand, 23 Rue des Palmiers 31400 Toulouse-France
e-mail: clegrand@froict 81
GERMANY
G.Burde, J.W.Goethe Universität, Fachbereich Mathematik
Robert Mayer-Str. 6-10, 6000 Frankfurt am Main-Germany
S.Wang, Ruhr Universität, Institut fur Mathematik 4630 Bochum-Germany
or Math. Dept. Peking Universty, Bejing 100871, China
H.Zieschang, Fakultät fur Mathematik, Ruhr- Universität,
Bochum, Postf.10 21 48, 4630 Bochum 1 Germany
e-mail: marlene.schwarz@ Ruba.Ruhr-Uni.Bochum.dbd.de
M. Hagelberg, Ruhr Universtät, Fakultät fur Mathematik
Postfach 102148,4630 Bochum 1 Germany

e-mail: Hagel@ math 16. Mathematik. Uni- Bielefeld.de
M.Hahne, Ruhr Universität, Felderhal 28 W-4030 Rativsen 1
Bochum-Germany
A.Kalesowsky, Ruhr Universitat,Auf der Papenburg 17-21 ZI.512
W-4630 Bochum Germany
O. Karalashvili, Ruhr Universtät, Fakultät fur Mathematik
postf. 10 21 48, 4630 Bochum 1 Germany
or Georgia, 380077 Tbilisi, A. Kasbegi ave. 29a, app. 52
U.Keil, Suntumer Str.86, 4630 1, Bochum Germany
M.Kriener, University of Tübingen, Uhlandstr. 16 7400 Tubingen
W-Germany e-mail: mmisa 01@ convex.zdv. uni-tvebingen de
F.Opitz, Niersteiner Str. 20, 600 Frankfurt Am Main 70 Germany
S.Rosebrock, J.W.Goethe Universität, Senckenberganlage 9
Didaktik d. Mathematik 6000 Frankfurt 11 Germany
e-mail: rosebrock@ mathematik. uni-frankfurt.dbp.de
PORTUGAL
M.E.Cesar de Sa, Seccao De Mathematica Faculdade de Ciencias
Universidade do Porto, Porto-Portugal
TURKEY
M.E.Bozhüyük, Atatürk Universitesi, Fen Edebiyat Fakultesi
Matematik Bolumu, Erzurum-Turkey
S. Akbulut, Ataturk Universitesi, Fen Edebiyat Fakultesi
Matematik Bolumu Erzurum-Turkey
Y.Altın, Yuzuncu Yil Universitesi, Fen Edebiyat Fakultesi
Matematik Bolumu Van-Turkey
I. Altintas, Yuzuncu Yil Yniversitesi, Fen Edebiyat Fakultesi
Matematik Bolumu, Van-Turkey
H.Aydın, Ataturk Universitesi, Fen Edebiyat Fakultesi
Matematik Bolumu, Erzurum-Turkey
H.Azcan, Anadolu Universitesi, Fen Edebiyat Fakultesi
Matematik Bolumu, Eskisehir-Turkey
A.Dane, Cumhuriyet Universitesi, Fen Ededbiyat Fakultesi
Mathematik Bolumu, Sivas-Turkey
E.Emir, Anadolu Universitesi, Fen Edebiyat Fakultesi
Matematik Bolumu, Eskisehir-Turkey
H.Ertugay, Ataturk Universitesi, Kazim Karabekir Egitim
Fakultesi, Matematik Bolumu, Erzurum-Turkey
A. Kacar, Ataturk Universitesi, Fen Edebiyat Fakultesi
Matematik Bolumu, Erzurum- Turkey
M. Kamali, Ataturk Universitesi, Fen Edebiyat Fakultesi
Matematik Bolumu, Erzurum-Turkey
F. N. Kiziloglu, Ataturk Universitesi, Kazim Karabekir Egitim
Fakultesi, Matematik Bolumu, Erzurum-Turkey

A.Kopuzlu, Ataturk Universitesi, Fen Edebiyat Fakultesi
Matematik Bolumu, Erzurum-Turkey
A. Magden, Ataturk Universitesi, Fen Edebiyat Fakultesi
Matematik Bolumu, Erzurum-Turkey
A.Küçük, Ataturk Universitesi, Fen Edebiyat Fakultesi
Matematik Bolumu, Erzurum-Turkey
M. Ozdemir, Ataturk Universitesi, Fen Ededbiyat Fakultesi
Matematik Bolumu, Erzurum-Turkey
B. Yildiz, Ataturk Universitesi, Fen Edebiyat Fakultesi
Matematik Bolumu, Erzurum-Turkey
UNITED KINGDOM
R.Fenn, University of Sussex, Matheatics Inst.,Falmer, Brighton
England BN1 9QH e-mail: rogerf@syma.susex.ac.uk.
S.Moran, Institute of Mathematics, University of Kent
Canterbury, Kent-U.K.
H.R.Morton, Dept. of Pure Mathematics, University of Liverpool
P.O.Box 147, Liverpool L69 3BX England
e-mail: h.r.morton@Liverpool.ac. UK.
P.R.Cromwell, University of Liverpool, Liverpool-England
I.M.James, Oxford University, Mathematical Institute. 24-29
ST. Giles, Oxford-England
C.Kearton, University of Durham,Department of Mathematical Sciences
South Road, Durham DH1 3LE, UK
e-mail Cherry.Kearton@durham.ac.uk
R.Rimanyi, University of Sussex, Math.Inst. Brighton England
or Dept. of Geometry, Elte, Rakoczi Ut 5, Budapest-1088 Hungary
J.O.Salkeld, University of Liverpool, Liverpool-England
G.Smith, University of Bath, School of Matematical Sciences
Claverton Down, Bath BA2 7AY England
e-mail: gcs@maths.bath.ac.uk
UNITED STATES
A.Casson, Math. Dept. University of California, Berkeley CA 94720, USA
e-mail casson@math.berkeley.edu
F.Delehan, Dept. Math. UCI, Irvine, CA 92717, USA
K.Ferguson, Dept. Math. UCI, Irvine, CA 92717, USA
e-mail kferguso@uci.edu
E.Flapan, Pomona College, Claremont,CA 91711, USA
J.Hempel, Rice University, Dept.of Mathematics, Wiess School of
Natural Sciences, P.O.Box 1892, Houston, Texas 77251 USA
R.Kirby, Department of Mathematics, University of California
Berkeley CA 94720 USA
R.Riley, State University of NY, Dept. of Matematical Sciences
P.O. Box 6000, Binghamton, New York 13902-6000 USA

NON-NATO COUNTRIES

AZERBAYCAN
A.Y.Aliev, Baku-122, Prosp. G.Zardabi 87, Apt.78 Azerbaycan
Ş.İ.İbrahimov Baku-12, Talybly Str.201,Apt.7 Azerbaycan
Y.C.Memmedov, Baku 122, Kievski Prosp. 9B, Apt.18 Azerbaycan
F.A.Tagiyev, Baku-125 8-TH MCR. 77, Apt.1 Azerbaycan
BULGARIA
Y.P.Mishev, Dept. of Matematics and Physics, Higher Institute of Forestry
KL. Okridski 10, 1156 Sofia Bulgaria
JAPAN
A.Kawauchi, Osaka City University, Dept. of Matematics Osaka,558,Japan
e-mail :d54454@jpnudpc.bitnet
T.Harikae, Kwansei Gakuin University, Vegahara, Nishinomiya,662 Japan
S.Kamada, Osaka City University,Sugimoto, Sumiyoshi Osaka,558, Japan
Y.Nakagawa, Yamaguchi Women's University, 3-2-1 Sakurabatake
Yamaguchi, 753 Japan
KIRGIZISTAN
M.A.Asankulova, Acedemy of Sciences, Bişkek-Kırgızistan
M.Ş.Batırkanov, Acedemy of Sciences,Bişkek-Kırgızistan
A.B.Bayzakov, Acedemy of Sciences, Bişkek-Kırgızistan
KOREA
G.T.Jin, Dept. of Mathematics,KAIST, Taejon. 305-701 Korea
e-mail:trefoil@math1.kaist.ac.kr
UKRAINE
A.Alexandrow, Kiev State University, Flat 26, 18/21 Kibalchich Street
Kiev 253139 Ukraine
I.Stukanev, Kiev State University, Inst.of Mathematics, Flat 74
9 Janvarsky Prosp. Belaya Tserkov 252 601 Kiev Ukraine
ACCOMPANYING PERSON
Mrs. Zieschang

KNOT PROJECTIONS AND KNOT COVERINGS

Prof. Dr. Mehmet Emin Bozhüyük
Atatürk Üniversitesi
Fen-Edebiyat Fakültesi
Matematik Bölümü
Erzurum-Turkey

ABSTRACT. In this paper, computer programs are given which draw knot diagrams, knot projections and representations of knot groups into the symmetric group of degree n.

1. Introduction

The act of drawing knot diagrams such as that of the shepherds knot in figure 1 and drawing knot projections such as figure 2 can be tedious and time consuming. The idea of using the computer programs in this paper is to speed up this process.

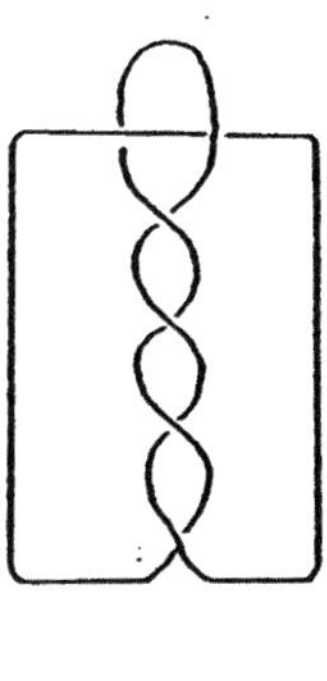

Figure 1: A Shepherd's knot

M.E. Bozhüyük (ed.), Topics in Knot Theory, 1–14.

2. The drawing programs.

2.1 Program G

This computer program draws random polygonal knot projections such as that shown in figure 2.

PROGRAM G

```
100 REM THIS PROGRAM DRAWS RANDOM GRAPHS.
110 REM IT IS WRITTEN BY MEHMET EMIN BOZHUYUK.
120 DIM A(100), DIM B(100)
130 LIBRARY "PLOTLIB***:TEK10"
140 DIN C(150)
150 PRINT "INPUT V, NO OF VERTICES"
160 INPUT V
170 CALL "LIMITS":C(),0,10,0,10
180 RANDOMIZE
190 FOR I=1 TO V
200 LET X=RND
210 LET Y=RND
220 LET A(I)=10*X
230 LET B(I)=10*Y
240 CALL "LINE":C(),A(I),B(I)
250 NEXT I
260 CALL "LINE":C(),A(1),B(1)
270 CALL "FINISH":C()
280 REM STOP FOR COPYING THE PICTURE.
290 LINPUT S$
300 GOTO 150
310 END
```

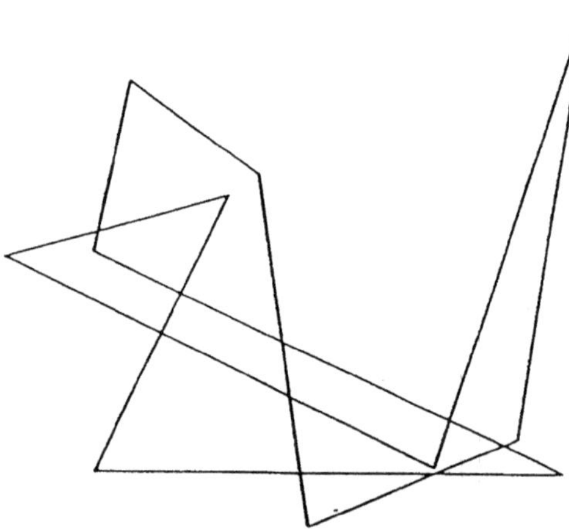

Figure 2: A random closed curve drawn by program G

2.2 Special cases

By specialising, the above method can be used to draw knots whose projection lies in a grid (see figure 3), is the closure of a braid (see figure 4), or is a plat, (see figure 5).

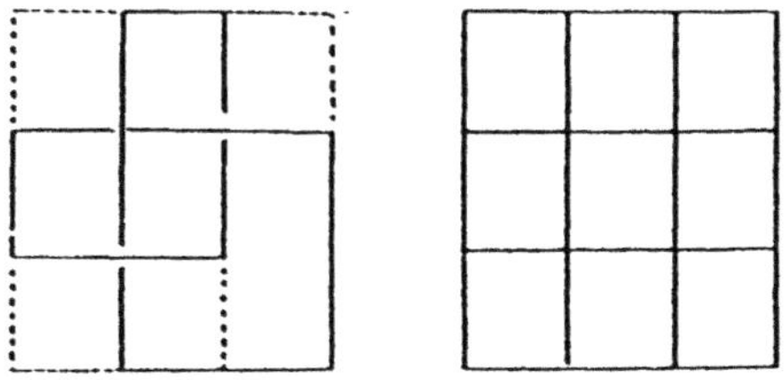

Figure 3: The trefoil drawn on a grid

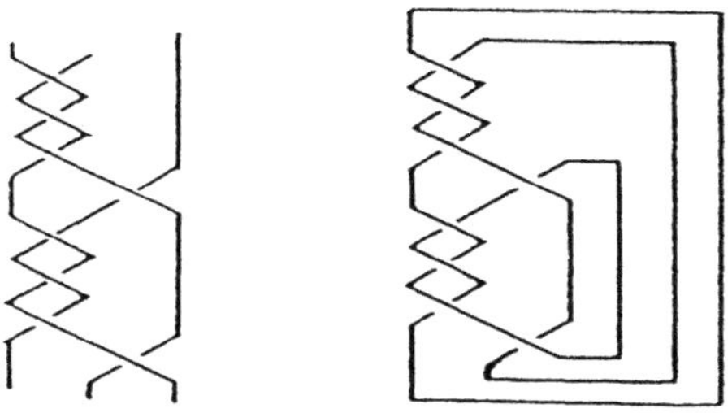

Figure 4: A closed braid

The closure of a braid is a knot or link and all knots and links can be obtained this way. See figure 4. Alternatively the braid can be closed as plat. See figure 5.

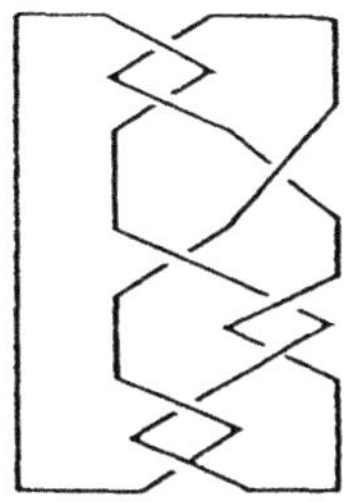

Figure 5: A plat

2.3 The plotting methods

The following computer program KNOT is a user interactive program. It takes the coordinates of the vertices of a graph and joins them successively. The programs named K, K2, K9 in [1] give the normal projections of the (3,4)-Turk's head knot, the closed braid form of the Figure Eight knot (also known as the (3,2) Turk's head knot) and (2,9) alternating torus knot respectively. The (2,n) alternating torus knots are also called the (2,n)-Turk's head knots. See Figures 6,7,8.

KNOT

```
100 REM THIS PROGRAM DRAWS KNOT PROJECTIONS.
110 REM IT IS WRITTEN BY MEHMET EMIN BOZHUYUK.
120 LIBRARY "PLOTLIB***:TEK10"
130 DIM C(150)
140 READ A,P,B,Q
150 DATA 0,6,0,12
160 CALL "LIMITS":C(), A,P,B,Q
170 REM THE INDEX I=10 LIFT, I=1 LINE
180 LET N=100
190 FOR J=1 TO N
200 INPUT X,Y,I
210 IF I=0 THEN 240
220 CALL "LINE":C(),X,Y
230 GOTO 200
240 CALL "LINE":C()
250 NEXT J
260 CALL "FINISH":C()
270 END
```

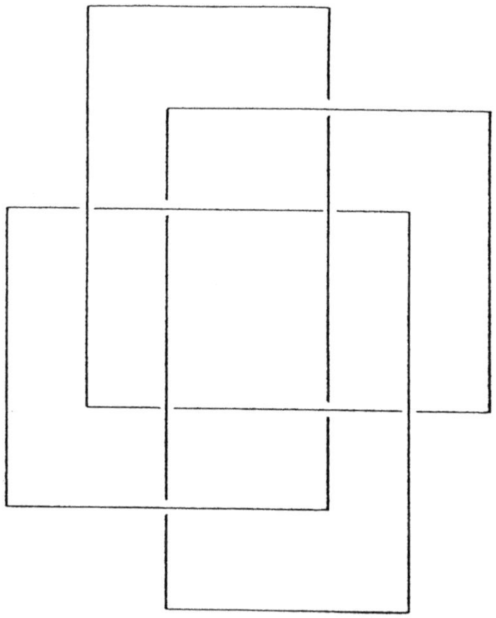

Figure 6: The (3,4)-Turk's head knot

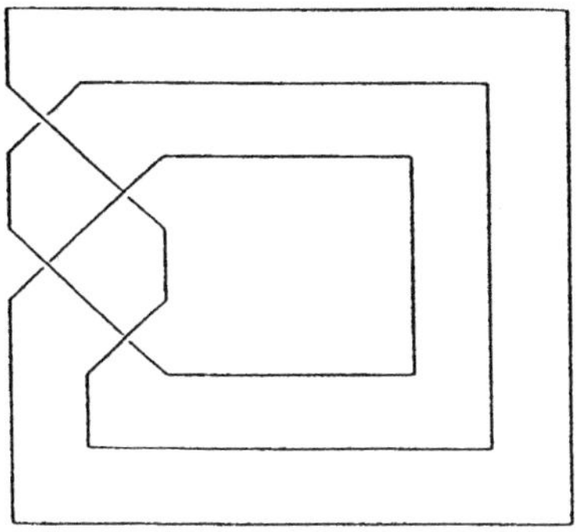

Figure 7: The (3,2)-Turk's head knot

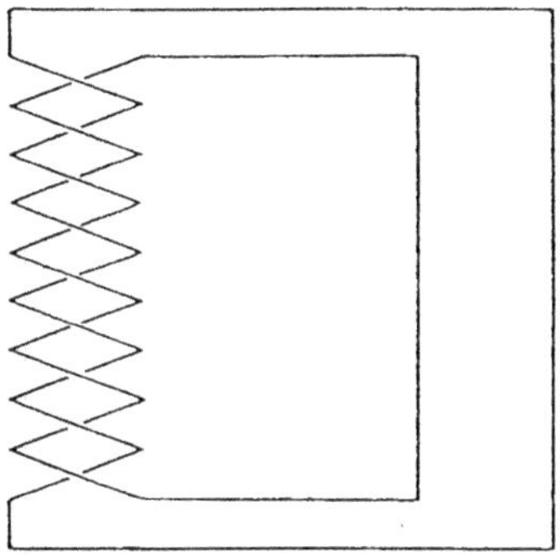

Figure 8: The (2,9)-Turk's head knot

2.4 Program star

This program draws regular projections of (2,2n+1)-Turk's head knots such as figure 9. Tom Kurtz taught me this program, [9].

STAR

```
100 REM THIS PROGRAM DRAWS REGULAR PROJECTIONS OF
110 REM (2,2N+1) TORUS KNOT WHEN YOU ENTER K=2.
120 LIBRARY "PLOTLIB***:TEK10"
130 DIM C(2000)
140 READ T,U,X,Y
150 DATA -1.3,1,3,-1,1
160 CALL "LIMITS":C(),T,U,X,Y
170 REM HERE IS THE PROGRAM.
180 PRINT "ENTER (N,K)"
190 INPUT N,K
200 CALL "STAR":C(),N,K
210 REM STOP FOR COPYING THE PICTURE.
```

```
220 LINPUT A
230 GOTO 180
240 CALL "FINISH":C()
250 END
260 SUN "STAR":C(),N,K
270 LET P1 = 3.14159265
280 FOR P=0 TO N
290 LET R=2*P1*P*K*/N
300 CALL "LINE":C(),COS(R),SIN(R)
310 NEXT P
320 CALL "LIFT":C()
330 SUBEND
```

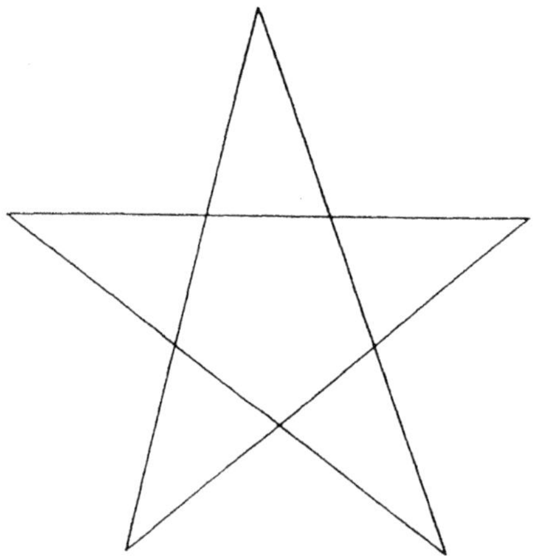

Figure 9: A projection of the (2,5)-torus knot

2.5 Program IZD

This program draws regular projections of (A,B)-Turk's head knots [1]. Where A,B are two relatively prime natural numbers. See Figure 10.

IZD

```
100 REM THIS PROGRAM DRAWS REGULAR PROJECTIONS
110 REM OF (A,B)-TURKS HEAD KNOTS.
120 REM IT IS WRITTEN BY MEHMET EMIN BOZHUYUK.
130 LIBRARY "PLOTLIB***:TEK10"
140 DIM C(170)
150 CALL "LIMITS":C(),-1,1,-1,1
160 CALL "SQUARE":C(),1
170 PRINT "ENTER A,B TWO RELATIVELY PRIME NUMBERS"
180 INPUT A,B
190 LET R=A/B
200 CALL "CLEAR":C()
210 FOR I=0 TO 2*3.1415*(A+.01) STEP .05
220 CALL "LINE":C(),COS(1)*(COS(R*I)+2)/3,SIN(I)*(COS(R*I)+2)/3
230 NEXT I
240 CALL "FINISH":C()
250 END
```

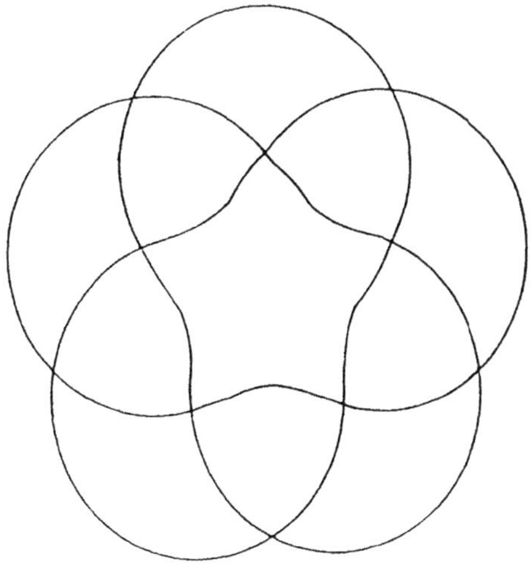

Figure 10: A projection of the (3,5)-Turk's head knot

2.6 Program Turk

This program draws (A, B)-Turk's head knots. Turk's head knots [8] can be presented by smooth functions. The positive integers A and B are called lead and bight respectively [7]. The (A, B)-Turk's head knot has $(A-1)B$ crossings. For instance, the $(3,4)$-Turk's head knot is known as 8_{18} in knot tables. In the projection there are $(A-1)$ concentric circles which carry the crossing points. The pattern of distribution of crossings can be determined. The angle between two consecutive crossings is π/B [1,3,4,5,6].

TURK

```
100 REM THIS PROGRAM DRAWS (A,B) TURK'S HEAD KNOT.
110 REM A AND B ARE TWO RELATIVELY PRIME NUMBERS.
120 DIM C(150)
130 LIBRARY "PLOTLIB***:TEK10"
150 CALL "LIMITS":C(),-1,1,-1,1
160 CALL "SQUARE":C(),1
180 PRINT "ENTER TWO RELATIVELY PRIME NUMBERS"
190 INPUT A,B
210 CALL "LIFT":C()
220 LET R=B/A
230 LET S=3.1415926/B
240 LET K=A-2
250 LET Q=1
260 FOR T=1 TO (A-1)*B
270 CALL "LIFT":C()
280 FOR I=Q*S +.04 TO (Q+FNS)*S -.04 STEP .08
290 LET X=COS(I)*(COS(R*I)+2)/3
300 LET Y=SIN(I)*(COS(R*I)+2)/3
310 CALL "LINE":C(),X,Y
320 NEXT I
330 FOR I=I+.008 TO (Q+P2)*S-.04 STEP .008
340 CALL "LINE":C(),X,Y
```

```
350 NEXT I
360 LET Q=Q+P2
400 NEXT T
410 CALL "FINISH":C()
420 PRINT "(";STR$(A);",";STR$(B);") TURKS KNOT"
440 LINPUT Z$
450 GO TO 180
460 DEF FNS
470 IF K<>0 THEN 510
480 LET FNS=2
490 LET K=A-2
500 GO TO 530
510 LET FNS=1
520 LET K=K-1
530 IF K<>0 THEN 570
540 LET FNS=FNS+2
550 LET K=A-2
560 GO TO 590
570 LET FNS=FNS+1
580 LET K=K-1
590 LET P2=FNS
600 FNEND
620 END
```

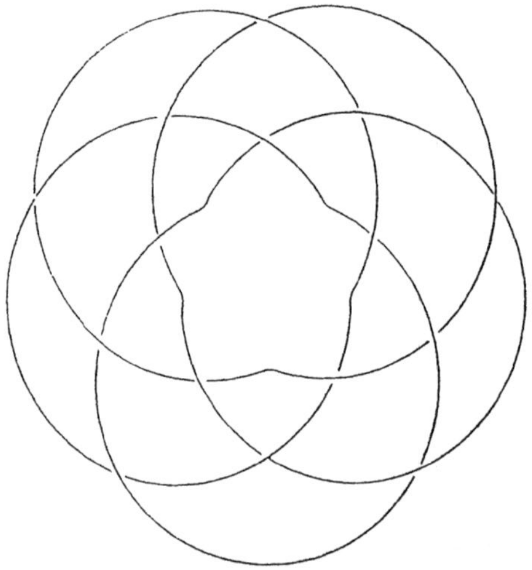

Figure 11: A (4,5)-Turk's head knot.

3. Representations of knot groups into the symmetric group

The computer program HOMO finds the representations of groups (especially knot or link groups) in the symmetric group of a given degree. We compare all these representations and if we find any differences we can conclude that the groups and hence the knots or links are distinct. For instance the Figure Eight knot does not have a representation onto S_3 but the trefoil does. Another way of saying this is that the Figure Eight knot is not 3-colourable whilst the trefoil is. Hence these two knots are distinct.

HOMO

```
90 REM THIS BASIC PROGRAM DETERMINES ALL
95 REM REPRESENTATIONS OF A GROUP GIVEN
100 REM BY ITS GENERATORS AND DEFINING
101 REM RELATIONS IN A SYMMETRIC GROUP
102 REM OF A GIVEN DEGREE, PROVIDED THAT
103 REM THE LENGTH OF THE RELATIONS ARE THE SAME.
104 REM IT PROVIDES ALSO DATA TO THE USER TO CALCULATE
105 REM PRESENTATIONS OF THE SUBGROUP (STABILIZER)
106 REM CORRESPONDING TO THESE REPRESENTATIONS. THIS
107 REM PROGRAM IS WRITTEN BY MEHMET EMIN BOZHUYUK,
108 REM AT DARTMOUTH COLLEGE OF USA IN FEBRUARY 1975.
109 DIM R(10,10)      'RELATION MATRIX
110 DIM H(10,10)      'HOMEOMORPHISM MATRIX
120 LET G=2      'NUMBER OF GENERATORS
130 LET R=1      'NUMBER OF RELATIONS
140 LET L=6      'THE LENGTH OF THE RELATIONS
150 MAT READ R(R,L)
160 DATA 1,2,1,4,3,4      'RELATIONS OF THE GROUP
170 REM GATHER INFORMATION ON V
175 REM FROM THE USER.
180 PRINT "ENTER V, THE DEGREE"
185 PRINT "OF THE SYMMETRIC GROUP"
190 INPUT V
200 IF V<=10 THEN 240
210 PRINT "V<=10, PLEASE ENTER SMALLER V"
220 REM FOR V>10, USER MAY CHANGE LINE 110.
230 GOTO 180
240 RESET # 0
250 CALL "PERM" :H(,),R(,),1,G,V,R,L
260 REM GOTO BEGINNING TO ENTER OTHER V'S.
270 GOTO 180
280 END
290 REM THIS RECURSIVE SUBPROGRAM
295 REM CREATES HOMEOMORPHISMS.
300 SUB "PERM" :H(,),R(,),N,G,V,R,L
310 IF N>G THEN 520
320 GOSUB 340
330 GOTO 860
340 LET Q=Q+1
350 IF Q>V THEN 460
360 LET H(Q,N)=0
370 LET A=H(Q,N)=H(Q,N)+1
380 FOR F=1 TO Q-1
390 IF H(F,N)=A THEN 420
400 NEXT F
410 GOSUB 340
420 IF H(Q,N)<V THEN 370
430 LET Q=Q-1
440 RETURN
450 REM THIS LOOP FINDS IMAGES
455 REM OF GENERATORS' INVERSE.
460 FOR F=1 TO V
470 LET H(H(F,N))
480 NEXT F
490 CALL "PERM" :H(,),R(,),N+1,G,V,R,L
500 LET Q=Q-1
510 RETURN
520 REM THIS LOOP CHECKS IF
525 REM EACH RELATOR IS MAPPED ONTO
530 REM IDENTITY. IF SO THEN IT
```

```
535 REM PRINTS THE MATRIX H(,).
540 FOR Q=1 TO V
550 LET X=Q
560 FOR T=1 TO R
570 FOR U=1 TO L
580 LET X=H(X,R(T,U))
590 NEXT U
600 IF X<>Q THEN 860
610 NEXT T
620 NEXT Q
630 PRINT
640 FOR Q=1 TO V
650 FOR F=1 TO 2*G
660 PRINT TAB(15);
670 PRINT USING "- +", H(Q,F);
680 NEXT F
690 PRINT
700 NEXT Q
710 REM AFTER H(,) IS PRINTED,
715 REM THIS LOOP CREATES INDI-
720 REM CES OF DEFINING RELATIONS OF STABILIZER.
730 PRINT
740 FOR Q=1 TO V
750 LET X=Q
760 FOR F=1 TO R
770 PRINT TAB(15)
780 PRINT USING "- +", Q:
790 FOR U=1 TO L
800 LET X= H(X,R(F,U))
810 PRINT USING "- +",X;
820 NEXT U
830 PRINT
840 NEXT F
850 NEXT Q
860 SUBEND
```

3.1 Example

The group of the Trefoil is $|x, y : xyx = yxy|$ and its representations in the symmetric group of order six are given here.

```
HOMO              25 Apr 75 21:31

INPUT V=NO OF SYMBOLS, G=NO OF GENERATORS
? 3,2
**********
1     1     1     1
2     2     2     2
3     3     3     3
1     1     1     1     1     1     1
2     2     2     2     2     2     2
3     3     3     3     3     3     3
**********
1     1     1     1
3     3     3     3
2     2     2     2
```

```
1  1  1  1  1  1  1
2  3  2  3  2  3  2
3  2  3  2  3  2  3
**********
1  2  1  2
3  1  3  1
2  3  2  3
1  1  2  3  3  2  1
2  3  3  2  1  1  2
3  2  1  1  2  3  3
**********
1  3  1  3
3  2  3  2
2  1  2  1
1  1  3  2  2  3  1
2  3  1  1  3  2  2
3  2  2  3  1  1  3
**********
2  1  2  1
1  3  1  3
3  2  3  2
1  2  3  3  2  1  1
2  1  1  2  3  3  2
3  3  2  1  1  2  3
**********
2  2  2  2
1  1  1  1
3  3  3  3
1  2  1  2  1  2  1
2  1  2  1  2  1  2
3  3  3  3  3  3  3
**********
2  3  2  3
1  2  1  2
3  1  3  1
1  2  2  1  3  3  1
2  1  3  3  1  2  3
3  3  1  2  2  1  3
**********
2  2  3  3
3  3  1  1
1  1  2  2
1  2  3  1  3  2  1
2  3  1  2  1  3  2
3  1  2  3  2  1  3
**********
```

3	3	2	2			
1	1	3	3			
2	2	1	1			
1	3	2	1	2	3	1
2	1	3	2	3	1	3
3	2	1	3	1	2	3

****

3	1	3	1			
2	3	2	3			
1	2	1	2			
1	3	2	2	3	1	1
2	2	3	1	1	3	2
3	1	1	3	2	2	3

****

3	2	3	2			
2	1	2	1			
1	3	1	3			
1	3	3	1	2	2	1
2	2	1	3	3	1	2
3	1	2	2	1	3	3

****

3	3	3	3			
2	2	2	2			
1	1	1	1			
1	3	1	3	1	3	1
2	2	2	2	2	2	2
3	1	3	1	3	1	3

3.2 Example

For the Figure Eight knot we have the following data.

HOMO 23 Apr 75 17:22

INPUT THE KNOT PARAMETERS (A,B)? 3,2
INPUT THE NUMBER OF GENERATORS? 4
DIMENSION OF SYMMETRIC GROUP? 3
****

2	2	2	2	3	3	3	3
3	3	3	3	1	1	1	1
1	1	1	1	2	2	2	2
1	2	3	2	1			
1	2	1	3	1			
1	2	3	2	1			
1	2	1	3	1			
2	3	1	3	2			
2	3	2	1	2			
2	3	1	3	2			
2	3	2	1	2			
3	1	2	1	3			

3	1	3	2	3
3	1	2	1	3
3	1	2	2	3

****

3	3	3	3	2	2	2	2
1	1	1	1	3	3	3	3
2	2	2	2	1	1	1	1
1	3	2	3	1			
1	3	1	2	1			
1	3	2	3	1			
1	3	1	2	1			
2	1	3	1	2			
2	1	2	3	2			
2	1	3	1	2			
2	1	2	3	2			
3	2	1	2	3			
3	2	3	1	3			
3	2	1	2	3			
3	2	3	1	2			

3.3 Coverings

A presentation of the fundamental group of covering spaces, from the computer data above, is obtained as follows. We consider, for instance, $G = |x, y : xyx\overline{y}\,\overline{x}\,\overline{y}|$ and the matrix

	x	y	$\overline{x}$	$\overline{y}$
1	2	1	2	1
2	1	3	1	3
3	3	2	3	2

This means $h(x) = h(\overline{x}) = (12)$ and $h(y) = h(\overline{y}) = (23)$. Namely, $h : G \to S_3$ is an onto homomorphism.

The matrix

1	2	3	3	2	1	1
2	1	1	2	3	3	2
3	3	2	1	1	2	3

means $h(r) = 1$, where $r = xyx\overline{y}\,\overline{x}\,\overline{y}$. From this matrix by Fox's algorithm one obtains

$$\pi_1(\Sigma - L) = |x_1, x_3, y_1, y_3 : x_1x_3\overline{x_1}\,\overline{y_1}, x_1x_3\overline{y_3}\,\overline{x_3}, y_1x_1\overline{y_1}\,\overline{y_3}, x_3y_3\overline{y_1}\,\overline{y_3}|$$

where Σ is the three sheeted irregular branched covering space of S_3 branched over the Trefoil knot and L is the link obtained from the closed braid, σ_1^4. In this case Σ is a 3-sphere. For more details see [2].

References

[1]. **Bozhüyük M. E.** *Basit Kapali Uzay Eğrileri*, Knots. TÜBİTAK Research Project, TBAG-155 Ankara, 1975.
[2]. —— *On Covering Spaces of Trefoil Knot.* Jour. Fac. Sci. of KTU, 2,31-40, 1979.
[3]. —— *On 3-Sheeted Covering Spaces of (3,2)-Turk's Head Knot.* Jour. Fac. Sci. of KTU, 3,31-35, 1981.
[4]. —— *On 3-Sheeted Covering Spaces of (3,4)-Turk's Head Knot.* Coll. Math. Soc. Janos Bolyai, 23 Topology, p.175-180, Budapest, 1978.
[5]. —— *On 3-Sheeted Covering Spaces of (3,5)-Turk's Head Knot.* Coll. Math. Soc. Janos Bolyai, 41 Topology and Applications, p.119-124, Eger, 1983
[6]. —— *Dallanmış Örtü Uzayları Teorisi ve Üç Boyutlu Kürenin Bazı Türk Düğümleri Üzerinde Dallanmış Örtü Uzayları.* Profesorlük Tezi, Atatürk Üniversitesi, Erzurum, 1983.
[7]. **Crowell R.H. and Fox R.H.** *Introduction to Knot Theory.* Blaisdell-Ginn, N.Y., 1963.
[8]. **Fox R.H.** *A Quick Trip Through Knot Theory.* Topology of 3-Manifolds, Proc. Top. Ins. Prentice Hall, Englewood-Cliffs N.J., 1962 p.120-216.
[9]. **Kemeny J. and Kurtz T.** *Basic Programming.* Wiley, N.Y., 1971.

KNOTS AND KNOT SPACES

Gerhard Burde
Fachbereich Mathematik
Johann Wolfgang Goethe-Universität
Postfach 11 19 32
D-6000 Frankfurt am Main / BRD

ABSTRACT. In this survey on knot spaces the central theorems are presented such as the Gordon-Luecke complement result, Waldhausen's theorem, hyperbolic and Seifert fibred knots spaces and the Johannson-Jaco-Shalen torus decomposition.

1. Introduction.

Classic knot theory is concerned with equivalence classes by ambient isotopy of p.l. or smooth embeddings $S^1 \hookrightarrow S^3$ (knots) or $\coprod_{i=1}^{\mu} S_i^1 \hookrightarrow S^3$ (links). Usually S^3 and the components S_i^1 of the link carry an orientation. The image of a representative of a class of embeddings will also be called a knot K or a link $L = \{K_i\}$, $1 \leq i \leq \mu$.

The homeomorphism type of the complement $C = S^3 - L$ is the most natural invariant of a link; we prefer in most cases a compact version of the complement, $C = S^3 - U(L)$, where $U(L)$ is an open tubular neighbourhood of L in S^3 it serves as well as the genuine complement in our case of tame embeddings. The complements are called knot or link spaces.

The purpose of this lecture is to give a survey on results and methods concerning the classification of knots and links and their spaces, to study the relations between them and to investigate geometric properties of the knot spaces. The most powerful algebraic tool to accomplish these feats is the fundamental group of the knot space C. Since C is a bounded manifold, there is not only a "knot (link) group" $G = \pi_1 C$ assigned to C but a family $\{G_i\}$ of conjugate subgroups of G to every component K_i of L, $G_i \cong \pi_1(\partial U(K_i))$. A homeomorphism of complements induces an isomorphism of the link groups respecting the whole "group system $(G, \{G_i\})$". Moreover a special element of the "peripheral subgroup" $\pi_1(\partial(U))$ of a knot K is distinguished (up to conjugation) by the complement C: the simple closed curve ℓ (called the "longitude" of K) which satisfies $\ell \sim K$ in $U(K)$ and $\ell \sim 0$ in C. The knot K itself determines a "meridian" m, a simple closed curve on $\partial U(K)$ with intersection number $\text{int}(m, \ell) = +1$ which bounds a disk in $U(K)$. Choosing the point of

M.E. Bozhüyük (ed.), Topics in Knot Theory, 15–23.

intersection of m and ℓ as a base point for $\pi_1 C$ the pair (m, ℓ) represents a pair of distinguished elements of $\pi_1 C$ determined up to a common conjugation factor. Evidently such a pair (m_i, ℓ_i) can also be assigned to every component K_i of a link.

Let us complete this introduction by some geometric conventions. By $-K$ we denote the knot obtained from K by reserving its orientation. K is called "invertible" if K and $-K$ are equivalent, $K \approx -K$. K^* means the image of K by a reflection. A knot is called "amphicheiral", if $K \approx K^*$; else it is "cheiral". The same notions apply to links. Special attention has to be payed, of course, as to which components of the link are supposed to be inverted.

A projection of a link onto a plane is called "regular" or a "link diagram", if the singularities consist of a finite set of double points of a transversal type. A link is "alternating", if there exists an alternating diagram, meaning that overcrossings and undercrossings alternate when running along any component.

The product $L_1 \# L_2$ of two links L_1 and L_2 is a link which allows an embedded 2-sphere $S^2 \subset S^3$ meeting $L_1 \# L_2$ in two points P and Q, such that the two components L_1' and L_2' of $L_1 \# L_2 - \{P, Q\}$ define $L_1 \approx L_1' \cup \alpha$ and $L_2 \approx L_2' \cup \alpha$ with α a simple arc on S^2 connecting P and Q. $L_1 \# L_2$ depends on the choice of the two components chosen in L_1 resp. L_2 to be joined, but only on this; the product of two knots is well-defined. It is easy to see that $K \# K' \approx K' \# K$ and $K \# K_0 \approx K$ for the trivial knot K_0. A knot K is called "prime", if $K = K_1 \# K_2$ implies $K_1 \approx K_0$ or $K_2 \approx K_0$.

2. Complement and group.

The first relation of a fundamental character between group and complement of a knot is an outcome of Dehn's Lemma. Every knot K is homotopic to the trivial knot K_0 in S^3; it is even isotopic to K_0. By definition there exists an ambient isotopy between K and K_0 only if K itself is trivial. If one looks at C instead of S^3, choosing for K its longitude, we know $K \sim 0$ in C, but $K \simeq 1$ holds if and only if $K \approx K_0$. The latter assertion is a consequence of Dehn's Lemma. Thus knottedness is at the same time a sharp characterization of the difference between ambient isotopy and homotopy and between homotopy and homology.

Dehn's Lemma can be put into equivalent forms: a knot space is homeomorphic to a solid torus, if and only if the knot is trivial. Or, the knot group is cyclic (abelian), if and only if K is trivial.

The general case is taken care of by an application of a famous theorem of Waldhausen [WA 68].

Theorem: *The complements C and C' of two non split links $L = \{K_i\}$ and $L' = \{K_i'\}$ are homeomorphic, if and only if there is an isomorphism*

$$h_{\#} : \pi_1 C \to \pi_1 C', \quad h_{\#}|\{\pi_1 \partial U(K_i)\} \to \{\pi_1 \partial U(K_i')\},$$

of the group systems (peripheral isomorphism).

In this case there is a homeomorphism $h : C \to C'$ which induces $h_{\#}$. Clearly the links L and L' will be equivalent if and only if h maps the meridians m_i of L onto the meridians m_i' of L'.

One might ask the question: are there non-peripheral isomorphisms between link groups? The answer is "yes", even in the case of knots. The following example was provided by Fox [Fox 52]: the groups of $K\#K$ and $K\#K^*$ are isomorphic, but the complements C and C' are in the cheiral cases (i.e. for K a trefoil) not homeomorphic. To prove this one may use a suitable class of representations of the knot group G of the square of a trefoil. There is a well characterized class $[\varphi]$ of homomorphisms of G mapping meridians onto glide reflections of a Euclidean plane E^2. $\varphi(\ell)$ is a non-trivial translation, $\varphi(\ell^*) = -\varphi(\ell)$, ℓ, ℓ^* longitudes of the trefoil K and its mirror image K^*. If there was a homeomorphism $h : C \to C'$, there would be a peripheral isomorphism $h_{\#}$ and a commutative diagram:

$$\begin{array}{ccc} \pi_1 C & \xrightarrow{h_{\#}} & \pi_1 C' \\ \varphi \searrow & & \swarrow \varphi' \\ & \mathrm{Isometr}(E^2) & \end{array}$$

$\mathrm{id} \neq \varphi(\ell^2) = \varphi_0' h_{\#}(\ell^2) = \varphi'(\ell\ell^*) = \mathrm{id}.$ leads to a contradiction.

The fact that this counterexample is based on composite knots is significant: Whitten [WHI 86] proved: Prime knots have homeomorphic complements if their groups are isomorphic.

3. Complements and knots.

Whether knots determine their complements was a question for over eighty years. The question was answered affirmatively by Gordon and Luecke [GL 89]. We will give a description of the problem and an idea of the ingenious proof.

It was early observed that the complement of a proper link can not determine the link: let one component of the link be trivial, then its complement is a solid torus V. A second component – inside V – suffers drastic changes as a knot under homeomorphisms of V.

The margin between a knot space and a knot is best described by a procedure called "Dehn surgery", a device used by Dehn to construct 3-manifolds from knots in S^3. Let meridian m and longitude ℓ of a knot K generate $H_1(\partial C) \simeq Z \oplus Z$. Every simple closed curve $\not\sim 0$ on ∂C is of the form $\alpha m + \beta \ell$, $(\alpha, \beta) \neq (0,0)$ a pair of coprime integers; they can be parametrized by the rationals including ∞, $\{\frac{\alpha}{\beta}\} = \mathbb{Q} \cup \{\infty\}$. A Dehn-$\frac{\alpha}{\beta}$-surgery along $K \subset S^3$ yields a 3-manifold $K(\frac{\alpha}{\beta}) = \underset{\alpha m + \beta \ell = m^*}{C \cup V^*}$ where V^* is a solid torus with meridian m^*; it is called non-trivial, if $\beta \neq 0$. The question whether a knot space determines its knot may now be restated in the following way: Can a non-trivial Dehn surgery along $K \subset S^3$ yield a 3-sphere? A stronger (algebraic) version of this question is: can it yield a homotopy 3-sphere $\tilde{S}^3$? Put algebraically: can one trivialize a knot group by adding a relator $m^\alpha \ell^\beta = 1$, $\beta \neq 0$? As we pointed out before the first question has the answer "no" by the work of Gordon and Luecke. In the second case the conjecture is also "no"; some classes of knots are known to allow no such surgery – which asserts that they "have property P". The first trivial observation is that one may assume $\alpha = 1$, since $H_1(K(\frac{\alpha}{\beta}))$ is cyclic of order $|\alpha|$.

Fig. 1

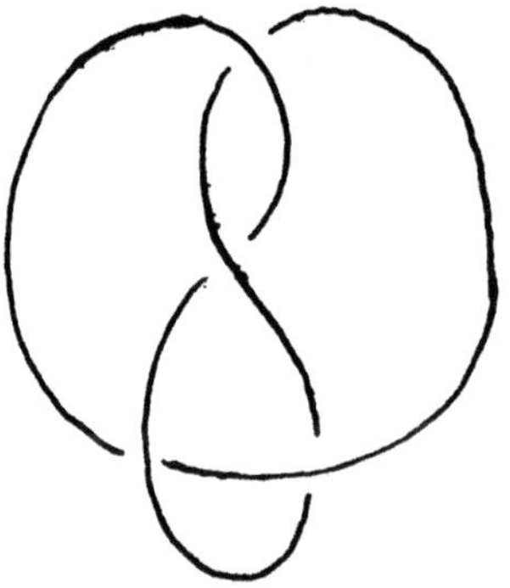

Fig.2.

The possibility of a counterexample for the property P conjecture is further restricted by a result called the "cyclic surgery theorem" [CGLS 87]: $\pi_1 K(\frac{1}{\beta})$ is cyclic (including trivial) implies $\beta = \pm 1$, if K is not a "torus knot" (a simple closed curve on the boundary of an unknotted solid torus in S^3). Surgery on torus knots are well known, [MO]. Nevertheless, the property P problem $K(1) \cong \tilde{S}^3$ is still open. As an example of Dehn surgery one obtains $K_0(\frac{\alpha}{\beta}) \cong L(\alpha, \beta) \cong \underset{\alpha m + \beta \ell = m^*}{V \cup V*}$ the lens spaces for the trivial knot K_0. For further use we observe that this yields a description of a so-called "punctured" lens space, that is, a lens space minus an open 3-cell e^3. We choose $e^3 = V^* - \partial V^* - D^*$, with $\partial D^* = m^*$, D^* a meridional disk of V^*. The punctured lens space then is $\underset{\alpha m+\beta \ell=\partial D^*}{V \cup D^*}$.

Now for the idea of the Gordon-Luecke proof: Let K, K^* be the two knots with homeomorphic complements E, E^* in $S^3 = E \cup \underset{m+\ell=m^*}{V^*}$, K^* the core of V^*. We think of K and K^* as "2m-plats" consisting of 2m-braids with arcs joining neighbouring upper and lower ends, Fig. 1. A certain technical condition is imposed on the plats which we omit. A braid has a natural height function and its level surfaces in E resp. E^* are planar surfaces S resp. S^* whose boundaries are meridional curves for K resp. K^*. Two such planar surfaces in a certain position are chosen. The homeomorphism $f : E^* \to E$ takes S^* to $f(S^*) = S'$ and $S \cap S'$ can be completed as well in S as in S' to a graph Γ resp. Γ' (with some additional structure) whose vertices correspond to the meridional boundary curves of S resp. S'. By a very tricky and subtle analysis of Γ and Γ' a so-called "Scharlemann-circle" is found which then yields a punctured lens space in the form $\underset{\alpha m+\beta \ell=\partial D^*}{V \cup D^*}$, $|\alpha| > 1$, in S^3. This is impossible since $\pi_1 S^3 = 1$.

4. Geometry of knot spaces.

Let V be a tubular neighbourhood of the trivial knot $K_0 \subset S^3$, and V' its complement. $V - K_0$ can be fibred by parallels of the simple closed curve $\alpha m + \beta \ell$, and the same holds for $V' - K_0'$, K_0' the core of V'. Putting back K_0 and K_0' into S^3 gives us the "Seifert fibrations" [SEI 33] of S^3 with K_0 and K_0' "exceptional" fibres of order $|\beta|$ and $|\alpha|$. 3-manifolds (compact) which can be fibred in this way (Seifert fibre spaces) with a (finite) number of exceptional fibres of arbitrary order were studied by Seifert and others [OVZ 67].

The links whose complements are Seifert fibre spaces are well known: Apart from a special case, the key-ring link, consisting of a trivial component and a number of meridional curves to it, such a link has as components any set of fibres of some $\alpha - \beta$-fibring of S^3, [BM]. Algebraically this class of links is characterized by the fact that the link group has a non-trivial centre.

Another type of fibring – transversal in a way to this – was discovered by Stallings: If and only If the commutator subgroup in the case of a knot is finitely generated, C is fibred over the circle S^1 – the fibres being Seifert surfaces of minimal genus spanning the knot. (In case of a link G' must be replaced by a larger subgroup, the kernel of a canonical homomorphism of the link group G onto Z.) One has $C = S \times I/\sim$, S = Seifert surfaces, and an identification

$$h(x, 0) = (x, 1)$$

by the holonomy $h : S \times 0 \to S \times 1$.

Of course, $S \times I$ carries two fibration, one with fibre S and one with fibre I. In $C = S \times I/\sim$ the second one leads to a codimension two foliation of C by lines or circles. Evidently it is a fibration by circles, if the holonomy h is of finite order $h^k = \text{id}$, in fact, then C is a Seifert fibre space and $\pi_1 C = G$ has a non trivial centre. The last property suffices for the fact: If $Z(G) \neq 1$, then G' is finitely generated and there is a Stallings fibration. Moreover a power of $h_\#$ is an inner automorphisms of G' and by a theorem of Nielsen h can be isotoped to be periodic.

It may be remarked at this juncture that knot complements always admit a codimension one foliation – generalizing the Stallings fibration. This was proved by Gabei [J. Diff. Geom. 18 (1983)]. Cantwell and Colon were able to give a geometric description of such "taut" foliations for special cases such as twist knots.

A completely different kind of geometry on a knot space was discovered by Riley [RIL 75]. He explicitly described a hyperbolic structure on certain (open) knot complements $S^3 - K$. W.Thurston then found a general theorem:

> **Theorem:** *The complement of a "simple" knot which is not a torus knot is a hyperbolic manifold of finite volume.*

A knot is "simple", if it is not a satellite, and a satellite is a knot K' contained in the tubular neighbourhood $U(K)$ of a non-trivial knot K such that K' is $\neq K$ and not already contained in a 3-cell $e^3 \subset U(K)$. A special case of a satellite is a composite knot $K_1 \# K_2$ as we shall presently see. Torus knots, for instance, are simple, likewise the four knot 4_1.

The hyperbolic structure of $S^3 - 4_1$ can be seen directly [THU 82]: Let an open unit 3-ball represent a Kleinian model of the hyperbolic space H^3. We place a double pyramid (two 3-simplices with a common 2-face) inside H^3 with the five vertices on the boundary of the 3-ball (the sphere of infinity of H^3). Now,

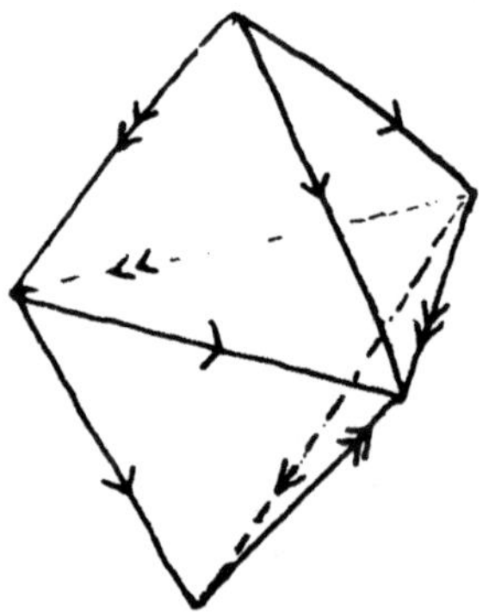

Fig.3

identify the 2-faces of the pyramid by non-Euclidean motions in such a way that equally arrowed edges coincide. It can be shown directly that the result is a 3-manifold homeomorphic to the complement $S^3 - 4_1$. It inherits a hyperbolic structure from H^3 which is the universal covering of $S^3 - 4_1$. The group of the four knot thus turns out to be a discrete subgroup of PSL(2, $\mathbb{C}$), the group of isometries of H^3, with a fundamental domain of finite volume.

By the so-called rigidity theorem of Mostow the hyperbolic structure is – up to isometries – uniquely determined by the link type. Several powerful invariants can be derived from the hyperbolic structure. Each component of the link determines a "cusp" of C, a neighbourhood of an ideal point of the form $T \times [0, 1)$, T a torus surface. It is most easily described in the Poincaré model of H^3 in the upper 3-space $\mathbb{R}^3$, $H^3 = \{(x, y, z) | z > 0\}$. Taking ∞ as the ideal point, the points $z \geq a > 0$ form a "horoball" at ∞, $z = a$ its "horosphere".

The horoball projects under the covering $H^3 \to C$ onto the cusp; the horosphere minus ∞ onto the torus T. Of course, the preimage of the cusp may consist of several horoballs. The pattern of horoballs – normalized in a certain fashion – gives rise to a set a very effective invariants. For instance, hyperbolic mutant knots are known to have complements of the same hyperbolic volume [RU 87], but can be told apart by their horoball patterns [AHW 91]. Another tool is a unique decomposition of a hyperbolic complement C into ideal simplices (or convex cells) such as in the case of the four knot. This makes it possible to determine the symmetries of the link completely.

These two geometric structures, Seifert fibring and hyperbolic metric, in a way govern link spaces. Johannson [JOH 79] Jaco and Shalen [JS79] proved a remarkable theorem:

There is a family of tori in the complement C of any K, disjoint and non-parallel which are incompressible in C and not boundary parallel (essential). These tori decompose C into pieces which are either Seifert fibre spaces or contain no essential tori (are atoroidal) and therefore, by Thurston's theorem, carry a hyperbolic structure. (A knot space is atoroidal if and only if the knot is not simple.) The important point is that a minimal system of tori satisfying these conditions is unique up to isotopy. As an example of the situation look at a composite knot $K = K_1 \# K_2$. Let S^2 be a separating 2-sphere meeting K in two points. $S^2 - U(K) = A$ is an "essential" annulus, and $\partial U(K) - S^2$ consists of two annuli A_1 and A_2 belonging to K_1 and

K_2 respectively. There are two essential tori $T_1 = A \cup A_1$ and $T_2 = A \cup A_2$ in $C = S^3 - U(K)$,

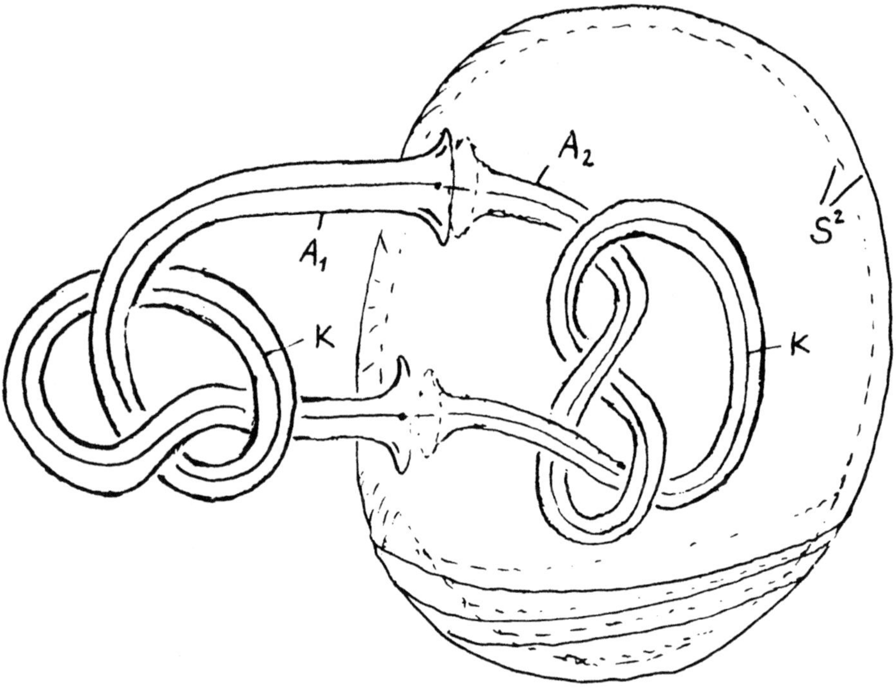

Fig.4

The geometrization of knot spaces has been effectively used to be obtain results in knot theory which had escaped prior attemps. An important tool fo hyperbolic pieces is Mortow's rigidity theorem which allows homeomorphisms to be replaced by hyperbolic isometries. This was employed for instance to determine symmetries of knots.

References

[**AHW91**] C. Adams, M. Hildebrand and J. Weeks, *Hyperbolic invariants of knots and links*, Trans Am. Math. Soc. 326, 1(1991), 1-56

[**BM70**] G. Burde and K. Murasugi, *Links and Seifert fibre spaces*, Duke Math. J. 37 (1970), 89-93

[**CGLS 87**] M. Culler, C. McA. Gordon, J. Luecke, and P.B. Shalen, *Dehn surgery on knots*, Annals of Math. 125 (1987), 237-300

[FOX 52] R.H.Fox, *On the complementary domains of a certain pair of inequivalent knots*, Indag. Math. 14 (1952), 37-40

[GL 89] C.McA. Gordon and J. Luecke, *Knots are determined by their complements*, Bull. Amer. Math. Soc. 20, No. 1(1989), 83-87

[JS 79] W.H.Jaco and P.B. Shalen, *Seifert fibre spaces in 3-manifolds*, Mem. Amer. Math. Soc. No.2 (1979).

[JOH 79] K. Johannson, *Homotopy equivalence of 3-manifolds with boundaries*, L.N. No 7761, (1979).

[OVZ 67] P. Orlik, E.Vogt, H. Zieschang, *Zur Topologie gefaserter dreidimensionaler Mannigfaltigkeiten*, Topology 6 (1967), 49-64.

[RIL 75] R. Riley, *Discrete parabolic representations of link groups*, Mathematika 22 (1975), 141-150.

[RU87] D. Ruberman, *Mutations and volumes of knots in S^3*, Invent. Math. 90 (1987), 189-215

[SEI 33] H. Seifert, *Topologie dreidimensionaler gefaserter Räume*, Acta Math. 60 (1933), 147-238.

[THU 82] W.P. Thurston, *The geometry and topology of 3-manifolds*, (preprint).

[WA 68] F. Waldhausen, *On irreducible 3-manifolds which are sufficiently large*, Ann. of Math. 87 (1968), 56-88.

[WHI 86] W. Whitten, *Knot complements and groups*, Topology 25 (1986).

KNOT GROUPS

Gerhard Burde
Fachbereich Mathematik
Johann Wolfgang Goethe-Universität
Postfach 11 19 32
D-6000 Frankfurt am Main / BRD

ABSTRACT. This lecture on classical knot and link groups concentrates – after some remarks on general properties – on representations of these groups and invariants derived from them. Metabelian and higher-step-metabelian representations are considered and homomorphisms into hyperbolic isometries.

Introduction.

The group of a knot K or link L is defined as the fundamental group of the complement C of K or L in S^3. What is generally known of these groups? The following properties are valid for any knot group G:

(1) G is finitely presentable
(2) $G/G' \simeq Z$, G' = commutator subgroup
(3) $G = \overline{\langle m \rangle}$ = normal closure of one element
(4) $H_2(G) = 0$

The last condition can be verified, if one knows that G has "defect one", this means that there is a presentation with n generators and $n-1$ relators. Thus these properties are easily established by looking at a Wirtinger presentation of G. Kervaire [KE 65] proved that (1)-(4) characterize knot groups in higher dimensions, $G = \pi_1(S^n - K^{n-2})$, $K^{n-2} \cong S^{n-2}$, for $n \geq 5$. For $n = 4$ some questions are still open – even after the affirmation of the topological Poincaré conjecture: There are groups satisfying Kervaire's conditions that are not knot groups and have defect > 1. On the other hand a 2-twist spin of a trefoil has a group with defect > 1, so that defect $= 1$ can not be substituted for $H_2 = 0$. In the classical case $n = 3$ a group satsifying the Kervaire conditions need not be a knot group. Examples were given of such groups which possess a non–symmetric Alexander polynome.

Recently S. Rosebrock [RO 91] constructed the following example:

$$G = \langle S_1, S_2, \ldots, S_8 | S_2^{-1}S_3^{-1}S_1S_3, S_3^{-1}S_4S_2S_4^{-1}, S_4^{-1}S_5^{-1}S_3S_5, \\ S_5^{-1}S_6S_4S_6^{-1}, S_6^{-1}S_7^{-1}S_5S_7, S_7^{-1}S_8S_6S_8^{-1}S_8^{-1}S_1^{-1}S_7S_1 \rangle.$$

M.E. Bozhüyük (ed.), Topics in Knot Theory, 25–31.

It satisfies (1)-(4), and it has the Alexander polynomial of a knot: $x^6 - 7x^5 + 20x^4 - 29x^3 + 20x^2 - 7x + 1$. This group is shown to be hyperbolic, hence, it is not a knot group. So, the symmetry of the Alexander polynomial as a fifth condition does not suffice to characterize classical knot groups.

A characterization in terms of presentations was given by Artin for link groups: let ζ be an n–string braid automorphism, then a group G is the group of a classical link, if and only if it admits a presentation

$$G = \langle S_1, \ldots, S_n | \zeta(S_1) = S_1, \ldots, \zeta(S_n) = S_n \rangle$$

Waldhausen proved [WA 68] the solvability of the word problem in link groups. Recently Sela [SEL 91] showed that knot groups even have a solvable conjugacy problem. The arguments appear to carry over to link groups. Some classes of link groups are quite well known – i.e. link groups with a non–trivial centre [BM 70]. But generally the groups are rather inaccessible – in the plain meaning of the word. The strength of the group as an invariant is great, so it is natural to look at homomorphic images of the group (representations) in the hope to find those more tractable and to extract from them calculable invariants.

§ 1 Metabelian representations.

Abelian images of a knot group are cyclic (property (2)) and carry no information on the individual knot. Metabelian ones are plentiful and have been studied since the beginning of knot theory.

Every metabelian image of a knot group G is a semidirect product $Z \ltimes A$, Z cyclic, A Abelian; the homomorphism $G \to Z \ltimes A$ factors through $G \to G/G'' \cong Z \ltimes H_1(C_\infty)$, or, in the case of a finite cyclic group, through $G \to Z_n \ltimes H_1(\overline{C}_n)$. Here C_∞ denotes the infinite cyclic covering and $\overline{C}_n$ the n–fold branched cyclic covering of C. As a module over $\mathbb{Z}Z$ the group $H_1(C_\infty)$ is known as the "Alexander module" M of G.

Though a single representation $\rho : G \to H$ may yield interesting information on G and the knot K, it is often desirable to obtain knowledge on the whole set of homomorphisms into a fixed group H. We call a representation trivial, if its image group is Abelian, and two representations ρ and ρ' equivalent, if they differ by an inner automorphism of H. The set $R(G, H)$ of equivalence classes of non-trivial homomorphisms $\rho : G \to H$, endowed with a suitable topology, is called a "representation space". As an example consider the group of orientation preserving similarities A of the complex plane $\mathbb{C}$. The elements of A are complex linear substitutions,

$$z \mapsto z' = az + b, \quad a \neq 0,\ b \in \mathbb{C}.$$

A representation $\rho : G \to A$ is determined by its values on a set $\{S_i\}$ of generators of G. We may assume $S_i \in \{m\}$, the conjugacy class in G of the meridian m. Put $\rho(S_i) : z \mapsto$

$a(z-b_i)+b_i$, $a \neq 1$. One may identify ρ with the set $\{a, b_1, b_2, \ldots, b_n\}$ of complex numbers. The representation space $R(G, A)$ consists of a 0-dimensional component for each root a of the Alexander polynomial $\Delta_1(t)$ which is not a root of the second Alexander polynomial $\Delta_2(t)$, and, in the genereal case, a component $\mathbb{C}^{k-1}$, if $\Delta_k(a) = 0$, $\Delta_{k+1}(a) \neq 0$.Thus there is a (partial) interpretation of Alexander polynomials via representation spaces.

The set $\{b_1, \ldots, b_n\}$, determined by $\rho : G \to A$, depends, of course, on the choice of the generators S_i. If K is periodic, one may select generators such that $\{b_1, \ldots, b_n\}$ shows the symmetry. Criteria for periodicity can derived from this [BU 78]. If, for instance, a knot allows a cyclic periodicity of order p, then, under certain assumptions concerning the $\Delta_k(t)$, one finds a set $\{b_1, \ldots, b_n\}$ with $b_\ell = \zeta b_j$ for some b_ℓ, b_j where ζ is a p-th root of unity. Since the b_i are in the splitting field of $\Delta_1(t)$ over $\mathbb{Q}$, so ζ must be.

Metabelian representations have been studied by Fox [FOX 70] and Hartley [HA 79]; they yield interesting invariants but cannot go beyond the Alexander module.

§ 2 Lifts of metabelian representations.

The metabelian representations of a knot group can be interpreted as lifts of the trivial (Abelian) ones.

$$\begin{array}{ccc} G & \longrightarrow & Z_n \ltimes H_1(\overline{C}_n) \\ & \searrow & \downarrow \\ & & Z_n \end{array}$$

One looks at the cyclic covering C_n determined by $\ker(G \to Z_n)$, and constructs a homomorphic image of G as an extension of $H_1(C_n)$ (or $H_1(\overline{C}_n)$ by Z_n using the operation induced by the covering transformations. This procedure can be iterated. The advantage gained over the metabelian representations by considering 3-step metabelian representations is the fact that the longitudes no longer are necessarily trivialized. They yield valuable "linkage invariants" which are very effective in detecting cheirality in knots, [REI 29], [BU 70], [HM 77]. They are also applicable in the much harder question of invertibility [HA 83]. We again illustrate the lifting process in a geometric example: If there is a prime p dividing $|H_1(\overline{C}_2)|$, then there is a dihedral (metacyclic) representation $\beta_p : G \to Z_2 \ltimes Z_p = D_{2p}$. Under certain conditions $R(G; D_{2p})$ consists of just one point. The elements of $\{m\}$ are mapped onto elements of order two in D_{2p}. We may represent them as reflections in lines of a Euclidian plane E^2 and regard β_p as a point of $R(G, B)$, B the group of motions of E^2. Then there exists a unique lift $\rho_p \in R(G, B)$ of β_p mapping the elements of $\{m\}$ on proper glide reflections. $\rho_p(\ell)$ will be a translation. ρ_p can be used to show that the complements of

$K\#K$ and $K\#K^*$ are not homeomorphic for a certain class of cheiral knots K, as pointed out in [Bu 92].

The knot invariants obtained from such representations by looking at the images of the longitudes can be interpreted as "linkage invariants": linking numbers of links covering the knot in certain coverings. In this contexts also Milnor's $\overline{\mu}$-invariants can be explained. They are linkage invariants of appropriate nilpotent covering spaces, [Mu 84].

§ 3 Hyperbolic isometries.

Metabelian representations – and consequently their lifts – do not always exist. Indeed, the necessary and sufficient condition for the occurrence of representations of this kind is the non-triviality of the Alexander polynomial, or equivalently, the fact that the commutator subgroup G' is not perfect $G' \neq G''$. Seifert [SE 34] found the first knot of this type, a (non-alternating) pretzel knot, and he used a representation of its group into the group of isometries of the hyperbolic plane to show that the knot itself was not trivial.

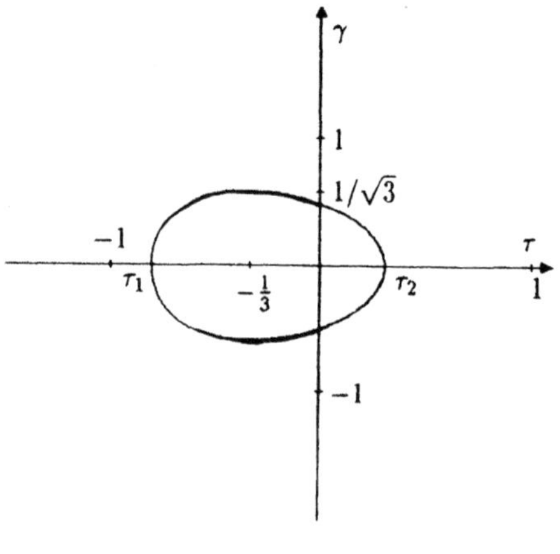

Fig. 1

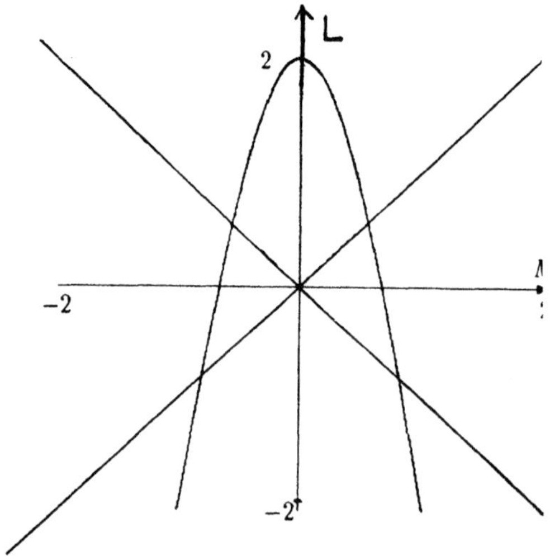

Fig. 2

Representations into this group had been used before by Reidemeister [REI 32] to classify alternating pretzel knots. Later Trotter used these representations to prove the existence of non-invertible knots [TR 64]. The group of the eleven-crossing Kinoshita-Terasaka-knot can be mapped onto groups of type $PSL(2,p)$, p a prime, as was shown in [MP 67]. This leads us to the group $PSL(2,\mathbb{C})$, the isometries of hyperbolic 3-space H^3, a group of fundamental significance in knot theory. R.Riley was the first to discover a hyperbolic structure on the complement of certain knots and hence faithful representation of the corresponding knot groups into $PSL(2,\mathbb{C})$. The general hyperbolization theorem for simple non-torus knots is due to W.Thurston. So representation spaces $R(G,\ PSL(2,\mathbb{C}))$ and $R(G,SL(2,\mathbb{C}))$ have been studied. (Every homomorphism $\rho : G \to PSL(2,\mathbb{C})$ of a knot group is known to lift to a homomorphism $\tilde{\rho} : G \to SL(2,\mathbb{C})$). Important papers on these topics are [CS 83], [CGLS 87] and [CCGLS 91]. $R(G,SU(2))$ was investigated explicitly for G the group of a torus knot or 2-bridge knot, [BU 90], [HEU 92]. The representation spaces could be completely described as plane real algebraic curves in special cases. As an example Fig. 1 shows $R(G,SU(2))$ for the four knot. Its group is generated by two meridional generators S and T. $\rho \in R(G,SU(2))$ is determined by two rotations $\rho(S)$ and $\rho(T)$ of 3-space through φ with an angle $\psi \neq 0$ between the axes. Fig. 1 shows the representation space in a (τ,γ)–coordinate plane, $\gamma = \operatorname{ctg}\frac{\varphi}{2}$, $\tau = \cos\psi$. The points $(\tau_i,0)$, $i = 1,2$, denote two dihedral representations, the two extreme points $(-\frac{1}{3},\pm\frac{1}{\sqrt{3}})$ correspond to tetrahedral representations. On the four arcs between dihedral and tetrahedral representations there is one representation satisfying the additional relator $m\ell$ (resp. $m\ell^{-1}$), that is, a (non-trivial) representation of $\pi_1(K(1))$ (resp. $\pi_1(K(-1))$).

This is closely connected with the fact that Casson's invariant of the four knot is -1.

There are other possibilities to extract information on knots and knot spaces from representation spaces $R(G,SL(2,\mathbb{C}))$. In [CS 83] incompressible surfaces (non-boundary-parallel) are constructed in a knot space using representation spaces, or rather spaces of characters of representations. One defines a map $\chi_g : R(G,SL(2,\mathbb{C})) \to \mathbb{C}$, $\chi_g(\rho) = \operatorname{tr}\rho(g)$, $g \in G$, $\rho \in R(G,SL(2,\mathbb{C}))$ and tr the trace. For the pair (m,ℓ) there is a well-defined map

$$\xi : \ R(G,SL(2,\mathbb{C})) \to \mathbb{C}^2$$

$$\rho \mapsto (\ \operatorname{tr}\rho(m),\ \ \operatorname{tr}\rho(\ell)).$$

The "curve-components" of $\operatorname{im}\xi \subset \mathbb{C}^2$ constitute the "character curve" which can be regarded as the zeroes of an invariant polynomial $A(M,L)$, $M = \operatorname{tr}\rho(m)$, $L = \operatorname{tr}\rho(\ell)$. This polynomial is known to be non-trivial for any hyperbolic knot [CGLS 87] and also for every torus knot [HEU 92]. Its computation, though, is rather difficult [CCGLS 91]. One may restrict oneself to $R(G,SU(2))$ and consider a plane real algebraic character curve

$$\xi : R(G,SL(2,\mathbb{C})) \to \mathbb{R}^2 .$$

As an example Fig. 2 shows this curve for the four knot. The polynomial is $A(M,L) = L - 2 - M^2(M^2 - 5)$. The intersections of this curve with diagonal lines correspond to the ± 1–Dehn surgeries on the four knot.

The study of the geometry of the space $R(G(\alpha,\beta),SU(2))$ for two bridge knots (α,β) implies in its first step the question of the reducibility of these spaces. Contributions to this question have been made in [BU 90], [RI 91],[HEU 92] and by Ohtsuki. In some cases the curves are known to be irreducible and in others they are known to split. Also some results have been obtained concerning the genus of the curves.

I have concentrated in this survey on representations of knot and link groups which allow in some way a systematic approach – at least for a class of knots or links. In these days of powerful computers information on a specific group can, of course, be obtained by searching for homomorphisms of the group into, say, permutation groups S_n for such n the computer can manage. This has, with good reason, been done to establish knot tables and to provide material for examples.

References

[BM 70] G. Burde and K. Murasugi, *Links and Seifert fibre spaces*, Duke Math. J. 37 (1970), 89-93

[BU 70] G. Burde, *Darstellungen von Knotengruppen und eine Knoteninvariante*, Abh. Math. Sem. Univ. Hamb. 35 (1970), 107-120

[BU 78] G. Burde, *Ueber periodische Knoten*, Arch. Math. 30 (1978), 487-492

[BU 90] G. Burde, SU(2)–*representation spaces for two-bridge knot groups*, Math. Ann. 288 (1990), 103-119

[BU 92] G. Burde, *Knots and knot spaces*, this volume

[CCGLS 91] D. Cooper, M. Culler, H. Gillet, D.D. Long and P.B. Shalen, *Plane curves associated to character varieties of 3-manifolds*, preprint

[CGLS 87], M. Culler, McA. Gordon, J. Luecke and P.B. Shalen, *Dehn surgery on knots*, Ann. of Math. 125 (1987) 237-300

[CS 83] M. Culler, and P.B. Shalen, *Varieties of group representations and splittings of 3-manifolds*, Annals of Mathematics, 117 (1983), 109-146

[HA 79] R. Hartley, *Metabelian representations of knot groups*, Pacific J. Math. 82 (1979), 93-104

[HA 83] R. Hartley, *Lifting group homomorphisms*, Pacific J. Math. 105 (1983), 311-320

[HM 77] R. Hartley and K. Murasugi, *Covering linkage invariants*, Canad. J. Math. 29 (1977), 1312-1339

[HEU 92] M. Heusener, *Darstellungsraeume von Knotengruppen*, Dissertation Frankfurt/Main (1992)

[KE 65] M. Kervaire, *Les noeds de dimensions supérieures*, Bull. Soc. Math. France 93 (1965), 225-271

[MP 67] W. Magnus and A. Peluso, *On knot groups*, Commun. Pure Appl. Math. 20 (1967), 749-770

[MU 84] K. Murasugi, *Nilpotent coverings of links and Milnor's invariant*, Proc. Sussex Conf. Low-dim.Top. (1982)

[REI 29] K. Reidemeister, *Knoten und Verkettungen*, Math. Z. 29 (1929), 713-729

[REI 32] K. Reidemeister, *Knotentheorie*, Ergebn. Math. Grenzgeb. Bd. 1 (1932)

[RI 91] R. Riley, *Algebra for Heckoid groups*, preprint (1991)

[RO 91] S. Rosebrock, *On the realization of Wirtinger presentations as knot groups*, preprint J.W.Goethe-Universitaet Frankfurt/Main

[SE 34] H. Seifert, *Ueber das Geschlecht von Knoten*, Math. Ann. 110 (1934), 571-592

[SEL 91] Z. Sela, *The conjugacy problem for knot groups*, preprint

[TR 64] H.F. Trotter, *Non-invertible knots exist*, Topology 2 (1964), 341-358

[WA 68] F. Waldhausen, *The word problem in fundamental groups of sufficiently large irreducible 3-manifolds*, Ann. of Math. 88 (1968), 272-280

AN INTRODUCTION TO SPECIES AND THE RACK SPACE

Roger Fenn
Math Dept Sussex University Brighton BN1 9QH UK
Colin Rourke
Brian Sanderson
Math Dept Warwick University Coventry CV4 7AL UK

ABSTRACT. Racks were introduced in [FR]. In this paper we define a natural category like object, called a species.* A particularly important species is associated with a rack. A species has a nerve, analogous to the nerve of a category, and the nerve of the rack species yields a space associated to the rack which classifies link diagrams labelled by the rack. We compute the second homotopy group of this space in the case of a classical rack. This is a free abelian group of rank the number of non-trivial maximal irreducible sublinks of the link.

Introduction

In this paper we continue the investigation into racks which was started in [FR]. Here we shall introduce a natural classifying space associated to a rack, which we call the **rack space**. Further results on this space will be found in [FRS].

This space is the realisation of a semi-cubical complex, and indeed the natural structure is cubical, rather than simplicial as is usually the case for classifying spaces. Investigating the formal structure of this space and its relationship with other classifying spaces has led us to formulate the concept of a **species**. A species is analogous to a category; it has vertices and edges (analogous to objects and morphisms in a category), but instead of composition (which can be regarded as given by preferred triangles of morphisms) it has *preferred squares* of edges. There is a concept of **mutation** between species, analogous to functors between categories, and a natural semi-cubical **nerve** of a species, analogous to the usual semi-simplicial nerve

* We use the term species here in a topological sense. It should not be confused with the same term used in combinatorics.

M.E. Bozhüyük (ed.), Topics in Knot Theory, 33–55.

of a category. A rack X defines a species $\mathcal{S}(X)$ with a single vertex and with X the set of edges. The preferred squares are of the following type.

$$\begin{array}{ccc} 0 & \xrightarrow{a^b} & 0 \\ \uparrow b & \circlearrowleft & \uparrow b \\ 0 & \xrightarrow{a} & 0 \end{array}$$

The **rack space** BX of X is the realisation of the nerve $N\mathcal{S}(X)$ of $\mathcal{S}(X)$. The 1–cells of BX are elements of X and 2–cells are the preferred squares.

The preferred square given above can be pictured as part of a link diagram with arcs labelled by a, b and a^b.

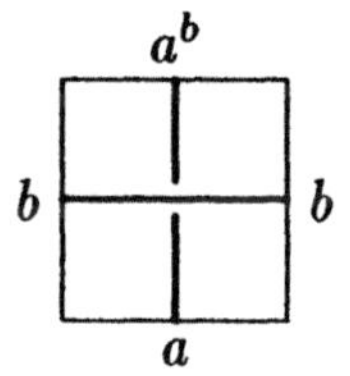

Diagram of a typical 2–cell of the rack space

Thus, given a map of a surface M in BX, we can use transversality [BRS chapter 7] to construct a link diagram in M labelled by elements of X. Similar considerations apply to higher dimensional manifolds, and it follows that BX is the classifying space for (cobordism classes of) link diagrams labelled by X.

If T is the trivial rack with one element then BT can be identified with $\Omega(S^2)$. There is a map $\Omega(S^2) \to BO$ where O is the infinite orthogonal group and it follows that for any rack X there is a map $BX \to BO$. This leads to the concept of an X–oriented manifold, namely a manifold with a lifting of the stable normal bundle over $BX \to BO$, and a consequent generalised (co)-homology theory. There is a geometric interpretation of this theory which is close to the geometric interpretation of Mahowald oriented manifolds given in [S].

There are also connections with classifying spaces for groups and crossed modules: BX naturally maps to $BAs(X)$ (the classifying space of the associated group of X) and there is an extension to give a classifying space for a G–rack, which has a natural map to the loop space of the Brown–Higgins classifying space of the associated crossed module [FR section 2 and BH].

The rack space can be used to define invariants of knots and links. There are two useful ways in which this can be done: Firstly, we can consider the rack space of the fundamental rack of the link. Any invariant of this space (for example a bordism group) is then an invariant of the original link. There is usually a geometric interpretation of such invariants: for example the bordism group $\Omega_n(BX)$ can be interpreted, by transversality, as cobordism classes of codimension one diagrams in n-manifolds labelled by X (or equivalently having a homomorphism of the fundamental rack to X). Secondly we can fix a particular rack X (for example the trivial rack on n elements) and then consider, for a particular link L in say S^3, obstructions to the existence of a homomorphism of the fundamental rack in X (i.e. obstructions to representing L by a particular map $S^2 \to BX$). Having obtained a particular representation, we can then consider its cobordism class i.e. the element of $\pi_2(BX)$ determined. In the case of the trivial rack on n elements, the representation divides the link into n disjoint sublinks and the cobordism class yield the linking numbers of these sublinks with each other.

The bulk of the results outlined above will be given in a future paper by the authors [FRS]. In this paper we shall give the basic definitions and calculate $\pi_2(BX)$ when X is the rack of a classical link.

1. Definitions and Examples of Racks

In this section we give a brief introduction to the theory of racks. More details may be found in [FR].

We consider sets X with a binary operation which we shall write exponentially

$$(a, b) \mapsto a^b.$$

There are several reasons for writing the operation exponentially.

(1) The operation is unbalanced and should be thought of as an action, i.e. think of a^b as meaning *the result of b acting or operating on a.*
(2) In group contexts exponention signifies conjugation. A group with conjugation is one of the principal examples of a rack — indeed this was the source for one strand of the earlier work on racks [CW]. A rack is an algebraic object which has just *some* of the properties of a group with conjugacy as the operation.
(3) Finally, and most conveniently, exponential notation allows brackets to be dispensed with, because there are standard conventions for association with exponents. In particular

$$a^{bc} \text{ means } (a^b)^c \text{ and } a^{b^c} \text{ means } a^{(b^c)}.$$

1.1 Definition *Racks*
A **rack** is a non-empty set X with a binary operation satisfying the following two axioms:

Axiom 1 Given $a, b \in X$ there is a unique $c \in X$ such that $a = c^b$.

Axiom 2 Given $a, b, c \in X$ the formula

$$a^{bc} = a^{cb^c}$$

holds. We call this formula the **rack identity (first form)**.

Several consequences flow from these axioms.

The first axiom implies that, for each $b \in X$ the function f_b given by $f_b(x) = x^b$ is a bijection of X to itself, and this fits with the idea that the operation is a (right) action of X on itself.

We shall write $a^{\bar{b}} = f_b^{-1}(a)$ for the element c given by the axiom, but notice that $a^{\bar{b}}$ is *a single symbol for an element of* X it is not suggested that $\bar{b}$ is itself an element of X; however the notation is suggestive (and intended to be) because now $a^{b\bar{b}} = a^{\bar{b}b} = a$ for all $a, b \in X$. Thus if we identify $\bar{b}$ with b^{-1} then we can give a meaning to any expression of the form x^w where $w = w(a, b, \ldots)$ is a word in $F(X)$ the free group on X, namely the result of repeatedly acting on x by $f_a, f_a^{-1}, f_b, f_b^{-1}$ etc. The word w is again *not to be regarded as an element of* X, but as an **operator** on X. Shortly, we shall formalise this by introducing the **operator group**.

The rack identity is a right self-distributive law as can be seen if we temporarily use the notation $a \cdot b$ for a^b:

$$(a \cdot b) \cdot c = (a \cdot c) \cdot (b \cdot c).$$

Thus both axioms are equivalent to the statement that *right multiplication is an automorphism.* The rack identity can be restated in more elegant and mnemonic form if we use the notation introduced above. Substituting $d = a^c$ in the rack identity and then changing d back to a gives the alternative form:

Axiom 2′ Given $a, b, c \in X$ the formula

$$a^{b^c} = a^{\bar{c}bc}$$

holds. This is the **rack identity (second form)**.

In other words b^c operates like $\bar{c}bc$, which makes clear the connection between the rack operation and conjugacy in a group.

The Operator and Associated Groups

In expressions such as $a^{b\bar{c}}$ we refer to a as being at **primary** level and $b, \bar{c}$ as at **operator** level. The second form of the rack identity makes clear that we do not need any "higher" level operators. Expressions involving repeated operations can always be resolved into one of the form a^w where $a \in X$ is at the primary level and w, lying in the free group $F(X)$ on X, is at the operator level.

In this way we have an action by the group $F(X)$ on X. In general if G acts on X, written $(a, g) \mapsto a \cdot g$ and if $\partial : X \to G$ is a map satisfying

$\partial(a\cdot g) = g^{-1}\partial(a)g$ then X has the structure of a rack given by $a^b := a\cdot\partial(b)$. In many situations this is the most convenient method of describing the rack operation. The similarity with crossed modules should be clear.

There is a useful concept of a rack with an *explicit* group action: We define a G–**rack** to be a set X with a G-action (which we write $(x,g) \mapsto x\cdot g \in X$, $g \in G$) and a function $\partial : X \to G$ satisfying the G–**rack identity**

$$\partial(a \cdot g) = g^{-1}\partial a g \quad \text{for all} \quad a \in X, g \in G.$$

If we interpret the operation of a rack as conjugation (i.e. read a^w as $w^{-1}aw$ then we obtain a group $As(X)$ called the **associated group**. More precisely let $As(X) = F(X)/K$ where K is the normal subgroup of $F(X)$ generated by the words $a^b b^{-1}a^{-1}b$ where $a, b \in X$. So $As(X)$ is the biggest quotient of $F(X)$ with the property that, when considered as a rack via conjugation, the natural map from $F(X)$ to $As(X)$ is a rack homomorphism.

Given a rack homomorphism $f : X \to Y$, then there is an induced group homomorphism $f_\sharp : As(X) \to As(Y)$; thus we have an **associated group functor** As from the category of racks to the category of groups.

Note that a G–rack is a plain rack if we ignore or forget about the explicit G-action. Conversely, there is a natural way to regard a plain rack as a G–rack by taking $G = As(X)$ (the associated group) with ∂ the natural map. Thus we can regard the category of racks as a subcategory of the category of G–racks and then the forgetful functor is a retraction of the larger category onto the smaller.

To make operators more precise we define **operator equivalence** by:

$$w \equiv z \iff a^w = a^z \text{ for all } a \in X$$

where $w, z \in F(X)$.

The equivalence classes form the **Operator Group** $Op(X)$ which could also be defined as $F(X)/N$ where N is the normal subgroup

$$N = \{w \in F(X) \mid w \equiv 1\}.$$

In general the operator group is a quotient of the associated group, cf. example 5 below.

Examples of operator equivalence
Since $b^{a^a} = b^{\bar{a}aa}$ (by the rack identity) $= b^a$ for all $a, b \in X$. We have

$$a^a \equiv a \quad \text{for all} \quad a \in X.$$

More generally if a^{a^n} means $a^{aa\ldots a}$ (n repeats) then $a \equiv a^{a^n}$.

In terms of operator equivalence, the rack identity can again be restated:

Axiom 2″ Given $a, b \in X$ we have

$$a^b \equiv \bar{b}ab.$$

This is the **rack identity (third form)**.

Examples of Racks

Example 1 *The Conjugation Rack*
Let G be a group, then conjugation in G i.e. $g^h := h^{-1}gh$ defines a rack operation on G. This makes G into the **conjugation rack.**

The operator group in this rack is the group of inner automorphisms of G and the orbits are the conjugacy classes. Given $g, h \in G$ then $g \equiv h$ if and only if gh^{-1} is in the centre of G.

Example 2 *The Dihedral Rack*
Any union of conjugacy classes in a group forms a rack with conjugation as operation. In particular let R_n be the set of reflections in the dihedral group D_{2n} of order $2n$ (which we regard as the symmetry group of the regular $n-$gon). Then R_n forms a rack of order n, with operator group D_{2n}, called the **dihedral** rack of order n.

Example 3 *The Reflection Rack*
Let P, Q be points of the plane and define P^Q to be P reflected in Q (i.e. $2Q - P$ in vector notation).

It is elementary to show that this is a rack operation. This example can be generalised by replacing the plane by any geometry with point symmetries satisfying certain general conditions (see Joyce [J] for details). Examples include the natural geometries of S^n and $\mathbf{RP}^n$. Interesting subracks of these latter racks are given by the action of Coxeter groups on root systems, cf. example 7 below.

Example 4 *The Alexander Rack*
Let Λ be the ring of Laurent polynomials $\mathbf{Z}[t, t^{-1}]$ in the variable t. Any $\Lambda-$module M has the structure of a rack with the rule $a^b := ta + (1-t)b$.

For example, letting M be the plane and the action of t multiplication by -1, yields the reflection rack of example 4.

The Quandle Condition

All the above examples have satisfied the identity

$$a^a = a \text{ for all } a \in X,$$

which we call the **quandle** condition We shall call a rack satisfying the quandle condition a **quandle rack** or **quandle.** The term quandle is due to Joyce [J].

Example 5 *The Cyclic Rack*
Here is a finite rack which is not a quandle:

The **cyclic rack of order** n, consists of the residues modulo n, $C_n = \{0, 1, 2, \ldots n-1\}$ with operation $i^j := i + 1 \bmod n$ for all $i, j \in C_n$.

This is an example of a rack for which the operator group ($\mathbf{Z}_n$ in this case) is a definite quotient of the associated group ($\mathbf{Z}$ in this case).

Example 6 *The Free Rack*
The **free rack** $\mathrm{FR}(S)$ on a given set S is defined, as a set, to be $S \times F(S)$. We write the pair (a, w) as a^w. Thus

$$\mathrm{FR}(S) = \{a^w \mid a \in S, w \in F(S)\}$$

and the rack operation is defined by

$$(a^w)^{(b^z)} = a^{w\bar{z}bz}.$$

Axiom 1 of definition 1.1 is easy to check whilst for the rack identity notice

$$(a^w)^{b^z} = a^{w\bar{z}bz} \equiv \overline{w\bar{z}bz}\ a\ w\bar{z}bz = \bar{z}\bar{b}z\ \bar{w}aw\ \bar{z}bz \equiv \overline{b^z}a^w b^z$$

which is the third form of the identity (axiom $2''$).

The operator group is $F(S)$ whilst the set of orbits is in bijective correspondence with the elements of S and all stabilizers are trivial.

Example 7. *Coxeter racks.*
Let $(\ ,\ )$ be a symmetric bilinear form on $\mathbf{R}^n$. Then, if S is the subset of $\mathbf{R}^n$ consisting of vectors $\mathbf{v}$ satisfying $\mathbf{v}.\mathbf{v} \neq 0$, there is a rack structure defined on S by the formula

$$\mathbf{u}^{\mathbf{v}} = \mathbf{u} - \frac{2(\mathbf{u}, \mathbf{v})}{(\mathbf{v}, \mathbf{v})}\mathbf{v}.$$

Geometrically, this is the result of reflecting $\mathbf{u}$ in the hyperplane $\{\mathbf{w} | (\mathbf{w}, \mathbf{v}) = 0\}$.

If we multiply the right-hand side of the above formula by -1, then the result geometrically is reflection in the line containing $\mathbf{v}$. In this case the formula

$$\mathbf{u}^{\mathbf{v}} = \frac{2(\mathbf{u}, \mathbf{v})}{(\mathbf{v}, \mathbf{v})}\mathbf{v} - \mathbf{u}.$$

defines a quandle structure on S.

Now a **root system** is precisely a finite subrack of S which is closed under multiplication by -1 (i.e. closed under *both* rack operations), and then the operator group is the corresponding Coxeter group.

Example 8. *The Fundamental Rack of a Link.*
This is the most important rack of all and is the *raison d'être* of the whole theory. A complete description with proofs can be found in [FR]. A (codimension two) **link** is defined to be a codimension two embedding $L : M \subset Q$ of one manifold in another. We shall assume that the embedding is proper at the boundary if necessary, that M is non-empty, that Q is connected and that M is **transversely oriented** in Q. I.e. we assume that each normal disc to M in Q has an orientation which is locally and globally coherent.

The link is said to be **framed** if there is a cross section (called a **framing**) $\lambda : M \to \partial N(M)$ of the normal disk bundle. Denote by M^+ the image of

M under λ. We call M^+ the **parallel** manifold to M. For simplicity we shall always assume that the link is framed.

We consider homotopy classes Γ of paths in $Q_0 = \text{closure}(Q - N(M))$ from a point in M^+ to a base point. During the homotopy the final point of the path at the base point is kept fixed and the initial point is allowed to wander at will on M^+.

The set Γ has an action of the fundamental group of Q_0 defined as follows: let γ be a loop in Q_0 representing an element g of the fundamental group. If $a \in X$ is represented by the path α define $a \cdot g$ to be the class of the composite path $\alpha \cdot \gamma$.

We can use this action to define a rack structure on Γ. Let $p \in M^+$ be a point on the framing image. Then p lies on a unique meridian circle of the normal circle bundle. Let m_p be the loop based at p which follows round the meridian in a positive direction. Let $a, b \in X$ be represented by the paths α, β respectively. Let $\partial(b)$ be the element of the fundamental group determined by the loop $\overline{\beta} \cdot m_\beta \cdot \beta$. (Here $\overline{\beta}$ represents the inverse path to β and m_β is an abbreviation for $m_{\beta(0)}$ the meridian at the initial point of β.) The **fundamental rack** of the framed link L is defined to be the set $\Gamma = \Gamma(L)$ of homotopy classes of paths as above with operation

$$a^b := a \cdot \partial(b) = [\alpha \cdot \overline{\beta} \cdot m_\beta \cdot \beta].$$

(If L were an unframed link then we could define its **fundamental quandle.** The definition is very similar. Let $\Gamma_Q = \Gamma_Q(L)$ be the set of homotopy classes of paths from the boundary of the regular neighbourhood to the base point where the initial point is allowed to wander during the course of the homotopy over the *whole* boundary. The rack structure on Γ_Q is similar to that defined on Γ.)

1.2 Proposition *The fundamental rack of a link satisfies the axioms of a rack.* □

Note that if G denotes the fundamental group $\pi_1(Q_0)$ then the set Γ is in fact a G–rack.

1.3 Proposition *The associated group of the fundamental rack $\Gamma(L)$ of a link L can be naturally identified with the fundamental group $\pi_1(Q - L)$ provided Q is simply connected.* □

Orbits and stabilizers

Since a rack is a set X with an action of $F(X)$ (or its quotients $Op(X)$, $As(X)$) we can use all the language of group actions in the context of racks. In particular X splits into disjoint **orbits** and each element has a **stabilizer** (in $F(X)$ or $Op(X)$ or $As(X)$) associated with it. The orbits in the fundamental G–rack of a link are in bijective correspondence with the components of the link and the next lemma identifies the corresponding stabilizers.

An element of the fundamental group represented by a loop of the form $\overline{\alpha} \cdot \gamma \cdot \alpha$ where γ lies in $\partial N(M)$ and α represents the element $a \in \Gamma$ is called a-**peripheral.** The set of a-peripheral elements forms the a-**peripheral subgroup.** If γ lies in the subset M^+ then the class of $\overline{\alpha} \cdot \gamma \cdot \alpha$ is called a-**longitudinal** and the set of a-longitudinal elements forms the a-**longitudinal** subgroup.

1.4 Lemma *With the notation above, the stabilizer of a in the fundamental group is the a-longitudinal subgroup.* □

2. Species

In this section we consider the notion of a **species.** This is a mathematical object loosely analogous to a category. A species is a directed graph, in other words it has vertices and directed edges in analogy to the objects and morphisms of a category, but instead of composition (which can be regarded as given by preferred *triangles* of morphisms) it has preferred *squares* of edges. However a species in its most primitive form is a much more basic notion than a category and so species are more numerous.

Definition An **oriented square** in a directed graph is a diagram of edges a, b, c, d which can be represented in two ways:

$$\begin{array}{ccc} C & \xrightarrow{c} & D \\ \uparrow b & \circlearrowleft & \uparrow d \\ A & \xrightarrow{a} & B \end{array} \qquad \text{or} \qquad \begin{array}{ccc} B & \xrightarrow{d} & D \\ \uparrow a & \circlearrowright & \uparrow c \\ A & \xrightarrow{b} & C \end{array}$$

The orientation being represented by the symbol $\circlearrowright$ or $\circlearrowleft$.

We shall sometimes refer to the edge a in *either* of the above repesentations as the **base** of the square.

Definition: A **species** in its most primitive form consists of:

S1. A directed graph Γ,

S2. A collection of oriented squares in Γ called **preferred squares.**

In addition a species may satisfy any or all of the following extra axioms:

S3. *Identity Squares:* Any loop (i.e. edge with the same start and end-point) in Γ may be designated as an **identity loop.** However we will never allow two identity loops based at the same vertex. We shall usually use the notation e_A for the identity loop at the vertex A. If identity loops exist then squares such as those illustrated below are preferred.

$$\begin{array}{ccc} A & \overset{a}{\rightarrow} & B \\ \uparrow e_A & \circlearrowleft & \uparrow e_B \\ A & \overset{a}{\rightarrow} & B \end{array} \qquad\qquad \begin{array}{ccc} B & \overset{e_B}{\rightarrow} & B \\ \uparrow a & \circlearrowleft & \uparrow a \\ A & \overset{e_A}{\rightarrow} & A \end{array}$$

Definition **A species with identities** is a species equipped with identity loops at *each* vertex.

S4. *Composition laws:*

Suppose that Γ is in fact a category, so that edges (morphisms) can be composed, then there are two composition laws which may be satisfied by a species:

Horizontal Composition: Preferred squares a, b, c, d and a', b', c', d' with $d = b'$ may be composed horizontally to form the preferred square aa', b, cc', d' with base aa'.

Vertical Composition: Preferred squares a, b, c, d and a', b', c', d' with $c = a'$ may be composed vertically to form the preferred square a, bb', c', dd' with base a.

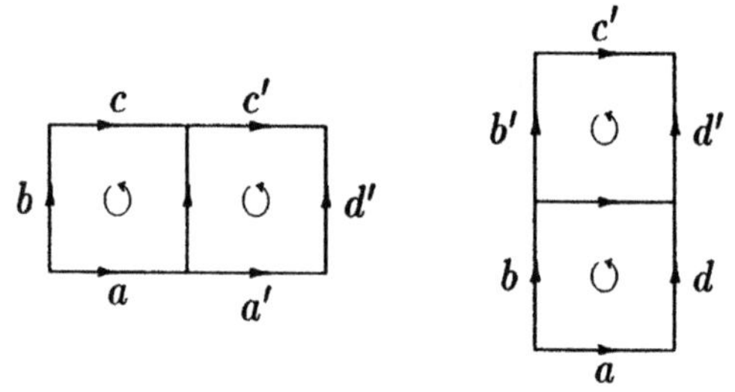

Composing Squares

S5. *The Vertebrate Laws:*

V1 Given edges $a: A \to B$ and $b: A \to C$ then there are unique c, d, D so that

$$\begin{array}{ccc} C & \overset{c}{\rightarrow} & D \\ \uparrow b & \circlearrowleft & \uparrow d \\ A & \overset{a}{\rightarrow} & B \end{array}$$

is preferred. Notice that the other edges are determined by two binary operations $c = a^b$ and $d = b_a$.

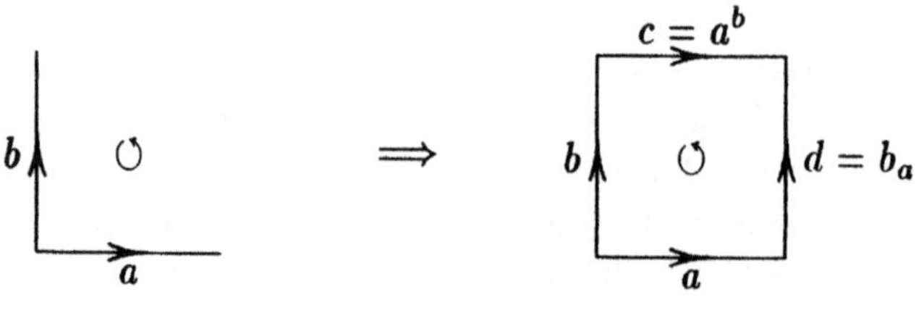

V2 In the following diagram, if the three squares centre, south and west are given and preferred then the diagram can be completed as shown so that the squares; outside, east and north are preferred. The outside square has the anticlockwise orientation.

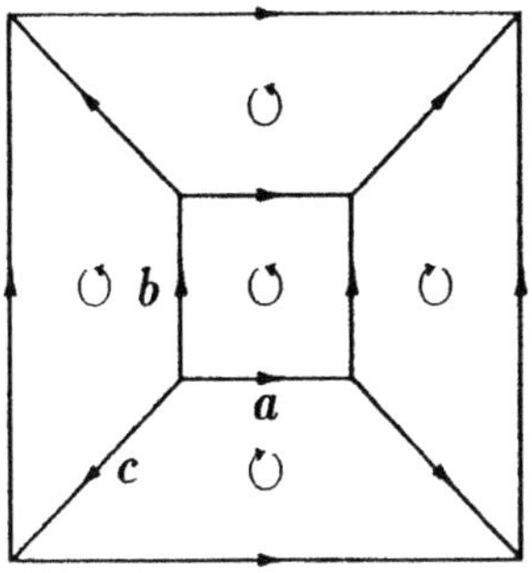

Given V1, an equivalent statement is that the edges a, b, c determine the entire diagram of preferred squares.

Definition A species is said to be **vertebrate** if it satisfies both vertebrate laws.

In a vertebrate species the *skeleton* of a square or cube (i.e. the edges a, b or a, b, c) has a unique completion to form the final square or cube, and this is why we have chosen this terminology.

Remark The two vertebrate axioms are loosely analogous to the composition and associativity laws of a category. The analogy can be sharpened by replacing our chosen skeleta, which are co-original edges (morphisms), by consecutive edges, as in the category axioms. This leads to an alternative theory of vertebrate species which, in the key example of racks (see example 4 below), coincides with theory given here. With some misgivings we have decided not to pursue this alternative approach. This is because the formulae for the boundary maps in the nerve (see section 3) are considerably simpler in the co-original approach.

2.1 Lemma *The binary operations of a vertebrate species satisfy the axioms:*

(a) $a^{bc_b} = a^{cb^c}$

(b) $a_{bc_b} = a_{cb^c}$

(c) ${a_b}^{c_b} = {a^c}_{b^c}$.

Proof We check these equalities by looking at the coincidental labelling on the following diagram.

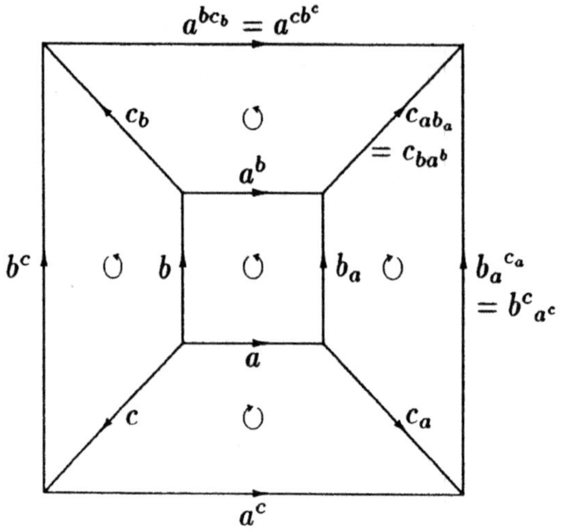

□

Remark If one of the operations is trivial (for example if $a_b = a$ for all a), then these laws reduce to the rack law for the other operation. Thus the three laws define a natural algebraic object which might be called a **birack**. This object, like the rack itself, has an application to links. A link diagram has a fundamental birack obtained by labelling arcs, and using both operations at crossovers. One operation determines the change of label on the understring and the other on the overstring. We shall explore these ideas further in a later paper.

Examples

1. Any category C determines a directed graph $\Gamma(C)$ with vertices the objects of C and edges the morphisms of C. Let $Spec(C)$ be the species with underlying graph $\Gamma(C)$, whose preferred squares are the commutative squares in C. This species satisfies both the identity and composition laws above.

2. Let C be a skeletal category with pushouts. Then C determines a vertebrate species with underlying graph $\Gamma(C)$ and preferred squares the pushout squares in C. This species satisfies all the extra axioms given above.

3. Let C be a category with one object (a monoid). Then C determines a vertebrate species with both operations **trivial** i.e. $a^b = a$ and $b_a = b$ for all a, b. This species again satisfies all the extra axioms above. But note that in this example the preferred squares are not necessarily commuting squares.

4. The Rack Species

This is the key example. A rack X defines a species $\mathcal{S}(X)$ with a single vertex and with X the set of edges. The preferred squares are of the following type.

$$\begin{array}{ccc} 0 & \xrightarrow{a^b} & 0 \\ \uparrow b & \circlearrowright & \uparrow b \\ 0 & \xrightarrow{a} & 0 \end{array}$$

5. A rack X determines a second species $\mathcal{S}_X(X)$ by taking as vertices the set X, and edges $a \xrightarrow{b} a^b$, and preferred squares of the following type.

$$\begin{array}{ccc} a^c & \xrightarrow{b^c} & a^{bc} = a^{cb^c} \\ \uparrow c & \circlearrowright & \uparrow c \\ a & \xrightarrow{b} & a^b \end{array}$$

The species $\mathcal{S}_X(X)$ covers the species $\mathcal{S}(X)$. Both are vertebrate with the binary operation a^b being the rack operation and the other operation b_a being trivial.

6. A G-rack $\partial : X \to G$ determines a species $\mathcal{S}_G(X)$, by taking as vertices the set G, and edges $g \xrightarrow{a} g^a$, for $a \in X$, where g^a is defined to be $(\partial a)^{-1} g$. The preferred squares are of the following type.

$$\begin{array}{ccc} g^b & \xrightarrow{a^b} & g^{ab} \\ \uparrow b & \circlearrowright & \uparrow b \\ g & \xrightarrow{a} & g^a \end{array}$$

7. There are two similar species associated to a G-rack defined by redefining g^a to be $g\partial a$ and $(\partial a)^{-1} g \partial a$ respectively.

8. The Cube Species

The n**-cube** I^n is the subset $[0,1]^n$ of $\mathbf{R}^n$. The 2^n **vertices** of I^n are $\{(\epsilon_1, \epsilon_2, \ldots, \epsilon_n)\}$ where $\epsilon_i = 0, 1$. It is often convenient to regard a vertex as a subset of $\{1, 2, \ldots, n\}$ as follows: $A \subset \{1, 2, \ldots, n\}$ corresponds to the vertex with $\epsilon_i = 1$ if and only if $i \in A$.

A **face** of I^n is a subcube defined by setting some of the coordinates equal to either 0 or 1. In particular the $2n$ $(n-1)$-faces $\partial_i^\epsilon I^n$, where $i \in \{1, 2, \ldots, n\}$ and $\epsilon \in \{0, 1\}$, are defined by setting $x_i = \epsilon$. The 1-faces are **edges** and have two vertices given in subset notation by A and $A \cup \{i\}$

where $i \notin A$. Faces have standard orientation given by their coordinate structure. In particular edges are oriented $A \to A \cup \{i\}$ and 2–faces are oriented

$$\begin{array}{ccc} A\cup\{j\} & \to & A\cup\{i\}\cup\{j\} \\ \uparrow & \circlearrowleft & \uparrow \\ A & \to & A\cup\{i\} \end{array}$$

where $i < j$.

Thus I^n becomes a species by taking its vertices and edges as the vertices and edges of the species and 2–faces as preferred squares, oriented as above. We shall use the notation I^n_{Spec} for this species, abbreviating it to I^n when no possibility of confusion can arise.

Mutations

Definition A **mutation** between species is a map which takes vertices to vertices, directed edges to directed edges and preferred squares to preferred squares. Mutations between species are analogous to functors between categories. There is thus a (large) category of species and mutations.

A species S determines an associated category $Cat(S)$. The objects of $Cat(S)$ are just the vertices of S The morphisms of $Cat(S)$ are generated by the edges of S with the relations which follow from insisting that preferred squares commute. Mutations become functors.

We have already seen in example 1 above that a category C determines a species $Spec(C)$. The functions $Spec$ and Cat can be seen to be adjoint in the sense of the following lemma:

0.1 Lemma *Given a species S and a category C then the set of mutations* $\mathrm{Mut}(S, Spec(C))$ *is in bijective correspondence with the set of functors* $\mathrm{Fun}(Cat(S), C)$. □

The category □

The category □ is the model category for semi-cubical sets. Recall that a **face** of the N–cube I^n is a subcube defined by setting some of the coordinates equal to either 0 or 1.

Let $p \le n$ and let J be a p–face of I^n. Then there is a canonical isometric identification $\gamma: I^p \to J$ defined by preserving the order of the coordinates and orientation of edges. We call the composition

$$\lambda = \mathrm{inc} \circ \gamma: \mathrm{I^p} \to \mathrm{I^n}$$

a **face map**.

Definition The category □ is the category whose objects are the n–cubes I^n for $n = 0, 1, \ldots, n$ and whose morphisms are the face maps.

The face map defined by the $(n-1)$–face $\partial_i^\epsilon I^n$ is denoted $\delta_i^\epsilon : I^{n-1} \to I^n$, and is given by:

$$\delta_i^\epsilon(\mathbf{x})_j = \begin{cases} x_j, & \text{if } j < i, \\ \epsilon, & \text{if } j = i, \\ x_{j-1}, & \text{if } j > i. \end{cases} \tag{1}$$

Note that the effect of δ_i^ϵ on the vertex A (in subset notation) is to add one to numbers $\geq i$, and to insert i if $\epsilon = 1$.

The following relations hold.

$$\delta_i^\epsilon \delta_{j-1}^\omega = \delta_j^\omega \delta_i^\epsilon \quad 1 \leq i < j \leq n \quad \text{and} \quad \epsilon, \omega \in \{0,1\}$$

Any face map λ can be uniquely written $\delta_{i_k}^{\epsilon_k} \cdots \delta_{i_1}^{\epsilon_1}$ where each $\epsilon_j \in \{0,1\}$ and we can assume either $i_1 < \cdots < i_k$, or $i_k \leq \cdots \leq i_1$.

There is an isomorphic copy of $\square$ given by replacing I^n by I^n_{Spec}. The same formulae, above, define the face maps (mutations) in this copy. This copy has a particularly simple definition:

2.3 Lemma *$\square$ is isomorphic to the category which has for objects I^n_{Spec}, $n = 1, 2, \ldots$ and morphisms all mutations $I^p \to I^n$ for $p \leq n$.* $\square$

We shall normally not distinguish between the two isomorphic copies of $\square$.

3. The Nerve of a Species and the Rack Space

In this section we describe the natural space built out of cubes which is associated to a species. The key example is the rack space. We also give the basic classifying properties of these spaces.

Semi-cubical sets and their realisations

A **semi-cubical set** or **$\square$–set** is a functor $X: \square^{op} \to \mathit{Sets}$. We write X^n for $X(I^n)$, λ^* for $X(\lambda)$ and in particular ∂_i^ϵ for $X(\delta_i^\epsilon)$.

(This use of ∂_i^ϵ is consistent with the previous use — as a particular face of I^n — because I^n can be regarded as a $\square$–set with X^p the set of p–faces of I^n and λ^* given by repeated application of ∂_i^ϵ's.)

A $\square$–**map** between $\square$–sets is a natural transformation.

The **realisation** $|X|$ of a $\square$–set X is given by making the identifications $(\lambda^* x, t) \sim (x, \lambda t)$ in the disjoint union $\coprod X^n \times I^n$, where I^n has its topological meaning.

The nerve of a species

Now let S be a species. The **nerve** of S, denoted $N(S)$, is the semi-cubical set defined by

$N(S)^{(n)} = \text{Mut}(I^n, S)$, the set of mutations from the cube species to S.

$\lambda^*(f) = f \circ \lambda$ where $\lambda: I^p \to I^n$ is a face map (mutation).

Key Example *The Rack Space*
Let X be a rack and let $\mathcal{S}(X)$ be the vertebrate species associated to X and described by example 4 in the previous section. We shall write BX for the realisation $|N\mathcal{S}(X)|$. This is the **rack space.**

As a cubical set the rack space has a particularly simple description analogous to the bar construction for groups. Indeed this is true for any vertebrate species. Let X be a vertebrate species then $|N\mathcal{S}(X)|$ has one cube for each sequence of length n, $(x_1, x_2, \ldots, x_n) \in X^n$ and the face maps are given by:

$$\partial_i^0(x_1, \ldots, x_n) = (x_1, \ldots, x_i, x_{i+1}, \ldots, x_n),$$

$$\partial_i^1(x_1, \ldots, x_n) = ((x_1)^{x_i}, \ldots, (x_{i-1})^{x_i}, (x_{i+1})_{x_i}, \cdots, (x_n)_{x_i}) \quad \text{for} \quad 1 \le i \le n.$$

For details here see [FRS].

3.1 Lemma *The fundamental group $\pi_1(BX)$ is isomorphic to $As(X)$, the associated group of X.*

Proof Since $S(X)$ has only one vertex the fundamental group of the 1-skeleton is the free group $F(X)$. Adding the 2-cells is equivalent to adding the relations $a^b = b^{-1}ab$ which gives the required result. □

We now investigate the homotopy group $\pi_2(BX)$. To do this we first find a combinatorial representation of $\pi_2|N(S)|$ for a general species S. Many of the concepts such as labelled diagrams can be generalised to give descriptions of the higher homotopy groups $\pi_n|N(S)|$ where $n > 2$ and of the bordism groups $\Omega_n|N(S)|$. We shall investigate these ideas in [FRS].

Labelled Diagrams

A **diagram** on a surface is a collection of oriented circles in general position such that at each crossing one of the arcs is regarded as the overcrossing arc and the other arc as the undercrossing arc.

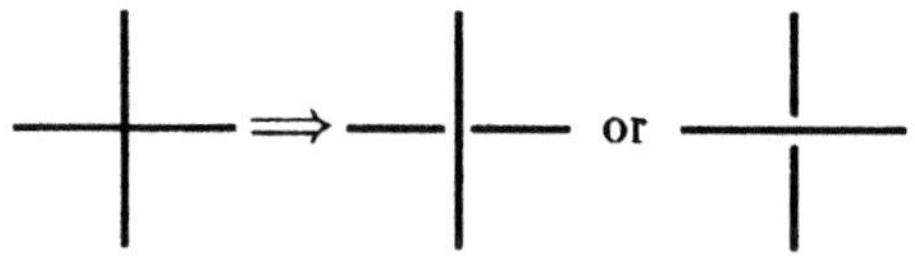

We say that a diagram is **labelled** by a species if each arc of the diagram is transversely oriented by an edge of the species,

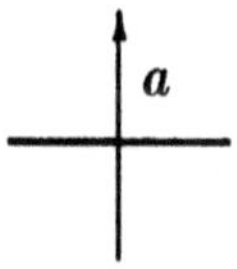

and at each crossing the resulting square of edges is preferred. In these diagrams the arcs will be indicated by heavier print than the transverse labelling.

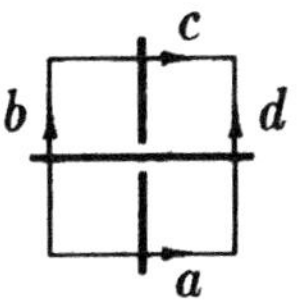

Labelling by a Preferred Square at a Crossing

Note that in the diagram the *base* a of the preferred square is parallel to the *overcrossing* arc, and this determines the orientation of the preferred square.

For an oriented diagram in S^2 we adopt the convention illustrated below for the transverse orientations of arcs.

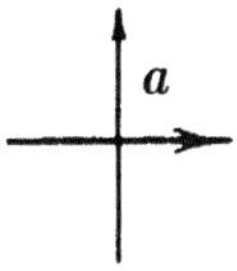

There is a strong connection between labelled diagrams and maps into the nerve.

Given a diagram D in a surface Σ labelled by the species S there is a map $f_D: \Sigma \to |N(S)|$ defined as follows. Thicken the diagram so that the surface is divided into three types of regions:

square regions about each crossing,
ribbon regions about the parts of arcs which don't lie in the square regions and
outer regions consisting of everything else.

For example the trefoil diagram gives rise to the following division with three square regions, six ribbon regions and five outer regions.

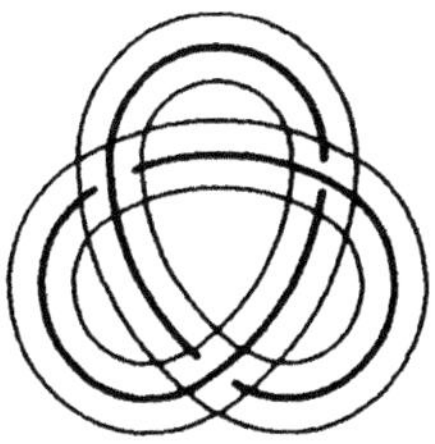

The Trefoil Knot

Now define f_D by sending each square to the corresponding square, projecting each ribbon to its transversely labelling edge and sending the outer regions to the corresponding vertex.

Conversely, given a map $f: \Sigma \to |N(S)|$, then by making f transverse to the 1 and 2–cells of the cubical complex $|N(S)|$ (see [BRS; chapter 7], [F]) we get a corresponding labelled diagram in Σ.

We now investigate the changes to the diagram which correspond to homotopy of the map in the nerve.

The first set of changes are familiar from knot theory. They are the Reidemeister Ω_2 and Ω_3 moves.

We say that an Ω_2 move is **legal** provided the diagram is labelled before and after the move, so that the labelling matches away from the move and both squares involved are preferred. The legal labels on the following picture show that the two preferred squares will necessarily be the same and have opposite orientation.

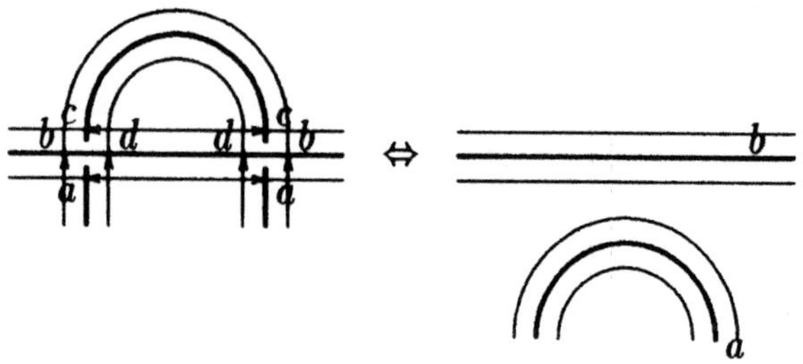

the Ω_2-move

Similarly, an Ω_3 move is legal if the diagram is labelled before and after the move, so that the labelling matches away from the move and all appropriate squares are preferred.

The following is an example of a legal Ω_3 move in which the labelling is by elements of a rack. The resulting homotopy is over a 3-cell in the rack space.

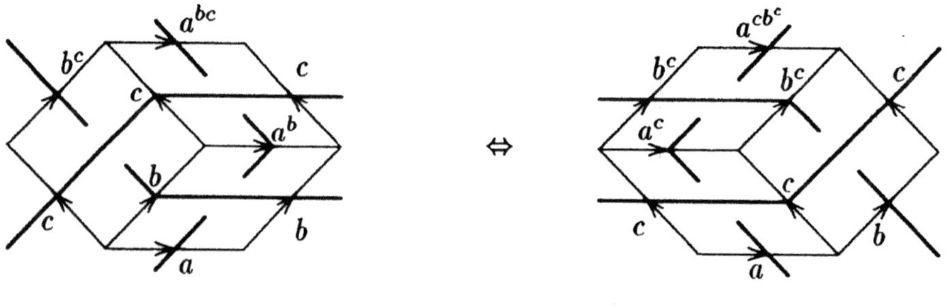

A Reidemeister Ω_3 move as a Homotopy over a 3-cell in the Rack Space

When the labels are from a general species, then it may not be possible to perform legal Ω_2 or Ω_3 moves, even when such moves are possible on the *unlabelled* diagram. However, when the species is a rack species, then this is always possible, indeed this is precisely what the rack laws imply, see [FR; section 4] and the remarks above corollary 3.4 below.

The other moves that we need are introduced in the next definition:

Definition **A cobordism by moves** between two labelled diagrams is a sequence of the following moves:

(1) Legal Reidemeister Ω_2 and Ω_3 moves.
(2) Introduction and deletion of unknotted and unlinked circle components in the diagram $D \Leftrightarrow D \cup \bigcirc$.
(3) A bridge move $\overset{a\longleftarrow}{\longrightarrow a} \Leftrightarrow a \hookleftarrow\!\supset \subset\!\hookrightarrow a$ between adjacent arcs with the same label and opposite orientations.

3.2 Proposition *Homotopy classes of maps $[\Sigma, |N(S)|]$ are in bijective corrrespondence with equivalence classes of diagrams in Σ labelled by S under cobordism by moves.*

Proof As remarked earlier, by using transversality ([BRS],[F]) we can homotope a map $f: \Sigma \to |N(S)|$ to be transverse to the 1 and 2–skeleta of $|N(S)|$ and produce a labelled diagram D in Σ such that $f = f_D$. Now suppose that f_D and $f_{D'}$ are homotopic. Then by relative transversality the corresponding diagrams are **cobordant** in the sense that there is a **labelled diagram** in $\Sigma \times I$ with boundary $D \times \{0\} \cup D' \times \{1\}$. By a labelled diagram in a 3–manifold we mean a self-transverse immersed surface which is transversely labelled by edges of S away from the double curves, and such that at double curves the transverse labellings form preferred squares which fit consistently together at triple points (thus forming cubes of $|N(S)|$). By looking at level sections of this cobordism in general position it is not hard to see that D and D' are cobordant by moves. (Bridge moves are saddles, maxima and minima correspond to introduction or removal of unknotted unlinked components, triple points correspond to Ω_3 moves, maxima or minima of double curves to Ω_2 moves.)

Conversely a cobordism by moves gives an ordinary cobordism which corresponds to a map of $\Sigma \times I$ in $|N(S)|$ by a construction similar to that for f_D given above. □

Remark The regions of a labelled diagram are mapped by the corresponding map to vertices of $|N(S)|$ and so can be regarded as labelled by vertices of S.

3.3 Corollary *$\pi_2|N(S)|$ is in bijective corrrespondence with equivalence classes of diagrams in S^2 labelled by S, such that the basepoint is labelled by the base vertex of S, under cobordism by moves away from the basepoint.*

We now turn to the case of the rack space. In this case the concept of a labelled diagram has an interpretation in terms of the fundamental rack of the link represented by the diagram. For simplicity consider a framed link L in S^3, and let X be a rack. A **labelling** of L by X means a homomorphism of the fundamental rack $\Gamma(L)$ to X. Such a homomorphism gives rise to a labelling of any diagram D corresponding to L by X i.e. the arcs are labelled by elements of X so that at crossings the labellings are consistent with the rack operation in X. But this is precisely the same as a labelling of D by the species $S(X)$. Conversely, any such labelling corresponds to a homomorphism. Thus, in this case, labelling is a property of the *link* not of the *diagram* and hence is invariant under isotopy i.e. under Ω_2 and Ω_3 moves, thus all such moves are automatically legal.

We say that labelled links L, L' are **labelled cobordant** if there is an embedded surface F in $S^3 \times I$ with boundary $L \times \{0\} \cup L' \times \{1\}$ and a homomorphism of the fundamental rack of F in $S^3 \times I$ to X extending the given homomorphisms of the fundamental racks of L and L'. By looking at the level sections of such a cobordism in general position we see that L and L' differ by isotopy, labelled bridge moves and the introduction or deletion of unknotted, unlinked components. I.e. diagrams for L and L' differ by cobordism by moves and the different concepts of cobordism all coincide.

Now by general position we can assume that the cobordism F misses $* \times I$ where $*$ is the basepoint of S^3, and hence the corresponding cobordism by moves can be assumed to miss the basepoint in S^2. We have proved:

3.4 Corollary *Let X be a rack. Then $\pi_2(BX)$ is in bijective correspondence with cobordism classes of links in S^3 labelled by X.*

4. The Second Homotopy Group of the Space of a Classical Rack

Let L be a tame classical link in S^3. We say that L is **non-split** or **irreducible** if no embedded 2-sphere in $S^3 - L$ divides the components of L into two non-empty subsets. In general a link can be written as a union $L = L_1 \cup \ldots \cup L_k$ where each L_i is a maximal irreducible sublink. We call the sublinks L_i the **blocks** of L. A block is said to be **trivial** if it is equivalent to the unknot with zero framing. The purpose of this section is to prove the following result.

4.1 Theorem *Suppose X is the fundamental rack of a tame classical link in S^3 and BX is the space of X. Then $\pi_2(BX) \cong \mathbf{Z}^p$ where p is the number of non-trivial blocks of L.*

Furthermore a basis of $\pi_2(BX)$ is given by diagrams representing these blocks.

Before proving the theorem we shall prove:

4.2 Lemma *Let X be any rack then the action of the fundamental group $\pi_1(BX)$ on $\pi_2(BX)$ is trivial.*

Proof We shall use the result from the last section that elements of $\pi_2(BX)$ can be represented as cobordism classes of diagrams in S^2 labelled by elements of the rack X.

We shall exhibit a cobordism between the diagram D and D^a where D^a is the diagram D in which every label has been acted upon by the rack element a. In words the cobordism proceeds as follows: introduce into D an unlinked unknotted circle labelled a. Now pass an arc of this circle over

D so that D is replaced by D^a. Now pass the arc under D^a to its original position and delete the circle. □

Remark A similar proof shows that BX is a **simple** space (π_1 acts trivially on π_n for each $n > 1$). For details see [FRS].

Proof of the theorem We shall deal first with the irreducible case.

Case 1: L is irreducible and non-trivial

We shall show that in this case $\pi_2(BX)$ is isomorphic to homotopy classes of maps from S^3 to itself. These classes are classified by degree which means that the group is, as claimed, the integers, $\mathbf{Z}$.

Given a map $f: S^3 \to S^3$, we can deform f to be transverse to a tubular neighbourhood N of L. Then the preimage is a link L' in S^3 and f determines a homomorphism of fundamental racks $\Gamma(L') \to \Gamma(L)$. In other words L' is labelled by X. Similarly a homotopy gives a labelled cobordism between labelled links.

There is thus a homomorphism from $[S^3, S^3]$ to cobordism classes of labelled links (labelled by X) which is $\pi_2(BX)$ by corollary 3.4. We shall now construct an inverse to this homomorphism. Let L' be any link labelled by X. Let N' be a tubular neighbourhood of L'. We shall construct the required map $f : S^3 \to S^3$ by first defining f on the tori $N' \to N$. Let D' be a diagram for L'. Pick an arbitary arc of D' and let $a \in X$ be its label. Let K' be the component of L' containing this arc. The label $a \in X$ will be the conjugate of a label of some component K of L. If we now carry on round the component K' until we return to our starting point the underpasses will spell out a word $w \in A(X)$. This word satisfies $a^w = a$ and so by lemma 1.4 is some power l^n of the framing longitude l of K. Note that n is well-defined since l has infinite order in $\pi_1(S^3 - L)$.

We can now define f on this component of $\partial N'$ to the corresponding component of ∂N by sending the standard meridian of the original arc to the meridian corresponding to a and by sending the longitude to l^n. Note that the degree of this map of surfaces is the integer n with appropriate choice of orientations.

Now extend the partially defined map over the interior of solid torus and in a likewise manner over the rest of the components of N'.

Consider now a cell complex subdividing the closure of the complementary space $S^3 - N'$. Since the fundamental group is generated by meridians and since the map respects their relations there is no obstruction to extending the map f over the 2-skeleton. Up to this point we have not used the fact that L is irreducible. Since this is the case the second homotopy group $\pi_2(S^3 - N)$ is trivial and there is no obstruction to defining f on closure$(S^3 - N') \to$ closure$(S^3 - N)$.

We can calculate the degree of f as follows. Pick any component T of ∂N and let $T'_1, \ldots, T'_k$ be the components of $\partial N'$ which get mapped to T.

Suppose that f has degree $n_1, \ldots, n_k$ on these components. Then $\deg f$ is the sum $n_1 + \ldots + n_k$.

We now observe that the degree is invariant under homotopy which is the same as invariance under (labelled) cobordism: a cobordism is equivalent to a cobordism by moves (see the last section) and it can be readily checked that the degree is invariant under each of the moves

Thus we have a well-defined inverse function $\pi_2(BX) \to [S^3, S^3]$ which completes the proof of case 1.

Case 2: L is irreducible and trivial

In this case the homomorphism $[S^3, S^3] \to \pi_2(BX)$ defined in case 1 is surjective by exactly the same argument (the map f can be constructed but its degree is not well-defined since l is zero in this case). But a generator of $[S^3, S^3]$ is represented by a single unknotted circle of framing zero which can be removed by a cobordism, hence $\pi_2(BX)$ is zero.

This completes case 2 and we now turn to the general case $L = L_1 \cup \ldots \cup L_k$ where each L_i is maximal irreducible. We need the following lemmas.

4.3 Lemma *Let M be an irreducible 3-manifold. Let M_0 be M with the interior of a k balls $B_1, \ldots, B_k$ removed. If $k = 1$ assume that $\pi_1(M)$ is non-trivial. Then $\pi_2(M_0)$ is generated as a $\pi_1(M_0)$ module by the spheres ∂B_i.*

Proof Let $\widetilde{M}$ and $\widetilde{M_0}$ be the universal covers of M and M_0 respectively. Then $\widetilde{M}$ can be obtained from $\widetilde{M_0}$ by filling in holes with copies gB_i of B_i one for each element g in $\pi_1(M_0)$, $i = 1 \ldots n$. Since M is irreducible $\pi_2(M) \cong H_2(\widetilde{M}) \cong 0$. Then using a Mayer–Vietoris sequence we see that as an abelian group $\pi_2(M_0) \cong H_2(\widetilde{M_0})$ has one generator for each pair (g, B_i) where $g \in \pi_1(M_0)$. □

4.4 Lemma *Let M be the connected sum $M = M_1 \sharp \ldots \sharp M_k$ of k irreducible 3-manifolds each with non trivial fundamental group. Then as a $\pi_1(M)$ module $\pi_2(M)$ is generated by the seperating spheres $S_1, \ldots, S_{k-1}$.*

Proof Let an element of $\pi_2(M)$ be represented by a map $f: S^2 \to M$ of the 2-sphere into M which we may assume is transverse to the seperating spheres. Consider an innermost disc D in S^2 which has boundary in the intersection of $f(S^2)$ and S_i say. Let D' be a (singular) disc in S_i which bounds ∂D. Then by the previous lemma the homotopy class of the sphere $D \cup D'$ is in the subgroup generated as a $\pi_1(M)$ module by the seperating spheres $S_1, \ldots, S_{k-1}$. By subtracting this element, we may perform a homotopy to remove this intersection curve. We can now argue by induction on the number of intersections. □

Returning now to the proof of the main theorem we attempt to construct as before a map f of the 3-sphere to itself. This time the obstructions to mapping in the 3-cells may be non zero. By the above lemma these can be made zero by introducing into the diagram D a number of disjoint copies of diagrams for the links L_i (possibly conjugated by an element of X). This means that the map f can be defined on the extended diagram. Indeed we may further extend the diagram so that f has degree zero. If we now make the resulting homotopy transverse to L we construct a cobordism between the original diagram D and and a number of disjoint copies of conjugated diagrams for the links L_i. By lemma 4.1, we can assume that these diagrams are labelled in a standard way (not by conjugates). Thus the diagrams for L_i form a generating set for π_2. Note that the diagrams for any trivial sublinks can be coborded away as in case 2.

Now no non-trivial linear combination of the remaining diagrams can be cobordant to zero because each component is labelled by elements of the rack which are distinct, so that no two arcs from different components can ever amalgamate by a bridge move. In particular no two distinct L_i can interact in their own annihilation. Moreover no non-trivial L_i can dissapear by the proof of case 1. Thus they form a basis for π_2 and the theorem is proved. □

References

[BH] **R.Brown and P.Higgins**, *The classifying space of a crossed complex*, Math. Proc. Cambridge Philos. Soc., 110 (1991) 95–120

[BRS] **S.Buonchristiano, C.Rourke and B.Sanderson**, *A geometric approach to homology theory*, London Math. Soc. Lecture Note Series, no. 18 C.U.P. (1976)

[CW] **J.Conway and G.Wraith**, correspondence, 1959

[F] **R.Fenn**, *Techniques of Geometric Topology*, London Math. Soc. Lecture Note Series, no. 57 C.U.P. (1983)

[FR] **R.Fenn and C.Rourke**, *Racks and links in codimension two*, to appear in J. of Knot Theory and its Ramifications, World Scientific

[FRS] **R.Fenn, C.Rourke and B.Sanderson**, *The origin of species: the rack space*, in preparation

[J] **D.Joyce**, *A classifying invariant of knots; the knot quandle*, J. Pure Appl. Alg., 23 (1982) 37-65

[S] **B.Sanderson**, *The geometry of Mahowald orientations* Algebraic Topology, Aarhus, 1978. Springer Lecture Notes in Mathematics, no. 763, 152-174

SOME REMARKS ON THE BRAID–PERMUTATION GROUP

Roger Fenn
Math Dept Sussex University Brighton BN1 9QH UK
Richárd Rimányi
Department of Geometry Eötvös Loránd University Budapest Rákóczi út 5. 1088 HUNGARY
Colin Rourke
Math Dept Warwick University Coventry CV4 7AL UK

ABSTRACT: A subgroup of the automorphism group of the free group is considered. This is the automorphism group of the free quandle. Various properties are demonstrated. A set of generators is given and also set of relations which show the close connection with both the classical braid and permutation groups. The elements of this group are pictured as braids generalised to allow some crossings to be "welded".

1. Introduction

In this paper we consider the subgroup of the automorphism group of the free group generated by the braid group and the permutation group. If $\{x_1, \cdots, x_n\}$ are the generators of the free group F_n then the braid subgroup B_n is the subgroup of $\mathrm{Aut}(F_n)$ generated by the elements σ_i, $i = 1, 2, \cdots, n-1$ given by

$$\begin{cases} x_i & \mapsto & x_{i+1} \\ x_{i+1} & \mapsto & {x_{i+1}}^{-1} x_i x_{i+1} \\ x_j & \mapsto & x_j \qquad j \neq i,\ i+1. \end{cases}$$

and the permutation subgroup consists of all permutations of the generators $\{x_1, \cdots, x_n\}$ with generators the transpositions τ_i, $i = 1, 2, \cdots, n-1$,

$$\begin{cases} x_i & \mapsto & x_{i+1} \\ x_{i+1} & \mapsto & x_i \\ x_j & \mapsto & x_j \quad j \neq i,\ i+1. \end{cases}$$

We will call this subgroup the **braid–permutation group** or BP_n for short. This group is clearly finitely generated and we shall list a finite set

M.E. Bozhüyük (ed.), Topics in Knot Theory, 57–68.

of relations. There are pictures for elements of BP_n analogous to the usual ones for braids but generalised to allow some crossings to be "welded".

2. Racks, quandles, free racks, free quandles

It is known that BP_n is isomorphic to $\mathrm{Aut}(FQ_n)$, the group of automorphisms of the free quandle of rank n, see [FR]. A quandle is an algebraic gadget intimately associated with a knot or link. Racks and quandles have been defined by many authors but we take our definitions from [FR] in 1991. A **rack** R is a set with a binary operation on it. This operation — which we will write exponentially — is subject to the following axioms:

(i) for all b and c in R there exists a unique $a \in R$ such that

$$a^b = c$$

(the element a will be denoted $a = c^{\bar{b}}$)

(ii) for all $a, b, c \in R$

$$a^{bc} = a^{cb^c}$$

where we have adopted the usual conventions for an operation written exponentially: $x^{yz} := (x^y)^z$ and $x^{y^z} := x^{(y^z)}$.

If Q is a rack and it satisfies the additional axiom

(iii) $a^a = a$ for all $a \in Q$

then it is called a **quandle.**

As usual in universal algebra the concept of free rack and free quandle can be defined.

Definition *Let S be a set. We call $FQ(S)$ $(FR(S))$ a* **free quandle** *(***free rack***) on the set S, if it satisfies the following two conditions:*

(1) $S \subset FQ(S)$ $(S \subset FR(S))$

(2) *whenever X is a quandle (rack) and $f : S \to X$ a function then there is a unique quandle (rack) homomorphism $\bar{f} : FQ(S) \to X$ $(FR(S) \to X)$ which is an extension of f.*

Let $F(S)$ denote the free group on the set S.

2.1 Theorem

(1) *Up to relative isomorphism there exists a unique free quandle and a unique free rack on S.*

(2) *$FQ(S)$ and $FR(S)$ can be constructed as follows. $FR(S)$ is the set $S \times F(S) = \{(a, w) = a^w | a \in S, w \in F(S)\}$ with the operation defined by:*

$$(a^w)^{b^u} := a^{w\bar{u}bu}$$

$FQ(S)$ is $FR(S)$ modulo the equivalence generated by $a^a = a$ for all $a \in S$.

Notation *If S is the finite set $\{x_1, \cdots, x_n\}$ then we will use the notation*

$$FQ_n = FQ(\{x_1, \cdots, x_n\})$$

for the free quandle on $\{x_1, \cdots, x_n\}$ and

$$F_n = F(\{x_1, \cdots, x_n\})$$

for the free group on $\{x_1, \cdots, x_n\}$.

Let $\partial : FQ_n \to F_n$ be defined by $\partial(a^w) = w^{-1}aw$. Then any automorphism $\phi : FQ_n \to FQ_n$ induces an automorphism $\phi_\sharp : F_n \to F_n$ such that the diagram below commutes.

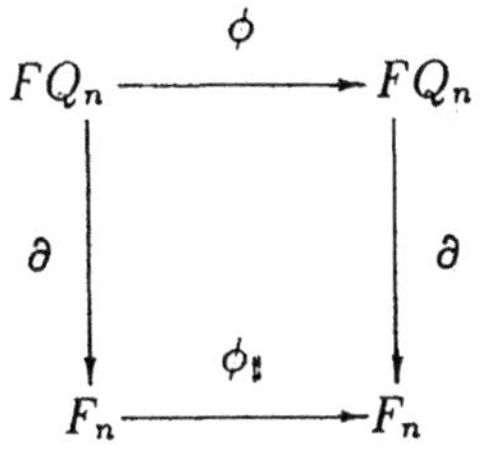

The correspondence $\phi \to \phi_\sharp$ embeds $\mathrm{Aut}(FQ_n)$ as a subgroup of $\mathrm{Aut}(F_n)$.

3. The map $B_n \to \mathrm{Aut}(FQ_n)$

Let γ be a fixed embedding of n points into D^2. Consider an embedding β of n disjoint line segments into $D^2 \times I$ so that the embedding of each line segment is monotone in the second coordinate (I) and the endpoints of the line segments are the image of the composition $\gamma_i : \{n \ \ points\} \to D^2 = D^2 \times \{i\} \subset D^2 \times I$ for $i = 1, 2$.

Figure 1

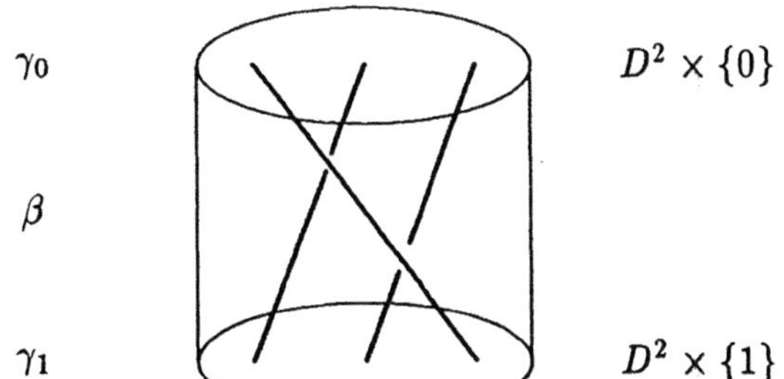

A braid β

We call n-braids the equivalence classes of the embeddings under relative isotopies of $D^2 \times I$ which preserve levels. Under the multiplication pictured in figure 2 the n-braids form the **braid group** on n strings called B_n.

Fig. 2

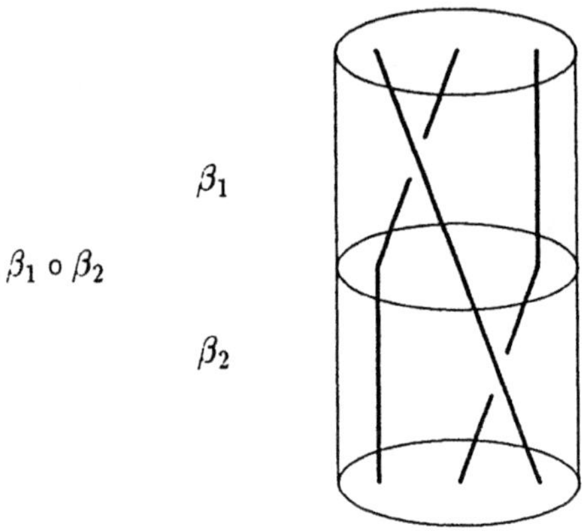

The braid β as the product $\beta_1\beta_2$

By abuse of language we will sometimes fail to distinguish between a braid and a representative β of that braid. This will cause no confusion because nothing we consider depends on which representative we take.

Recall that a fundamental quandle $\Gamma(\iota)$ can be defined for all embeddings ι of codimension two.

3.1 Lemma

(1) $\Gamma(\gamma) \cong FQ_n$

(2) $\Gamma(\beta) \cong FQ_n$

Instead of proving these results we will just give the isomorphisms that we will use through the paper without proving that they are isomorphisms. The formal proof is an easy exercise.

(1) It would be helpful at this stage for the reader to refer to figure 3. Assume the points $\{P_1, \ldots, P_n\}$ are positioned along the x-axis in D^2. Take a base point $(0,-1) = B \in \partial D^2 = S^1$, and let X be the antipodal point $(0,1)$.

Connect X with the n line segments : $l_i = \overline{P^i X}$. An element of $\Gamma(\gamma)$ is represented by an arc α connecting some P_k with B in $D^2 \setminus \{P_1, \cdots, P_n\}$. Going along α from P_k to B we can spell out a word w as follows: if we meet l_i from left to right then write down x_i, if we meet it from right to left write down x_i^{-1}. In this way we obtain a word w in the x_i's. Now the map $p : \Gamma(\gamma) \to FQ_n$ is $\alpha \mapsto x_k^w$.

For example in figure 3 we get $\alpha \mapsto x_1^{x_1 x_3^{-1}} = x_1^{x_3^{-1}} \in FQ_3$.

Figure 3

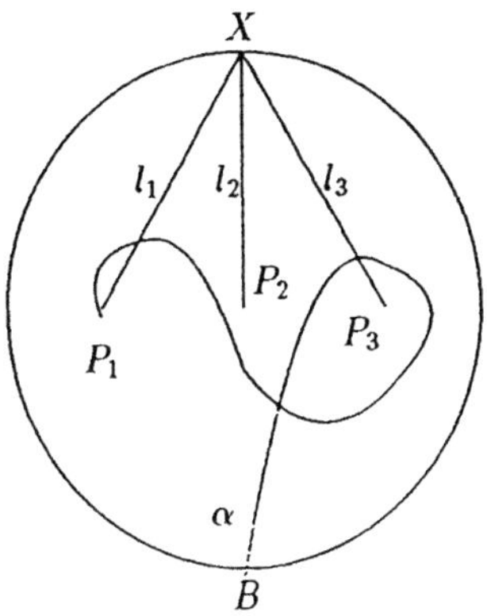

(2) Take a base point $B_i = B \times \{i\}$ for $i = 1, 2$. We are going to show two maps $q_i : \Gamma(\beta) \to \Gamma(\gamma)$, $i = 0, 1$, which are isomorphisms. If α is in $\Gamma(\beta)$ then it is represented by an arc from one of the line segments of the n-braid to B_i avoiding the image of β. Let this arc α "drop" down (or up) to $D^2 \times \{i\}$ by a homotopy of α which avoids the image of β. Of course the tail of the arc can wander on the braid during the homotopy — see the definition of the fundamental quandle in [FR]. This gives rise to an element in $\Gamma(\gamma)$.

Let us now take an n-braid β. Its embedding space is $D^2 \times I$. Let the embedding of its n top (bottom) endpoints into D^2 be called $\gamma_0 = \{P_1, \cdots, P_n\} \times 0$ $(\gamma_1 = \{P_1, \cdots, P_n\} \times 1)$.

Then from the isomorphisms described above we can construct an automorphism of FQ_n:

$$FQ_n \overset{p}{\cong} \Gamma(\gamma_0) \overset{q_0}{\cong} \Gamma(\beta) \cong \Gamma(\beta) \overset{q_1}{\cong} \Gamma(\gamma_1) \overset{p}{\cong} FQ_n \qquad (*)$$

The two $\Gamma(\beta)$'s in the above formula are taken with the different base points B_0 and B_1 respectively.The isomorphism between them is induced by the arc $\{B\} \times I$ (because we can clearly assume that $\{B\} \times I$ is disjoint from β).

The composition in $(*)$ (reading from left to right) is called $\Phi(\beta) : FQ_n \to FQ_n$. So $\Phi : B_n \to \mathrm{Aut}(FQ_n)$.

The map Φ is obviously a homomorphism and it is in fact injective. The injectiveness can be shown in a similar way as it is shown for free *groups* (instead of free quandles) in [BZ].

As an example one can see that the braid in figure 1 determines the automorphism: $x_1 \mapsto x_3$, $x_2 \mapsto x_1^{x_3}$, $x_3 \mapsto x_2^{x_3}$.

4. A presentation of the map Φ

Now we present an easy way to calculate the automorphism of FQ_n assigned to an n-braid. Let β be an n-braid and let us take a diagram of it. Let us orient all strings "downwards" (i.e. from the 0 level to the 1 level). Let us label the bottom ends of the strings by $x_1, \ldots, x_n$. Now we will label all of the arc components of the diagram with elements of FQ_n. Let us do it so that at a crossing like this below (figure 4) the relation $c = b^a$ is true.

Figure 4

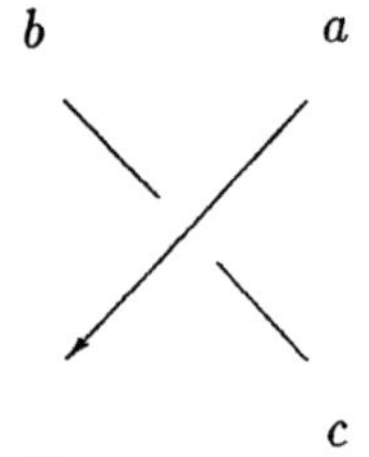

Labelling near a crossing

Consequently if the top ends of the strings are labelled with $w_1, \ldots, w_n \in FQ_n$ respectively then $\Phi(\beta)$ is the automorphism determined by $x_i \mapsto w_i$.

For example the braid represented by the diagram in figure 5 determines the automorphism given by

Figure 5

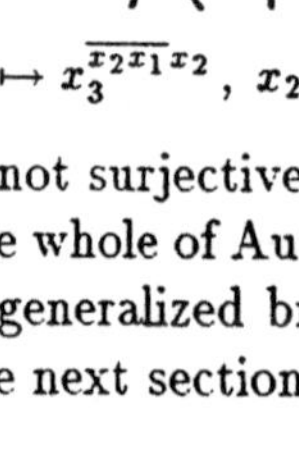

$$w_1 = x_3^{\overline{x_1^{x_2}}} = x_3^{\overline{x_2 x_1} x_2}$$
$$w_2 = x_2^{w_1} = x_2^{\overline{x_2} x_1 x_2 x_3 \overline{x_2 x_1} x_2}$$
$$w_3 = x_1^{x_2}$$

$$x_1 \mapsto x_3^{\overline{x_2 x_1} x_2}, \; x_2 \mapsto x_2^{\overline{x_2} x_1 x_2 x_3 \overline{x_2 x_1} x_2}, \; x_3 \mapsto x_1^{x_2}.$$

The map Φ is not surjective, so studying Φ itself is not enough to state anything about the whole of $\mathrm{Aut}(FQ_n)$. So the goal is extending the map Φ from a braid to a generalized braid diagram called a welded braid diagram. This is done in the next section.

5. The welded braid diagrams

A generalized — or **welded** — braid diagram is just like an ordinary braid diagram except that at a crossing of two strings we can ***weld*** the two strings together. An example of a welded braid diagram is given in figure 6.

Figure 6

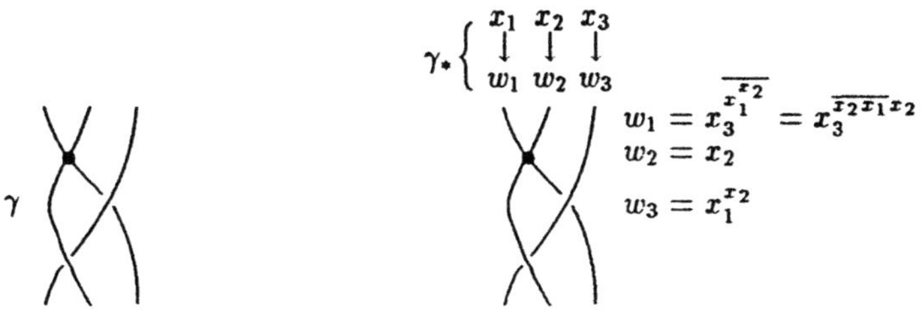

For reasons which will be clear later we do not regard a welded braid as a 2-dimensional projection of some 3-dimensional object.

Now we would like to assign an automorphism of FQ_n to an n-string welded braid diagram. Section 4 motivates the following definition.

First orient all strings downwards. Assign $x_1, \ldots, x_n$ to the bottom of the strings. Now change the welded braid diagram for a moment so that at a welded crossing the strings avoid each other without break (which can not be pictured in the plane.) Then let us take components of this and label them with the elements of FQ_n in the usual way (see figure 4). Now we can read off the automorphism of FQ_n as before: if w_i is assigned to the ith bottom endpoint then the automorphism is determined by $x_i \mapsto w_i$. For example the welded braid diagram in figure 6 determines the automorphism

$$\begin{cases} x_1 & \mapsto & x_3^{\bar{x}_2 x_1 x_2} \\ x_2 & \mapsto & x_2 \\ x_3 & \mapsto & x_1^{x_2}. \end{cases}$$

All welded braids can be built up from two types of atomic welded braids. The first type familiar from braid theory is σ_i which interchanges the i-th and $i+1$-th strands so that the left hand string croses over the right hand string as we go downwards. The second type τ_i replaces the crossing in σ_i by a weld. These two types define the automorphisms

$$\begin{cases} x_i & \mapsto & x_{i+1} \\ x_{i+1} & \mapsto & x_i^{x_{i+1}} \\ x_j & \mapsto & x_j \quad j \neq i,\ i+1. \end{cases}$$

and

$$\begin{cases} x_i & \mapsto & x_{i+1} \\ x_{i+1} & \mapsto & x_i \\ x_j & \mapsto & x_j \quad j \neq i,\ i+1. \end{cases}$$

respectively. (Compare with the formulæ in the introduction).

The most important result — which is the main reason for using welded braid diagrams — is the following.

5.1 Theorem *Every automorphisms of the free quandle can be obtained from a welded braid diagram.*

The proof depends on the Nielsen theory of *free groups* and can be found in [FR].

The question now arises: which welded braid diagrams determine the same element of $\mathrm{Aut}(FQ_n)$. One can make local changes to the welded braid diagram as pictured in figure 7. These changes do not change the assigned automorphism.

Figure 7

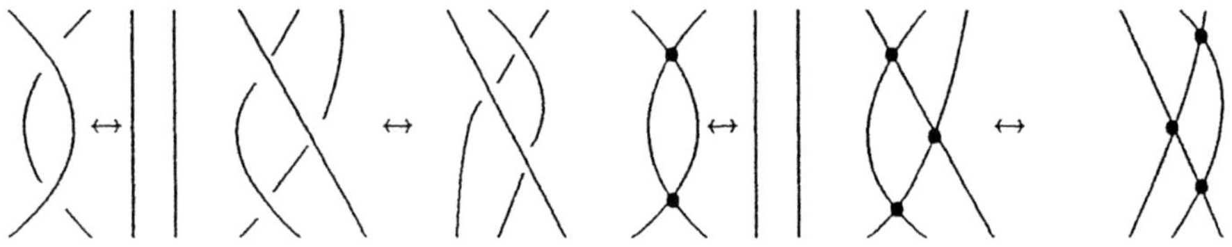

The first two types are the Reidemeister moves and one can easily check that the other two do not alter the assigned automorphism either. Actually, if we identify welded braid diagrams that differ only in the Reidemeister moves then we can embed the group of braids into the semigroup of welded braid diagrams. If we also identify those braids that differ in the second two moves then we can give the welded braid diagrams the structure of a group.

One can, however, see that there are some more local changes that do not change the assigned automorphism. In figure 8 we present two of them.

Figure 8

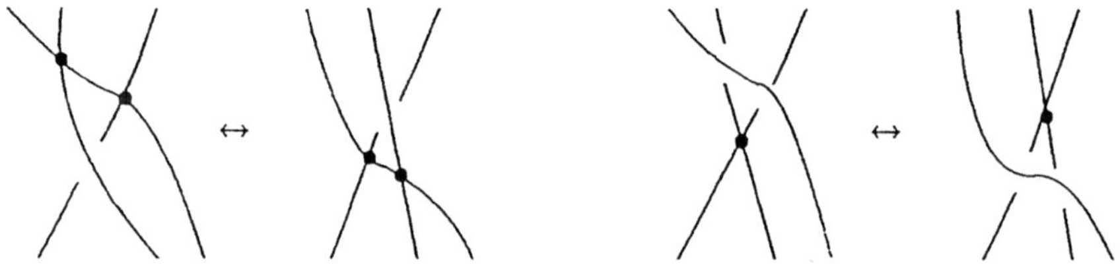

All of the moves seen up to now are similar to the Reidemeister moves, and therefore one could be forgiven for thinking the following: Welded braids can be defined as the 2-dimensional projection of 3-dimensional welded braids and two welded braid diagrams give the same automorphism of FQ_n if they are projections of equivalent welded braids — where the equivalence among 3-dimensional welded braids is some kind of natural equivalence, e.g. isotopy in 3-space.

This is not so and can be readily seen by realizing that the local change pictured in figure 9 **changes** the assigned automorphism.

Figure 9

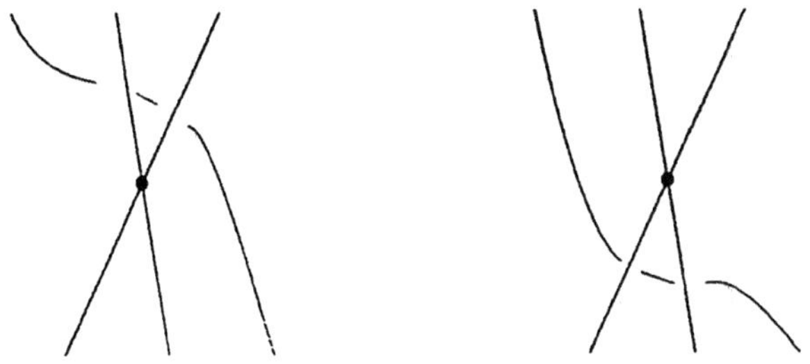

Two welded braids which do not induce the same element of $Aut\,FQ_n$.

It is as if there were a semi-infinite rod starting at each weld and pointing into the plane of the diagram preventing the arc behind from passing.

5.2 Theorem *Two welded braid diagrams determine the same automorphism of FQ_n iff they can be obtained from each other by a finite sequence of the local moves in figure 7 and 8.*

We can illustrate this by looking at the welded braid in figure 10.

Figure 10

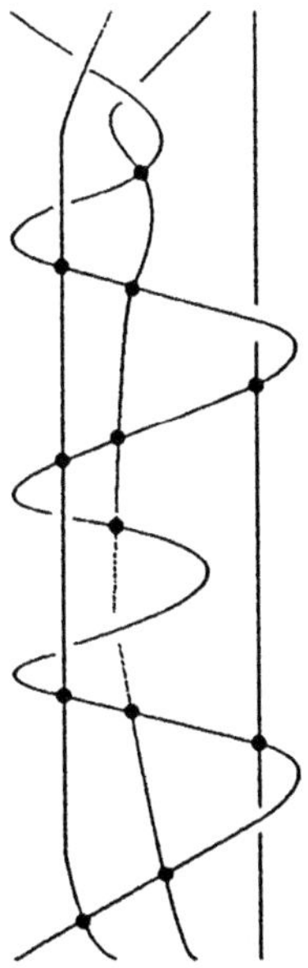

This induces the identity map of the free quandle on four generators and can be reduced to the identity braid by a series of moves illustrated in figure 11 below.

Figure 11

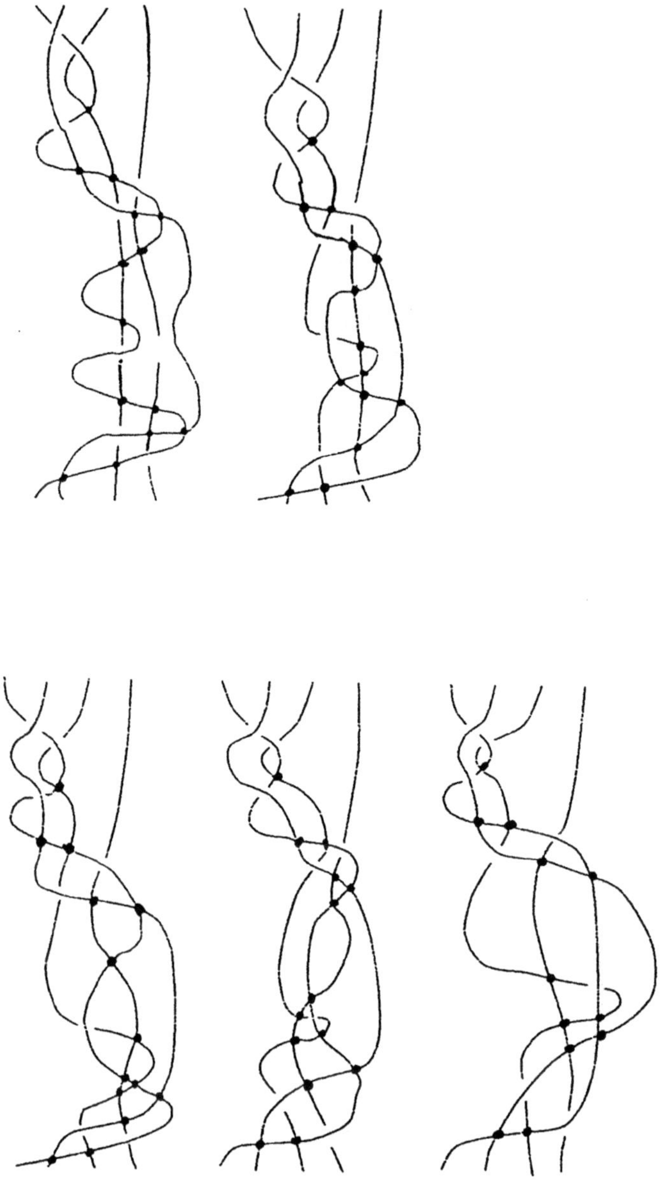

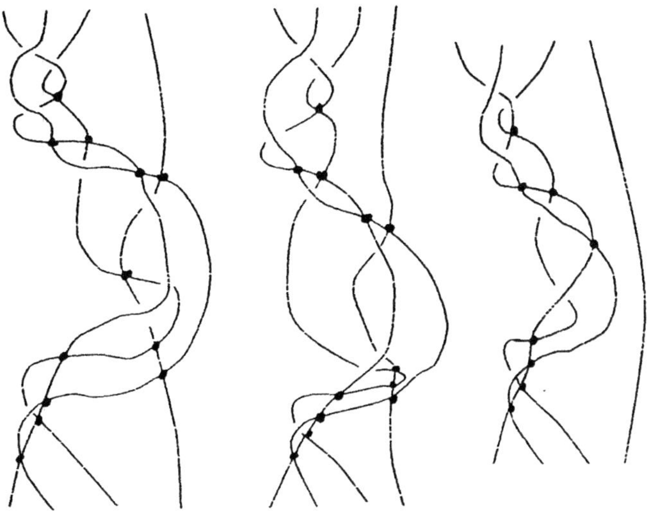

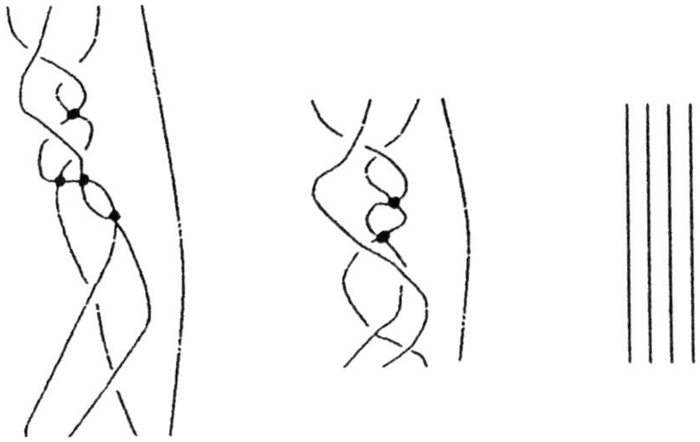

The proof of theorem 5.2 will be given in a later paper. Part of the proof depends on work by Krüger [K]. The following result is the algebraic formulation of the theorem and gives the promised presentation of the braid–permutation group. The presentation makes the connection with both the braid and permutation groups clear.

5.3 Theorem *The group $BP_n \cong Aut(FQ_n)$ has a presentation with generators σ_i, τ_i, $i = 1, \cdots, n-1$, and relations*

$$\begin{cases} \sigma_i\sigma_j & = \sigma_j\sigma_i \quad |i-j|>1 \\ \sigma_i\sigma_{i+1}\sigma_i & = \sigma_{i+1}\sigma_i\sigma_{i+1} \end{cases}$$

The braid relations

$$\begin{cases} \tau_i^2 & = 1 \\ \tau_i\tau_j & = \tau_j\tau_i \quad |i-j|>1 \\ \tau_i\tau_{i+1}\tau_i & = \tau_{i+1}\tau_i\tau_{i+1} \end{cases}$$

The permutation group relations

$$\begin{cases} \sigma_i\tau_j & = \tau_j\sigma_i \quad |i-j|>1 \\ \tau_i\tau_{i+1}\sigma_i & = \sigma_{i+1}\tau_i\tau_{i+1} \\ \sigma_i\sigma_{i+1}\tau_i & = \tau_{i+1}\sigma_i\sigma_{i+1} \end{cases}$$

The mixed relations

References

[BZ] **G.Burde, H.Zieschang**, *Knots*, de Gruyter Studies in Mathematics no. 5

[CM] **H.S.M.Coxeter, W.O.J.Moser**, *Generators and Relations for Discrete Groups*, Springer–Verlag, (1980)

[FR] **R.Fenn, C.Rourke**, *Racks and Links in Codimension Two*, to appear in the Journal of Knot Theory and its Ramifications

[K] **B.Krüger**, *Automorphe Mengen und die Artinschen Zopfgruppen*, Bonner Mathematische Schriften (1990)

INTRODUCTION TO TOPOLOGICAL IMITATIONS OF (3,1)-DIMENSIONAL MANIFOLD PAIRS

Akio KAWAUCHI
Department of Mathematics
Osaka City University
Osaka, 558, Japan

ABSTRACT. This article is an introduction to the topological imitation theory for (3,1)-manifold pairs, which propose a method how to construct a topologically close, new (3,1)-manifold pair from a given (3,1)-manifold pair. Some elementary new applications to knot theory are given.

0. Knot theory from a 4-dimensional aspect

We consider the equatorial 3-sphere S^3 in the 4-sphere S^4. Then it is well-known that every knot K in S^3 bounds a smoothly imbedded disk D in S^4, although in general int$D$ meets S^3. A knot K in S^3 is called a *slice* if K bounds a smoothly imbedded disk D in S^4 with int$D \cap S^3 = \emptyset$. The disk D is then called a *slice disk*. Note that the union $D \cup rD$ forms a 2-knot in S^4, where r denotes the standard involution on S^4 with $\mathrm{Fix}(r, S^4) = S^3$. The concept of a slice knot was introduced by R. H. Fox and J. W. Milnor (cf. R. H. Fox[**1**]). In this article, we shall concern a special kind of slice knot(cf. R. Kirby-P. Melvin[**12**], W. Brakes[**0**]).

DEFINITION: A knot K in S^3 is a *superslice* if K bounds a slice disk D such that the 2-knot $D \cup rD$ is trivial in S^4.

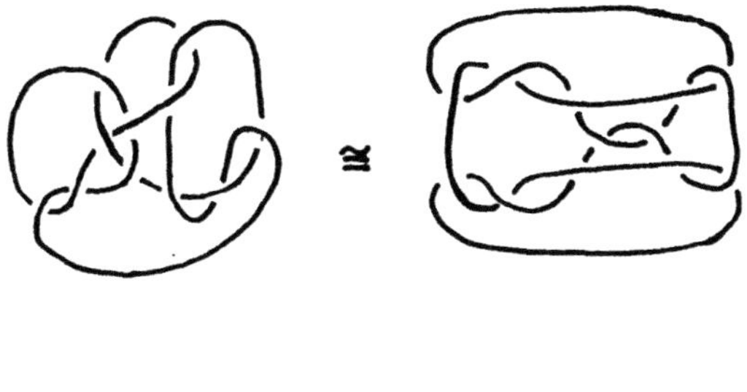

Fig. 1

The Kinoshita-Terasaka knot K_{KT} (See [**13**]), illustrated in Fig. 1 is the first example of a superslice (cf. S. Suzuki[**19**,10.22]), which has been known at latest in 1970 by R. H. Fox, F. Hosokawa, T. Yanagawa and others. See Section 2 for

M.E. Bozhüyük (ed.), Topics in Knot Theory, 69–83.

more recent historical notes. We shall replace in Lemma 2.1 the concept of a superslice knot with the existence of a certain map, called an *almost identical imitation*, so that a knot K in S^3 is a superslice if and only if for a trivial knot O in S^3 there is an almost identical imitation $q : (S^3, K) \to (S^3, O)$. The topological imitation theory, which we shall develop in this article treats a generalization of the concept of a superslice knot in terms of such a map. In Section 1, we describe basic concepts of topological imitation theory for manifold pairs (M, L) such that M is a smooth connected oriented 3-manifold and L is $\emptyset$ or a proper (possibly disconnected) oriented smooth 1-submanifold. In Section 2, we describe historical notes and main theorems. In Section 3, some elementary applications to knot theory are given.

1. The topological imitation theory

By a *manifold pair*, we mean a pair (M, L) such that M is a smooth connected oriented 3-manifold and L is $\emptyset$ or a proper (possibly disconnected) oriented smooth 1-submanifold. In particular, the manifold pair (M, L) is identified with the 3-manifold M when $L = \emptyset$ and called a (3,1)-*manifold pair* when $L \neq \emptyset$. A (3,1)-manifold pair (M, L) is called an r-component *link* if L consists of just r loop components, and an r-string *tangle* if L consists of just r arc components. A 1-component link is also called a *knot*. Let $I = [-1, 1]$.

DEFINITION: For a manifold pair (M, L), a smooth involution α on $(M, L) \times I$ is a *reflection* in $(M, L) \times I$ if:

(1) $\alpha((M, L) \times 1) = (M, L) \times (-1)$,
(2) The fixed point set, $\mathrm{Fix}(\alpha, (M, L) \times I)$ of α in $(M, L) \times I$ is a manifold pair.

The following definition contains an improvement of the definitions in [**5**], [**7**] and can be found in [**8**].

DEFINITION: For a manifold pair (M, L), let α be a reflection in $(M, L) \times I$.

(1) α is *standard* if $\alpha(x, t) = (x, -t)$ for all $(x, t) \in M \times I$,
(2) α is *normal* if $\alpha(x, t) = (x, -t)$ for all $(x, t) \in \partial(M \times I) \cup N(L) \times I$ for a tubular neighborhood $N(L)$ of L in M,
(3) α is *isotopically standard* if $f^{-1}\alpha f$ is standard for a diffeomorphism f of $M \times I$ which is isotopic to the identity by an isotopy keeping $\partial(M \times I) \cup N(L) \times I$ fixed for a tubular neighborhood $N(L)$ of L in M,
(4) α is *isotopically almost standard* if $L \neq \emptyset$ and α defines an isotopically standard reflection in $(M, L - a) \times I$ for each component a of L.

In the above definition and from now on, we understand that $N(L) = \emptyset$ when $L = \emptyset$.

DEFINITION: For a manifold pair (M, L), a smooth imbedding ϕ from a manifold pair (M^*, L^*) to $(M, L) \times I$ is a *reflector* of a reflection α in $(M, L) \times I$ if $\phi(M^*, L^*) = \mathrm{Fix}(\alpha, (M, L) \times I)$.

DEFINITION: The composite

$$q : (M^*, L^*) \xrightarrow{\phi} (M, L) \times I \overset{\text{projection}}{\longrightarrow} (M, L)$$

for a reflector ϕ of a reflection α in $(M, L) \times I$ is an *imitation*. Further, if α is normal, then q is a *normal imitation*, and if α is isotopically almost standard, then q is an *almost identical imitation*.

When $q : (M^*, L^*) \to (M, L)$ is an imitation (or a normal imitation or an almost identical imitation, resp.), we also call (M^*, L^*) an *imitation* (or a *normal imitation* or an *almost identical imitation*, resp.) of (M, L) with *imitation map* q. Clearly, an almost identical imitation implies a normal imitation. The following theorem is a reformation of several results in [**6**].

THEOREM 1.1. *For manifold pairs (M, L), (M^*, L^*), (M^{**}, L^{**}), we have the following:*

(1) *If $q : (M^*, L^*) \to (M, L)$ and $q^* : (M^{**}, L^{**}) \to (M^*, L^*)$ are normal (or almost identical, resp.) imitations, then there is a normal (or an almost identical, resp.) imitation $q^{**} : (M^{**}, L^{**}) \to (M, L)$,*
(2) *If $q : (M^*, L^*) \to (M, L)$ is a normal imitation and $p : (\tilde{M}, \tilde{L}) \to (M, L)$ is a finite covering, unbranched or branched along some components of L, then $p^* : (\tilde{M}^*, \tilde{L}^*) \to (M^*, L^*)$ is a covering and $\tilde{q} : (\tilde{M}^*, \tilde{L}^*) \to (\tilde{M}, \tilde{L})$ is a normal imitation in the following pullback diagram:*

$$\begin{array}{ccc} (\tilde{M}^*, \tilde{L}^*) & \xrightarrow{\tilde{q}} & (\tilde{M}, \tilde{L}) \\ {\scriptstyle p^*}\downarrow & & \downarrow{\scriptstyle p} \\ (M^*, L^*) & \xrightarrow{q} & (M, L), \end{array}$$

(3) *If $q : (M^*, L^*) \to (M, L)$ is a normal imitation, then there are tubular neighborhoods $N(L^*), N(L)$ of L^*, L in M^*, M, respectively such that q defines a diffeomorphism from $N(L^*)$ onto $N(L)$, denoted by q_N, and a degree one map from the exterior $E(L^*, M^*) = \mathrm{cl}(M^* - N(L^*))$ onto the exterior $E(L, M) = \mathrm{cl}(M - N(L))$, denoted by q_E, inducing an epimorphism $(q_E)_\# : \pi_1(E(L^*, M^*), x^*) \to \pi_1(E(L, M), x)$ whose kernel group is perfect, i.e., $H_1(\mathrm{Ker}(q_E)_\#; Z) = 0$.*

A sphere component S of ∂M is called an *n-pointed sphere* for a manifold pair (M, L) if $|S \cap L| = n$. Let $q : (M^*, L^*) \to (M, L)$ be an almost identical imitation. Then $L \neq \emptyset$ and the map $(M^*, L^* - a^*) \to (M, L - a)$ defined by q is homotopic to a diffeomorphism by a homotopy relative to $\partial M^* \cup N(L^* - a^*)$ for any components a^*, a of L^*, L with $q(a^*) = a$ and a tubular neighborhood $N(L^* - a^*)$ of $L^* - a^*$ in

M^*. Hence by a choice of the reflector ϕ used for the definition of q, we can identify M^* with M so that $q|\partial M$ =identity. From this reason, we write the almost identical imitation $q : (M^*, L^*) \to (M, L)$ as $q : (M, L^*) \to (M, L)$. If S is a 1-pointed sphere for (M, L), then the almost identical imitation map $q : (M, L^*) \to (M, L)$ is homotopic to a diffeomorphism by a homotopy relative to $\partial M \cup N(L^* - a^*)$ for a component a^* of L^* with $|S \cap a^*| = 1$ and a tubular neighborhood $N(L^* - a^*)$ of $L^* - a^*$ in M. If S is a 2-pointed sphere for (M, L), then we construct a (3,1)-manifold pair (M^+, L^+) from (M, L) by a *spherical completion*, i.e., adding a cone over $(S, S \cap L)$. Then the almost identical imitation $q : (M, L^*) \to (M, L)$ extends to an almost identical imitation $q^+ : (M^+, L^{*+}) \to (M^+, L^+)$ with $E(L^{*+}, M^+) \cong E(L^*, M)$. Thus, to obtain an almost identical imitation $q : (M, L^*) \to (M, L)$ with $E(L^*, M), E(L, M)$ non-diffeomorphic, we may assume without loss of generality that there are no n-pointed spheres for (M, L) with $0 \leq n \leq 2$. Such a manifold pair (M, L) is said to be *good*. Links in the 3-sphere S^3 and $r(\geq 2)$-string tangles in the 3-ball B^3 are typical examples of good (3,1)-manifold pairs. From any manifold pair (M, L), we can obtain a unique good manifold pair by spherical completions, which we denote by $(M, L)_\wedge$. Then any normal imitation $q : (M^*, L^*) \to (M, L)$ extends to a unique normal imitation $q_\wedge : (M^*, L^*)_\wedge \to (M, L)_\wedge$. Our main purpose is to construct many kinds of almost identical imitations $q : (M, L^*) \to (M, L)$ with $E(L^*, M)$ non-diffeomorphic to $E(L, M)$ for any good (3,1)-manifold pair (M, L).

2. Historical notes and main theorems

Early examples of imitations come from superslice knots. To state them, we observe the following lemma:

LEMMA 2.1. *A knot K in S^3 is a superslice if and only if there is an almost identical imitation $q : (S^3, K) \to (S^3, O)$ for a trivial knot O in S^3.*

PROOF: Since the 'if 'part is clear, it suffices to show the 'only if 'part. Let D be a slice disk for K such that $F = D \cup rD$ is a trivial 2-knot in S^4. Then $\mathrm{Fix}(r, (S^4, F)) = (S^3, K)$. Let V be a 3-ball in S^3 with $V \cap K = a_V$ a trivial arc in V. Remove from (S^4, F) an r-invariant bicollar of $(\mathrm{int}V, \mathrm{int}a_V)$ in (S^4, F), so that the resulting (4,2)-ball pair is diffeomorphic to $(B^3 \times I, E)$ with E invariant under the standard reflection α_B in $B^3 \times I$. Note that E is isotopic to $a_O \times I$ with a_O a trivial arc in B^3 by an ambient isotopy of $B^3 \times I$ keeping the boundary fixed. Then by uniqueness of a tubular neighborhood of $a_O \times I$ in $B^3 \times I$, we obtain from α_B an isotopically almost standard reflection in $(B^3, a_O) \times I$, which extends to an isotopically almost standard reflection α in $(S^3, O) \times I$ with $\mathrm{Fix}(\alpha, (S^3, O) \times I) \cong (S^3, K)$. Thus, there is an almost identical imitation $q : (S^3, K) \to (S^3, O)$. This completes the proof.

Since the Alexander polynomial of a superslice knot can be seen to be trivial by Theorem 1.1(3) and the Alexander polynomials of non-trivial knots with up to 10 crossings are non-trivial, we see that the Kinoshita-Terasaka knot K_{KT} in Fig. 1 which has 11 crossings is a non-trivial superslice with the smallest crossing number. In 1975 C. Gordon and D. W. Sumners [**2**] construct I-stably trivial (4,2)-ball pairs, that is, (4,2)-ball pairs (B^4, D) with diffeomorphism $(B^4, D) \times I \cong (B^4, B^2) \times I$ for a trivial (4,2)-ball pair (B^4, B^2). In particular, they showed:

THEOREM 2.2 ([**2**]). *A knot (S^3, K) with K the untwisted double of any slice knot bounds an I-stably trivial (4,2)-ball pair and hence is a superslice.*

In 1980 W. Brakes [**0**] considered some other superslices. In 1981 Y. Nakanishi [**15**] considered the K-T (=Kinoshita-Terasaka) tangle (B^3, T_{KT}), illustrated in Fig. 2 and showed the following:

LEMMA 2.3. *There is a normal imitation $q : (B^3, T_{KT}) \to (B^3, T_O)$ of a trivial 2-string tangle (B^3, T_O).*

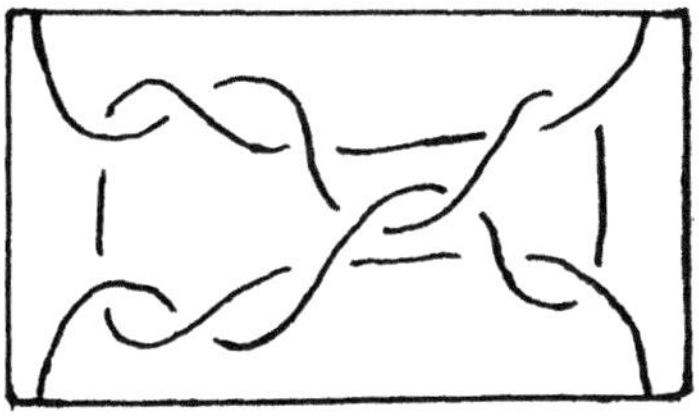

Fig. 2

Actually, we can take an almost identical imitation as q in this lemma, but we need a delicate argument like [**8**,§2]. This enables us to construct many normal imitations of any given link (See [**15**]) and, more generally, of any good (3,1)-manifold pair. To proceed this argument further, we use the concept of hyperbolic 3-manifolds.

DEFINITION: A (compact connected oriented) 3-manifold M is ***hyperbolic*** if we have the following (1) or (2):

(1) ∂M is $\emptyset$ or a union of tori and $\mathrm{int}M$ has a complete hyperbolic structure (that is, a complete Riemannian structure of constant curvature -1),
(2) The double $\mathrm{D}_1 M$ of M pasting along the non-torus components of ∂M has the property (1) when we regard $\mathrm{D}_1 M$ as M in (1).

A hyperbolic 3-manifold is a good 3-manifold. For a hyperbolic 3-manifold M, the ***volume*** $\mathrm{Vol}M$ and the ***isometry group*** $\mathrm{Isom}M$ of M are defined to be the hyperbolic volume $\mathrm{Vol}(\mathrm{int}M)$ and the hyperbolic isometry group $\mathrm{Isom}(\mathrm{int}M)$, respectively, if M is in (1), or $\mathrm{Vol}(\mathrm{int}(\mathrm{D}_1 M))/2$ and the quotient group of the group $\{f \in \mathrm{Isom}(\mathrm{int}\mathrm{D}_1 M) | f\tau = \tau f\}$ by the unique involutive isometry τ induced from the involution exchanging the two copies of M in $\mathrm{D}_1 M$, respectively, if M is in

(2). Vol$M$ and Isom$M$ (up to conjugations in Diff$M$) are known to be topological invariants of M(cf. [**20**]). By an argument in the tangle theory, T. Soma in [**18**] showed that the exterior $E(T_{KT}, B^3)$ has no incompressible torus. Using this fact and Lemma 2.3 and Myers gluing lemma in [**14**] and an observation by T. Kanenobu in [**4**], we have the following:

THEOREM 2.4. *For any link L in S^3, there is a normal imitation $q : (S^3, L^*) \to (S^3, L)$ such that $E(L^*, S^3)$ is hyperbolic.*

Even if the imitation in Lemma 2.3 is taken to be almost identical, it is difficult to establish the almost identical imitation version of Theorem 2.4 by using Lemma 2.3. To construct an almost identical imitation, we use the 2-string tangle (B^3, T'_{KT}) illustrated in Fig. 3. Let T_O^+ be a 1-string tangle in a 3-ball B^{3+} obtained from the trivial 2-string tangle (B^3, T_O) by adding a standard (3,1)-disk pair (D^3, D^1) as it is illustrated in Fig. 4.

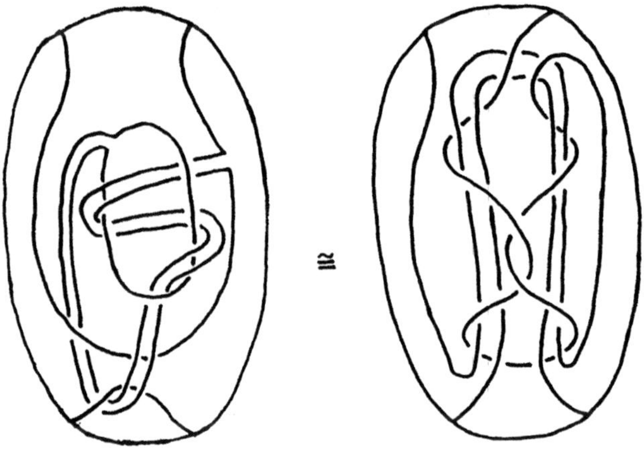

Fig. 3

LEMMA 2.5. *The tangle (B^3, T'_{KT}) has the following properties:*

(1) *T'_{KT} consists of an arc representing the Kinoshita-Terasaka knot K_{KT} and a trivial arc, and $E(T'_{KT}, B^3) \cong E(T_{KT}, B^3)$,*

(2) *There is a normal reflection α in $(B^3, T_O) \times I$ such that*

$$\mathrm{Fix}(\alpha, (B^3, T_O) \times I)) \cong (B^3, T'_{KT})$$

and the normal reflection α^+ in $(B^{3+}, T_O^+) \times I$ defined by α and the standard reflection in $(D^3, D^1) \times I$ is isotopically standard.

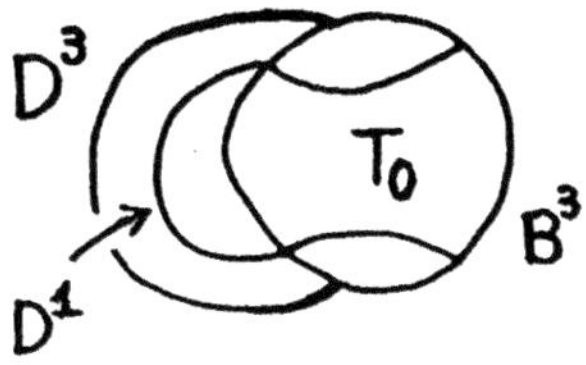

Fig. 4

The proof of Lemma 2.5 was given in [**7**] except an observation in [**8**,§2] which we need for an alternation of the definition of isotopically standard reflection. By combining Lemma 2.5 with an argument of Heegaard splitting, Myers gluing lemma [**14**] and an argument of T. Kanenobu [**4**], we have the following:

THEOREM 2.6. *For any good (3,1)-manifold pair* (M, L) *and any positive number* $C > 0$, *there is an almost identical imitation* $q : (M, L^*) \to (M, L)$ *such that* $E(L^*, M)$ *is hyperbolic with* $\mathrm{Vol}E(L^*, M) > C$.

For a good 3-manifold M and a trivial knot O in M, we apply Theorem 2.6 to obtain an almost identical imitation $q : (M, O^*) \to (M, O)$ such that $E(O^*, M)$ is hyperbolic with $\mathrm{Vol}E(O^*, M) > C$. By Thurston's argument [**20**] on hyperbolic Dehn surgery, there is a normal imitation $q_m : M_m^* \to M$ such that M_m^* is hyperbolic and $\mathrm{Vol}M_m^* > C$, obtained from q by the $1/m$-Dehn surgery along O^*, O for all integers m with $|m| \geq$ a constant. Hence we have the following:

COROLLARY 2.6′. *For any good 3-manifold* M *and a positive number* C, *there is a normal imitation* $q : M^* \to M$ *such that* M^* *is hyperbolic and* $\mathrm{Vol}M^* > C$.

The proof of Theorem 2.6 given in [**7**] has a weak point that the constructing pair (M, L^*) admits essential Conway spheres (i.e., spheres S meeting L with 4 points such that $S - L$ is incompressible and non-∂-parallel in $M - L$), so that any double covering space of M branched along L^* is never hyperbolic if it exists. Our next result is related to the hyperbolicity of a finite regular (branched) covering over (M, L^*). For this purpose, we introduce the concept of ***hyperbolic covering property***. Let (M, L) be a good manifold pair. Let $p : (\tilde{M}, \tilde{L}) \to (M, L)$ be any finite regular covering, unbranched or branched along a component union F_L of L such that $\tilde{M}$ is connected, which is referred to as any ***normal covering***. Let L_0 be a component union (possibly, $\emptyset$) of L such that $L_0 \supset L - F_L$. Let $\tilde{L}_0 = p^{-1}L_0$. By spherical completions, we obtain from $(\tilde{M}, \tilde{L}_0)$ a unique good manifold pair, denoted by $(\check{M}, \check{L})$, and from the covering $p|(\tilde{M}, \tilde{L}_0) : (\tilde{M}, \tilde{L}_0) \to (M, L_0)$ a unique (branched) covering, denoted by $\check{p} : (\check{M}, \check{L}) \to (M^+, L_0^+)$. This pair $(\check{M}, \check{L})$ is called a ***branch-missing good manifold pair*** of the normal covering $p : (\tilde{M}, \tilde{L}) \to (M, L)$.

Let $G(\tilde{M} \to M)$ denote the covering transformation group of the covering $p|\tilde{M} : \tilde{M} \to M$, which extends uniquely to an action on $(\check{M}, \check{L})$.

DEFINITION: A good manifold pair (M, L) has the *hyperbolic covering property* if $E(\check{L}, \check{M})$ is hyperbolic for all branch-missing good manifold pairs $(\check{M}, \check{L})$ of any normal covering $p : (\tilde{M}, \tilde{L}) \to (M, L)$.

Note that a good 3-manifold M has the hyperbolic covering property if and only if M is hyperbolic.

DEFINITION: A normal imitation $q : (M^*, L^*) \to (M, L)$ of a good manifold pair (M, L) with $E(L^*, M^*)$ hyperbolic is *rigid* if

$$\mathrm{Isom}E(\tilde{L}^*, \tilde{M}^*) \cong G(\tilde{M} \to M)$$

for the lift $p^* : (\tilde{M}^*, \tilde{L}^*) \to (M^*, L^*)$ of any normal covering $p : (\tilde{M}, \tilde{L}) \to (M, L)$ by q.

DEFINITION: A normal imitation $q : (M^*, L^*) \to (M, L)$ of a good manifold pair (M, L) such that (M^*, L^*) has the hyperbolic covering property is *J-rigid* for a positive integer J if we have all of the following (1)-(3):

(1) q is rigid,

(2) For any branch-missing good manifold pair $(\check{M}^*, \check{L}^*)$ of the lift $p^* : (\tilde{M}^*, \tilde{L}^*) \to (M^*, L^*)$ of any normal covering $p : (\tilde{M}, \tilde{L}) \to (M, L)$ of degree $\leq J$ by q,

$$\mathrm{Isom}E(\check{L}^*, \check{M}^*) \cong G(\tilde{M} \to M),$$

(3) Every normal covering $p^* : (\tilde{M}^*, \tilde{L}^*) \to (M^*, L^*)$ of degree $\leq J$ is the lift of a normal covering $p : (\tilde{M}, \tilde{L}) \to (M, L)$ by q.

Note that (2) is contained in (1) when $L = \emptyset$. We have the following:

THEOREM 2.7. *For any good (3,1)-manifold pair (M, L) and any positive integer J and any positive number C, there is a J-rigid almost identical imitation $q : (M, L^*) \to (M, L)$ such that (M, L^*) has the hyperbolic covering property and $\mathrm{Vol}E(L^*, M) > C$.*

This result is a combination result of the main result of [**8**] and [**10**,Lemma 1.3]. Combining it with Thurston's hyperbolic Dehn surgery argument [**20**], we obtain the following:

COROLLARY 2.7'. *For any good 3-manifold M and any positive integer J and positive number C, there is a J-rigid normal imitation $q : M^* \to M$ such that M^* is a hyperbolic 3-manifold and $\mathrm{Vol}M^* > C$.*

Our next purpose is to combine this topological imitation theory with a concept of mutation due to Ruberman [**17**] generalizing a certain kind of Conway mutation on a link, which is done in [**10**]. An involution ρ on a closed surface F is called a *symmetry* of F if the orbit space F/ρ is a 2-sphere. A good manifold pair (M', L') is an *e-mutation* of a good manifold pair (M, L) if there is a closed separating surface F of genus 2 in $\mathrm{int}E(L, M)$ such that (M', L') is obtained from (M, L) by cutting along F and regluing by a symmetry ρ of F. Then (M, L) is also an e-mutation of (M', L') and we can say without ambiguity that $(M, L), (M', L')$ are *e-mutative.*

DEFINITION: Two good manifold pairs $(M, L), (M', L')$ are *mutative* if there is a finite sequence of good manifold pairs $(M^{(m)}, L^{(m)}), m = 1, 2, ..., s$, with $(M^{(1)}, L^{(1)}) = (M, L)$ and $(M^{(s)}, L^{(s)}) = (M', L')$ such that $(M^{(m)}, L^{(m)})$ and $(M^{(m+1)}, L^{(m+1)})$ are e-mutative for all m.

Noting that a hyperbolic 3-manifold is unchanged under e-mutation associated with any symmetry of any *compressible* surface of genus 2 (See [**10**]), we see from a result of D. Ruberman in [**17**] the following:

LEMMA 2.8. *If $(M, L), (M', L')$ are mutative good manifold pairs and $E(L, M)$ is hyperbolic, then $E(L', M')$ is hyperbolic and $\mathrm{Vol}E(L, M) = \mathrm{Vol}E(L', M')$.*

For two normal imitations $q : (M^*, L^*) \to (M, L), q' : (M'^*, L'^*) \to (M, L)$ of a good manifold pair (M, L), we put the following two definitions:

DEFINITION: q, q' are *properly mutative* if $(\tilde{M}^*, \tilde{L}^*), (\tilde{M}'^*, \tilde{L}'^*)$ are mutative and $E(\tilde{L}^*, \tilde{M}^*), E(\tilde{L}'^*, \tilde{M}'^*)$ are non-diffeomorphic for the lifts $p^* : (\tilde{M}^*, \tilde{L}^*) \to (M^*, L^*)$, $p'^* : (\tilde{M}'^*, \tilde{L}'^*) \to (M'^*, L'^*)$ of any normal covering $p : (\tilde{M}, \tilde{L}) \to (M, L)$ by q, q'.

DEFINITION: q, q' are *J-properly mutative* for a positive integer J if q, q' are properly mutative and $E(\check{L}^*, \check{M}^*), E(\check{L}'^*, \check{M}'^*)$ are non-diffeomorphic for any branch-missing good manifold pairs $(\check{M}^*, \check{L}^*), (\check{M}'^*, \check{L}'^*)$ of the lifts $p^* : (\tilde{M}^*, \tilde{L}^*) \to (M^*, L^*), p'^* : (\tilde{M}'^*, \tilde{L}'^*) \to (M'^*, L'^*)$ of any normal covering $p : (\tilde{M}, \tilde{L}) \to (M, L)$ of degree $\leq J$ by q, q'.

We have the following:

THEOREM 2.9. *For any good (3,1)-manifold pair (M, L) and any positive integers J, N and any positive number C, there are J-rigid almost identical imitations*

$$q^{(n)} : (M, L^{*(n)}) \to (M, L), n = 1, 2, ..., 2^N,$$

such that $(M, L^{*(n)})$ has the hyperbolic covering property with $\mathrm{Vol}E(L^{*(n)}, M) > C$ and $q^{(n)}, q^{(n')}$ are J-properly mutative for all n, n' with $n \neq n'$.

COROLLARY 2.9$'$. *For any good 3-manifold M and any positive integers J, N and positive number C, there are J-rigid normal imitations $q^{(n)} : M^{*(n)} \to M, n = 1, 2, ..., 2^N$, with $M^{*(n)}$ hyperbolic 3-manifolds with $\mathrm{Vol}M^{*(n)} > C$ such that $q^{(n)}, q^{(n')}$ are properly mutative for all n, n' with $n \neq n'$.*

These results are proved in [**10**], where, further, certain equivariant versions together with a mutative reduction property on isometry groups are established.

3. Some elementary applications to knot theory

The following application suggests a topological imitation theory for a Seifert surface of a link:

APPLICATION 1. *If F is a compressible connected Seifert surface for any link L in S^3 with a non-∂-parallel compressible loop, then there are a new Seifert surface F^* for the same link L and a map $q : (S^3, F^*) \to (S^3, F)$ such that $q|(N(F^*), F^*) : (N(F^*), F^*) \to (N(F), F)$ is a diffeomorphism and $q|E(F^*) : E(F^*) \to E(F)$ is a normal imitation with non-diffeomorphic $E(F^*), E(F)$ for some regular neighborhoods $N(F^*), N(F)$ of F^*, F in S^3 and $E(F^*) = \mathrm{cl}(S^3 - N(F^*)), E(F) = \mathrm{cl}(S^3 - N(F))$.*

In Application 1, F, F^* have the same Seifert matrix with respect to any corresponding bases of $H_1(F^*; Z), H_1(F; Z)$ by q.

PROOF: Let D be a compression disk for F with ∂D a non-∂-parallel loop. A line bundle $D \times I$ of D in S^3 is regarded as a 2-handle attaching to F with $F \cap D \times I = (\partial D) \times I$. Let F' be the surface obtained from F by the surgery along $D \times I$, and E', the 3-manifold obtained from S^3 by splitting along F'. Consider a proper arc $a = p \times I, p \in \mathrm{int}D$, in E'. Except when F' is a disk, (E', a) is a good (3,1)-manifold pair and there is a normal imitation $q' : (E', a^*) \to (E', a)$ such that $E(a^*, E')$ is a hyperbolic 3-manifold with a large volume, so that $E(a^*, E')$ is non-diffeomorphic to $E(a, E')$. Let F^* be the surface obtained from F' by a 1-handle surgery along a^*. Then the normal imitation map q' defines a desired map $q : (S^3, F^*) \to (S^3, F)$. When F' is a disk, E' is a 3-ball and we have the same conclusion by considering a normal imitation $(S^3, k_a^*) \to (S^3, k_a)$ of the knot (S^3, k_a) obtained from (E', a) by spherical completion with $E(k_a^*, S^3)$ a hyperbolic 3-manifold with a large volume. This completes the proof.

APPLICATION 2. *For any $r(\geq 2)$-string tangle in a 3-ball B^3 and any 2-handle h^2 on B^3 attaching along an incompressible, non-∂-parallel loop in the sphere with $2r$ open disks removed, $\partial B^3 \cap E(t, B^3)$, there is an almost identical imitation $q : (B^3, t^*) \to (B^3, t)$ such that $E(t^*, B^3), E(t, B^3)$ are non-diffeomorphic and the extension $q^+ : (B^3 \cup h^2, t^*) \to (B^3 \cup h^2, t)$ of q by the identity on h^2 is homotopic to a diffeomorphism by a homotopy relative to $\partial(B^3 \cup h^2) \cup N(t^*)$.*

PROOF: Let $M = B^3 \cup h^2$ and $A = h^2 \cap \partial B^3$. Let a be a trivial arc in h^2, joining the two components $\partial h^2 - A$ so that $E(a, M) \cong B^3$. Then $(M, t \cup a)$ is a good (3,1)-manifold pair and we obtain an almost identical imitation q_M : $(M, t^* \cup a) \to (M, t \cup a)$ such that $E(t^* \cup a, M), E(t \cup a, M)$ are non-diffeomorphic. Identifying $E(a, M)$ with B^3, we see that q_M defines a desired almost identical imitation $q : (B^3, t^*) \to (B^3, t)$.

Let O^r denote the r-component trivial link. We consider a knot K and a spanning band b in S^3. Let $K(b)$ be the link obtained from K by surgery along b. Note that the component number of $K(b)$ is 2 or 1 according to whether the band b spans K with coherent or non-coherent orientation. Two such bands b, b' are said to be *equivalent* if there is a diffeomorphism f of S^3 with $f(K \cup b) = K \cup b'$. When $K(b) = O^2$, we can construct a unique ribbon 2-knot in S^4 from (K, b), denoted by $S(K, b)$. Y. Nakanishi and Y. Nakagawa [**16**] showed that there is a ribbon knot K which admits two inequivalent spanning bands b, b' with $K(b) = K(b') = O^2$. In fact, they showed that $S(K, b), S(K, b')$ are inequivalent. We have the following:

APPLICATION 3. *For any positive integer $N > 1$, there is a non-trivial superslice K^* which admits N inequivalent spanning bands $b_1, \ldots, b_N$ with $K^*(b_i) = O^2$ and $S(K^*, b_i)$, a trivial 2 knot,$i = 1, \ldots, N$.*

It is unknown whether or not the associated ribbon disks $D(K^*, b_i) \subset B^4, i = 1, \ldots, N$, are equivalent, although $\pi_1(B^4 - D(K^*, b_i)) \cong Z$ for all i. This question leads to a question asking whether or not the 2-knot $S_{ij} \subset S^4, i \neq j$, pasting $D(K^*, b_i) \subset B^4$ and $(K^*, b_j) \subset B^4$ along $K^* \subset S^3$ is (smoothly) trivial, although $\pi_1(S^4 - S_{ij}) \cong Z$.

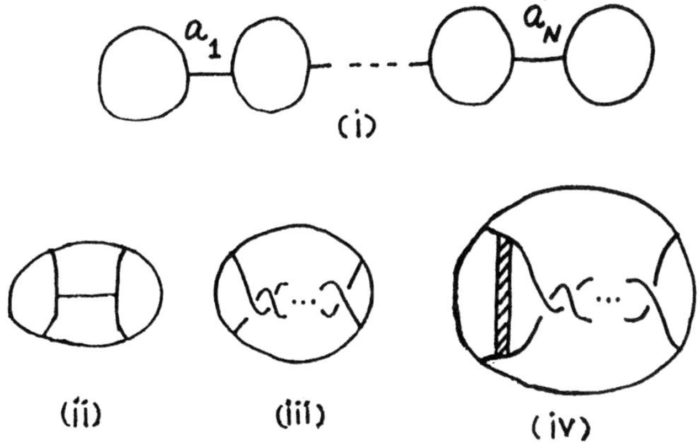

Fig. 5

PROOF: Let Γ be a graph in S^3, illustrated in Fig. 5(i). Let (M, L) be a good (3,1)-manifold pair obtained from (S^3, Γ) by removing an open 3-ball neighborhood of each degree 3 vertex of Γ. Let $q_M : (M, L^*) \to (M, L)$ be a 2-rigid almost identical imitation such that (M, L^*) has the hyperbolic covering property. Let $q_\Gamma : (S^3, \Gamma^*) \to (S^3, \Gamma)$ be the almost identical graph imitation obtained from q_M by spherical completions. Let V_i be a 3-ball neighborhood around the arc a_i, illustrated in Fig. 5(ii). We may consider that $q_\Gamma|(V_i, V_i \cap \Gamma^*) : (V_i, V_i \cap \Gamma^*) \to (V_i, V_i \cap \Gamma)$ is the identity. In particular, $a_i = q^{-1}(a_i)$. Let $q : (S^3, K^*) \to (S^3, K)$ be the almost identical imitation obtained from q by replacing the H-graph $V_i \cap \Gamma^* = V_i \cap \Gamma$, illustrated in Fig. 5(ii) with a 2-string braid with m_i full twists, illustrated in Fig. 5(iii). Then K is a trivial knot and hence K^* is a superslice for all m_i. Let b_i be a band in V_i spanning $V_i \cap K^* = V_i \cap K$, illustrated in Fig. 5(iv). By the property of almost identical imitation, we have $K^*(b_i) = K(b_i) = O^2$ for all i and all m_i. For the centerline $a_i'^*$ (or a_i',resp.) of the band b_i, regarded as a band spanning $K^*(b_i) = O^2$ (or $K(b_i) = O^2$, resp.), we see from the property of almost identical imitation that $a_i'^*$ is homotopic to a_i' by a homotopy relative to O^2. This implies that $S(K^*, b_i), S(K, b_i)$ are equivalent (cf. [**3**,Lemma 2.7]). Since $S(K, b_i)$ is trivial, $S(K^*, b_i)$ is trivial. Let $E = \mathrm{cl}(S^3 - \cup_{i=1}^N V_i)$. Note that the double covering space S_2^3 of S^3 branched along K^* is a union of the double covering space E_2 of E branched along along $E \cap K^*$ and the solid tori, T_i lifting $V_i, i = 1, ..., N$. Since (M, L^*) has the hyperbolic covering property, E_2 is hyperbolic. By Thurston's hyperbolic Dehn surgery [**20**], there is a positive integer m_0 such that $S_2^3, E_2^i = \mathrm{cl}(S_2^3 - T_i)$ are hyperbolic 3-manifolds and any isometry $E_2^i \to E_2^j$ preserves the unions of the cores of the attaching solid tori for all m_i's with $|m_i| \geq m_0$ and all i, j. For any such m_i's, suppose that $b_i, b_j, i \neq j$ are equivalent spanning bands for K^*. Then there is a diffeomorphism $E_2^i \to E_2^j$. By Mostow rigidity [**20**], there is an isometry of E_2 sending the end ∂T_i to the end ∂T_j. But, since q_M is 2-rigid, we see from Mostow rigidity that $\mathrm{Isom}E_2(\cong G(E_2 \to E))$ preserves each end of E_2, a contradiction. Hence any two bands of the b_i's are inequivalent spanning bands for K^*. This completes the proof.

For non-coherently spanning bands, we have a similar result as follows:

APPLICATION 4. *For any positive integer $N > 1$, there is a non-trivial superslice K^* which admits N inequivalent spanning bands $b_1, ...,\ b_N$ with $K^*(b_i) = O^1, i = 1, ..., N$.*

The Klein bottle $DM(K^*, b_i) \subset S^4$ which is the double of the associated Möbius band $M(K^*, b_i) \subset B^4$ is seen to be trivial for all i by a non-orientable version of [**3**, Lemma 2.7]. Hence $\pi_1(B^4 - M(K^*, b_i)) \cong Z_2$ for all i. However, it is unknown whether or not the Möbius bands $M(K^*, b_i) \subset B^4, i = 1, ..., N$ are equivalent and whether or not the Klein bottle $K_{ij} \subset S^4, i \neq j$, pasting $M(K^*, b_i) \subset B^4$ and $M(K^*, b_j) \subset B^4$ along $K^* \subset S^3$ is trivial, although $\pi_1(S^4 - K_{ij}) \cong Z_2$.

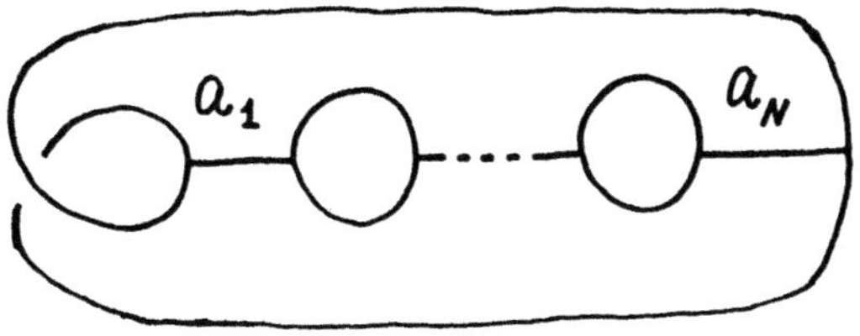

Fig. 6

PROOF: The proof is obtained from an argument parallel to the proof of Application 3 if we consider the graph Γ in S^3 illustrated in Fig. 6 in place of Fig. 5(i) and we take $\Sigma_{i=1}^{N} m_i = 0$ to assure that K is trivial.

Finally, we consider an unknotting operation on a knot K in S^3. We call a 3-ball B in S^3 a *place of unknotting operation on* a knot K if $K \cap B$ is a trivial 2-string tangle in B and one crossing change of the two strings of $B \cap K$ makes K a trivial knot. Two places B, B' of unknotting operations on K are said to be *equivalent* if there is a diffeomorphism f of S^3 such that $f(K) = K$ and $f(B) = B'$. We have the following result, answering a question of Y. Nakanishi and Y. Uchida asking whether there is an unknotting number one knot with two or more inequivalent places of unknotting operations:

APPLICATION 5. *For any positive integer $N > 1$, there is an unknotting number one superslice K^* with N inequivalent places of unknotting operations.*

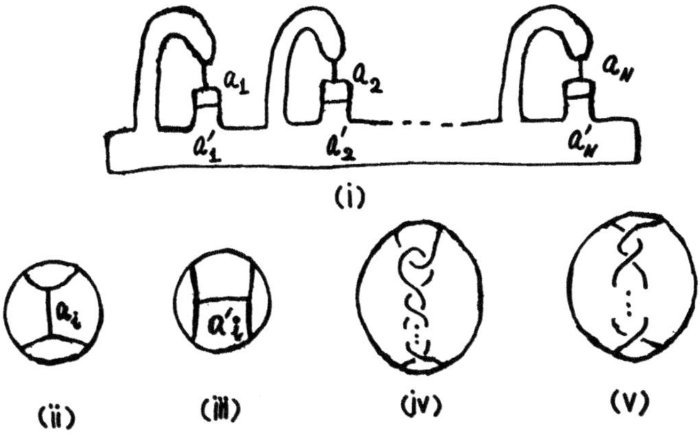

Fig. 7

PROOF: Let Γ be a graph in S^3 illustrated in Fig. 7(i). Let (M, L) be a good (3,1)-manifold pair obtained from (S^3, Γ) by removing an open 3-ball neighborhood of each degree 3 vertex of Γ. Let $q_M : (M, L^*) \to (M, L)$ be a 2-rigid almost identical imitation such that (M, L^*) has the hyperbolic covering property. Let $q_\Gamma : (S^3, \Gamma^*) \to (S^3, \Gamma)$ be the almost identical graph imitation obtained from q_M by spherical completions. Let B_i (or B_i',resp.) be a 3-ball neighborhood around the arc a_i (or a_i', resp.), illustrated in Fig. 7(ii) (or 7(iii), resp.). Let $V = \cup_{i=1}^N (B_i \cup B_i')$ and $E = \mathrm{cl}(S^3 - V)$. We may consider that $q_\Gamma | V \cap (S^3, \Gamma^*) : V \cap (S^3, \Gamma^*) \to V \cap (S^3, \Gamma)$ is the identity. In particular, $a_i = q^{-1}(a_i), a_i' = q^{-1}(a_i')$. Let $q : (S^3, K^*) \to (S^3, K)$ be the almost identical imitation obtained from q by replacing the H-graph $B_i \cap \Gamma^* = B_i \cap \Gamma$, illustrated in Fig. 7(ii) with a 2-string tangle with m_i full twists, illustrated in Fig. 7(iv) and the H-graph $B_i' \cap \Gamma^* = B_i' \cap \Gamma$, illustrated in Fig. 7(iii) with a 2-string braid of $-m_i$ full twists, illustrated in Fig. 7(v). Then K is a trivial knot and hence K^* is a superslice for all m_i's. By the property of almost identical imitation, we can consider each B_i as a place of unknotting operation on K^*. Note that the double covering space S_2^3 of S^3 branched along K^* is a union of the double covering space E_2 of E branched along $E \cap K^*$ and the solid tori, T_i, T_i' lifting $B_i, B_i', i = 1, ..., N$. Since (M, L^*) has the hyperbolic covering property, E_2 is hyperbolic. By Thurston's hyperbolic Dehn surgery [**20**], there is a positive integer m_0 such that $S_2^3, E_2^i = \mathrm{cl}(S_2^3 - T_i)$ are hyperbolic 3-manifolds and any isometry $E_2^i \to E_2^j$ preserves the unions of the cores of the attaching solid tori for all m_i's with $|m_i| \geq m_0$ and all i, j. For any such m_i's, suppose $B_i, B_j, i \neq j$ are equivalent places of unknotting operations on K^*. Then there is a diffeomorphism $E_2^i \cong E_2^j$. By Mostow rigidity [**20**], there is an isometry of E_2 sending the end ∂T_i to the end ∂T_j. But, since q_M is 2-rigid, we see from Mostow rigidity that $\mathrm{Isom} E_2 (\cong G(E_2 \to E))$ preserves each end of E_2, a contradiction. Hence K^* is an unknotting number one superslice with N inequivalent places of unknotting operations. This completes the proof.

There are three other recent applications of topological imitations. One is done for constructing links with the same skein polynomial close to the skein polynomial of any previously given link(see[**9**]). The second is done for constructing 3-manifolds with the same quantum invariant close to the quantum invariant of any previouly given 3-manifold(see[**11**]). The third is done for constructing knots along which the Dehn surgeries with a fixed slope produce the same manifold.

REFERENCES

[0] Brakes, W. (1980), *Property R and superslices*, Quart. J. Math. **31**, 263-281.

[1] Fox, R. H. (1962), *A quick trip through knot theory*, "Topology of 3-Manifolds and Related Topics," Prentice-Hall, pp. 120-167.

[2] Gordon, C. McA.-Sumners, D. W. (1975), *Knotted ball pairs whose product with an interval is unknotted*, Math. Ann. **217**, 47-52.

[3] Hosokawa, F.-Kawauchi, A. (1979), *Proposals for unknotted surfaces in four-spaces*, Osaka J. Math. **16**, 233-248.

[4] Kanenobu, T. (1986), *Hyperbolic links with Brunnian properties*, J. Math. Soc. Japan **38**, 295-308.

[5] Kawauchi, A. (1988), *Imitations of (3,1)-dimensional manifold pairs*, (in Japanese), Sugaku **40**, 193-204; Sugaku Expositions **2** (1989), 141-156.

[6] Kawauchi, A. (1989), *An imitation theory of manifolds*, Osaka J. Math. **26**, 447-464.

[7] Kawauchi, A. (1989), *Almost identical imitations of (3,1)-dimensional manifold pairs*, Osaka J. Math. **26**, 743-758.

[8] Kawauchi, A. (1992), *Almost identical imitations of (3,1)-dimensional manifold pairs and the branched coverings*, Osaka J. Math. **29**, 299-327.

[9] Kawauchi, A. (1992), *Almost identical link imitations and the skein polynomial*, "Knots 90," Walter de Gruyter, Berlin-New York, pp. 465-476.

[10] Kawauchi, A. (1992), *Almost identical imitations of (3,1)-dimensional manifold pairs and the manifold mutation*, revised preprint.

[11] Kawauchi, A. (1992), *Topological imitations of 3-manifolds and the quantum invariants*, preprint.

[12] Kirby, R. C.-Melvin, P. (1978), *Slice knots and Property R*, Invent. Math. **45**, 57-59.

[13] Kinoshita, S.-Terasaka, H. (1957), *On union of knots*, Osaka Math. J **9**, 131-153.

[14] Myers, R. (1983), *Homology cobordisms, link concordances, and hyperbolic 3-manifolds*, Trans. Amer. Math. Soc. **278**, 271-288.

[15] Nakanishi, Y. (1981), *Primeness of links*, Math. Sem. Notes Kobe Univ. **9**, 415-440.

[16] Nakanishi, Y.-Nakagawa, Y. (1982), *On ribbon knots*, Math. Sem. Notes Kobe Univ. **10**, 423-430.

[17] Ruberman, D. (1987), *Mutation and volumes of knots in S^3*, Invent. Math. **90**, 189-215.

[18] Soma, T. (1983), *Simple links and tangles*, Tokyo J. Math. **6**, 65-73.

[19] Suzuki, S. (1976), *Knotting problems of 2-spheres in 4-sphere*, Math. Sem. Notes Kobe Univ. **4**, 241-371.

[20] Thurston, W. P. (1978), *The Geometry and Topology of 3-Manifolds*, preprint.

A WILD VARIATION OF ARTIN'S BRAIDS

SIEGFRIED MORAN
The University
Canterbury
Kent CT2 7NF
England

ABSTRACT. We show how to derive a theory of (infinite) braids which includes the well known classical theory of braids due to Emil Artin. We use this theory to give a theory of (infinite) knots which includes the classical theory of knots. A vital role in the new theory is played by the topological free group which was first defined and studied by Graham Higman. This is only a preliminary attempt to extend the classical theory. As we make use of a somewhat intuitive approach to the subject, we leave a number of basic problems still outstanding. We assume that the reader is familiar with the theory of braids and knots as given in E. Artin [1], J.S. Birman [2] and S. Moran [5]. We also make use of the fact that the fundamental group of a certain region in $\boldsymbol{R}^2$ is a topological free group as given in G. Higman [4], H.B. Griffiths [3], J.W. Morgan and I.A. Morrison [6].

0. Let n be a fixed integer ≥ 1. Let ${}_{-n}B_n$ denote the group of braids with top ends $P_{-n}, \ldots, P_{-1}, P_0, P_1, \ldots, P_n$ and bottom ends $Q_{-n}, \ldots, Q_{-1}, Q_0, Q_1, \ldots, Q_n$. The latter will also be denoted by the integers $-n, \ldots, -1, 0, 1, \ldots, n$. Then we have the usual set of generators

$$\sigma_{-n},\ \sigma_{-n+1},\ \ldots,\ \sigma_{-1},\ \sigma_0,\ \sigma_1,\ \ldots,\ \sigma_{n-1}$$

for ${}_{-n}B_n$. Now there is a natural embedding of

$$ {}_{-n}B_n \quad \text{in} \quad {}_{-n-1}B_{n+1} \ .$$

So one has the group

$$\bigcup_{n\geq 1}\ {}_{-n}B_n\ ,$$

which we denote by ${}_{-\infty}B_{\infty}$. Further ${}_{-\infty}B_{\infty}$ is a topological group with

$$U_0 \supset U_1 \supset \ldots \supset U_m \supset \ldots$$

being the fundamental system of neighbourhoods of the unit element, where

$$U_m = <\sigma_i\ ;\ \ -m-2 \geq i \quad \text{or} \quad i \geq m+1>$$

for all $m \geq 0$. Clearly U_m consists of those braids whose $0, \pm 1, \ldots, \pm m$ strings are of the trivial form given in Figure 1 and no other strings overpass or underpass these strings.

M.E. Bozhüyük (ed.), Topics in Knot Theory, 85–106.

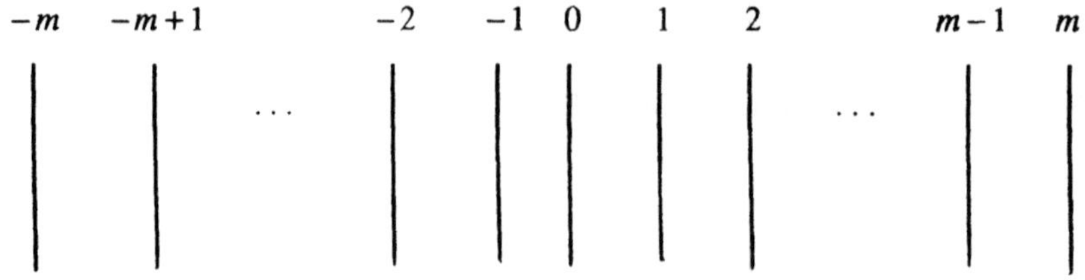

Figure 1. An element of U_m.

The following properties of this fundamental system of neighbourhoods are easily seen to hold:

(i) $\bigcap_{m \geq 0} U_m = \langle e \rangle$;

(ii) if σ belongs to ${}_{-\infty}B_{\infty}$ and U_m is given, then there exists a positive integer n so that

$$\sigma^{-1} U_n \sigma \subseteq U_m .$$

Note that the strings of an arbitrary element of ${}_{-\infty}B_{\infty}$ differ from the strings of the trivial braid e only for a finite number of strings.

We can now define the completion ${}_{-\infty}\hat{B}_{\infty}$ of ${}_{-\infty}B_{\infty}$ with respect to this fundamental system of neighbourhoods. A Cauchy sequence $\{\sigma^{(\nu)}\}_{\nu \geq 0}$ in ${}_{-\infty}B_{\infty}$ is a sequence of elements of ${}_{-\infty}B_{\infty}$ with the property that given a positive integer m there exists a positive integer $N = N(m)$ such that

$$U_m \, \sigma^{(\mu)} \, U_m \;=\; U_m \, \sigma^{(\nu)} \, U_m$$

for all $\mu, \nu \geq N$.

Two Cauchy sequences $\{\sigma^{(\nu)}\}$ and $\{\tau^{(\nu)}\}$ are said to be *equivalent* when given a positive integer m there exists a positive integer $N = N(m)$ so that

$$U_m \, \sigma^{(\nu)} \, U_m \;=\; U_m \, \tau^{(\nu)} \, U_m$$

for all positive integers $\nu \geq N$.

The collection of all equivalence classes of Cauchy sequences is denoted by ${}_{-\infty}\hat{B}_{\infty}$ with the product operation not always being defined by

$$\{\sigma^{(\nu)}\} \, . \, \{\tau^{(\nu)}\} = \{\sigma^{(\nu)} \, . \, \tau^{(\nu)}\} ,$$

since this operation is not always invariant under equivalence. This product will subsequently be referred to as the *convergent product*. The topological space ${}_{-\infty}\hat{B}_{\infty}$ has the following fundamental system of neighbourhoods of the unit element

$$\hat{U}_0 \supset \hat{U}_1 \supset \ldots \supset \hat{U}_m \supset \ldots ,$$

where $\hat{U}_m$ denotes the closure of U_m in ${}_{-\infty}\hat{B}_\infty$ for every m.

It is clear from the above construction that every element of ${}_{-\infty}\hat{B}_\infty$ can be expressed as an ordered product of braids of the form

$$\prod_{-\infty<j<\infty} \sigma_{i_j}^{\varepsilon_j} = \lim_{n\to\infty} \left[\prod_{-n<j<n} \sigma_{i_j}^{\varepsilon_j}\right],$$

where every ε_j is an integer and $|i_j| \geq m\ (>0)$ for all j such that

$$|j| \geq \text{some positive integer } N = N(m).$$

An element of ${}_{-\infty}\hat{B}_\infty$ will be called a *convergent* braid. We shall now concern ourselves mainly with another class of braids – sane braids – which will be introduced in the following geometric way.

1. We consider infinite braids which consist of strings starting at $P_{-\infty}, \ldots, P_{-1}, P_0, P_1, \ldots, P_\infty$ (with $P_{-\infty}$ or P_∞ or both may be possibly omitted) and each string terminating at some Q_j for a uniquely defined integer j. Each string is taken to be of bounded length and is made up of a finite number of straight line segments (including strings at $\pm\infty$). Further we assume that each P_i has the same z-coordinate and the same y-coordinate for all i, while

$$x\text{-coordinate of } P_i < x\text{-coordinate of } P_j$$

for $i < j$. We also insist that

$$d(P_i, P_{i+1}) \to 0 \quad \text{and} \quad d(P_{-i}, P_{-i-1}) \to 0$$

as $i \to \infty$ with

$$\lim_{i\to\infty} P_i \neq P_\infty \quad \text{and} \quad \lim_{i\to-\infty} P_i \neq P_{-\infty}.$$

We assume that a similar situation holds for the lower ends

$$Q_i\ (i = 0, \pm 1, \pm 2, \ldots).$$

The removal of all but a finite number of the strings of the braid is always assumed to result in an ordinary finite braid (one may of course have to renumber the ends of the braid). For each integer n, the n-th string is taken to be such that it overpasses and underpasses only a finite number of strings. In the case of the string at either $-\infty$ or ∞, it is assumed that as one traverses from one end of the string to the other end, there are only a finite number of variations from over (under) passing to under (over) passing. Further, except possibly at the initial stages, no string goes off to $-\infty$ or ∞ at any later stage. If one takes any horizontal section of the strings of a fixed braid, then the distance between neighbouring strings tends to zero as one goes either to the left or to the right along this section. Every braid we consider lies of course inside a finite right circular cylinder of $\boldsymbol{R}^3$. A braid of the above described form is called a *sane braid*. In fact more concisely one could say that each of its strings is made up of a finite number of finite line segments. The collection of all sane braids is denoted by ${}_{-\infty}B_\infty^c$. Every sane braid has a corresponding (1:1) mapping

$$\phi:\ S \to S/\{\pm\infty\},$$

where S is one of the following four sets

$$\{0, \pm 1, \pm 2, \ldots, \pm\infty\} \quad \text{or} \quad \{0, \pm 1, \pm 2, \ldots, \infty\} \quad \text{or}$$

$$\{0, \pm 1, \pm 2, \ldots, -\infty\} \quad \text{or} \quad \{0, \pm 1, \pm 2, \ldots\} .$$

This is determined by the fact that the i-th string terminates at Q_j for i in S.

Two sane braids are said† to be *equal* (or *string isotopic*) if there exists a continuous deformation of one braid onto the other so that at each stage of the deformation one has a sane braid. Further the removal of all but a finite number of strings of the first braid leaves one always with an ordinary braid which is string isotopic (under the above given deformation) to an ordinary braid which is obtained from the second braid given above by the removal of all but a finite number of its strings.

The above defined sane braids form a topological partial group ${}_{-\infty}B^{c}_{\infty}$ under the usual product operation of braids (called the *braid product*).

Note the following properties:

(a) the trivial braid is the unit element;

(b) the braid product of two sane braids both of which have a string at either $-\infty$ or ∞ is not defined;

(c) a sane braid which has a string at either $-\infty$ or ∞ does not have an inverse;

(d) if τ is a sane braid without strings at either $-\infty$ or ∞ and σ is any sane braid, then τ^{-1}, $\sigma\tau$, $\tau\sigma$

are sane braids.

So the collection of all sane braids without strings at either $-\infty$ or ∞ form a group under the operation of braid product. We denote this group by ${}_{-\infty}B^{cf}_{\infty}$. This is a topological group under a similar topology as that described in Section **0** above.

One can define as above the partial groups

$${}_{n}B^{c}_{\infty} \quad \text{and} \quad {}_{-\infty}B^{c}_{n}$$

in terms of the groups

$${}_{n}B_{\infty} \quad \text{and} \quad {}_{-\infty}B_{n} ,$$

where n is an integer.

2. We have the usual generating braids σ_i and σ_i^{-1}, which are given in Figure 2 respectively, where $i = 0, \pm 1, \pm 2, \ldots$ and the *height* of σ_i and σ_i^{-1} (that is $d(P_i, Q_i)$) tends to zero as $i \to \infty$ or $i \to -\infty$.

† It is not yet clear whether this is the appropriate definition.

Figure 2. The braids σ_i and σ_i^{-1}.

The following are examples of infinite braids which are constructed in terms of convergent braids – they may or may not belong to the above defined

$$_{-\infty}B^c_\infty \ , \quad _{-\infty}B^c_n \ , \quad _nB^c_\infty \ .$$

(i) $\sigma_1\sigma_2 \ldots \sigma_n \ldots$ is a convergent braid, has a picture of the form given in Figure 3 and hence is not considered here as $1 \to \infty$.

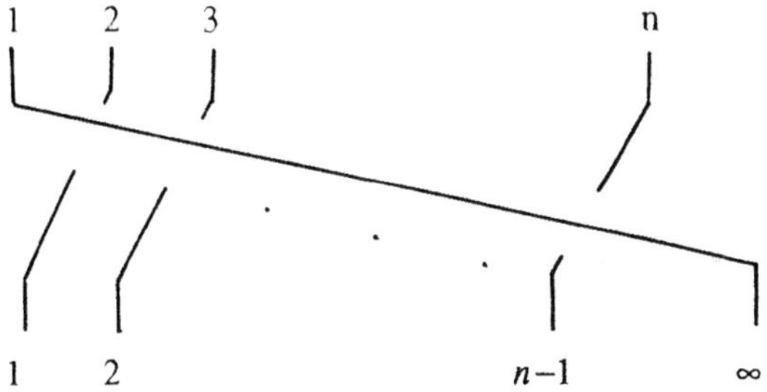

Figure 3. The braid $\sigma_1\sigma_2 \ldots \sigma_n \ldots$.

(ii) $\ldots \sigma_n \ldots \sigma_2\sigma_1$ is a convergent braid, belongs to $_1B^c_\infty$ and has a picture of the form given in Figure 4. The corresponding mapping is the "permutation" $(1\ 2\ 3 \ldots \infty)$.

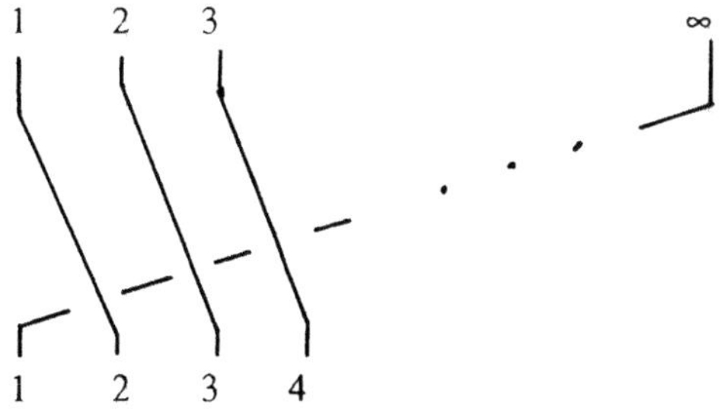

Figure 4. The braid $\ldots \sigma_n \ldots \sigma_2\sigma_1$.

(iii) $(\ldots\sigma_n\ldots\sigma_2\sigma_1)^2$ is not defined in ${}_1B^c_\infty$.

(iv) $\ldots\sigma_{-n}\ldots\sigma_{-2}\sigma_{-1}$ is a convergent braid, belongs to ${}_{-\infty}B^c_0$ and has a picture of the form given in Figure 5. The corresponding mapping is the "permutation" $(-1\ -2\ -3\ \ldots -\infty)$.

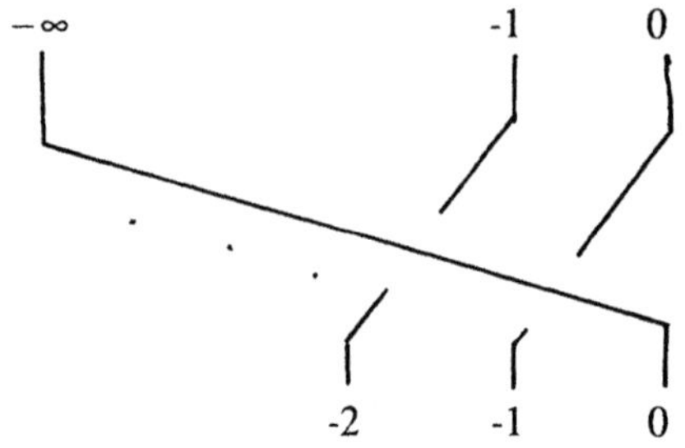

Figure 5. The braid $\ldots\sigma_{-n}\ldots\sigma_{-2}\sigma_{-1}$.

(v) $\ldots\sigma_n^3\ldots\sigma_3^3\sigma_2^3\sigma_1^3$ is a convergent braid but it does not belong to ${}_1B^c_\infty$. The string at ∞ does not consist of a finite number of straight line segments. It has a picture of the form given in Figure 6.

Figure 6. The braid $\ldots\sigma_n^3\ldots\sigma_3^3\sigma_2^3\sigma_1^3$.

(vi) $\ldots\sigma_{-n}\ldots\sigma_{-2}\sigma_{-1}\sigma_0\sigma_1\sigma_2\ldots\sigma_n\ldots$ is a convergent braid which does not belong to ${}_{-\infty}B^c_\infty$, since the corresponding mapping maps $-\infty$ to ∞.

(vii) $\sigma_1\sigma_3^3\sigma_5^5\ldots\sigma_{2n-1}^{2n-1}\ldots$ is a convergent braid which belongs to ${}_1B^c_\infty$. It has a picture of the form given in Figure 7 and (1 2) (3 4) . . . $(2n-1\ 2n)$ . . . is the corresponding mapping.

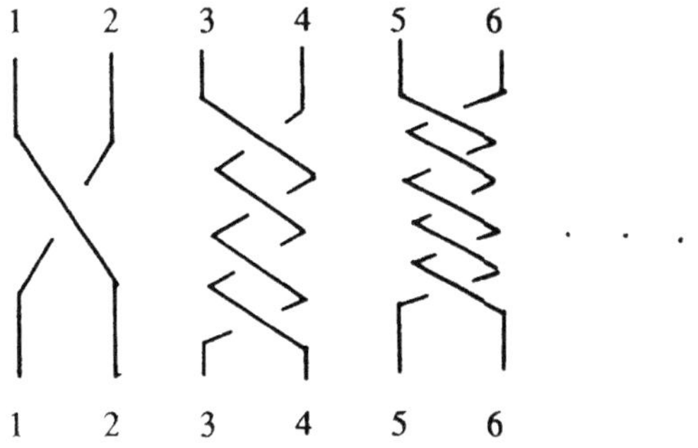

Figure 7. The braid $\sigma_1\sigma_3^3\sigma_5^5\ldots\sigma_{2n-1}^{2n-1}\ldots$.

(viii) $\sigma_1^2\sigma_2^2\sigma_3^2\ldots\sigma_n^2\ldots$ is a convergent braid which belongs to ${}_1B^c_\infty$. It has a picture of the form given in Figure 8. The corresponding mapping is the identity mapping.

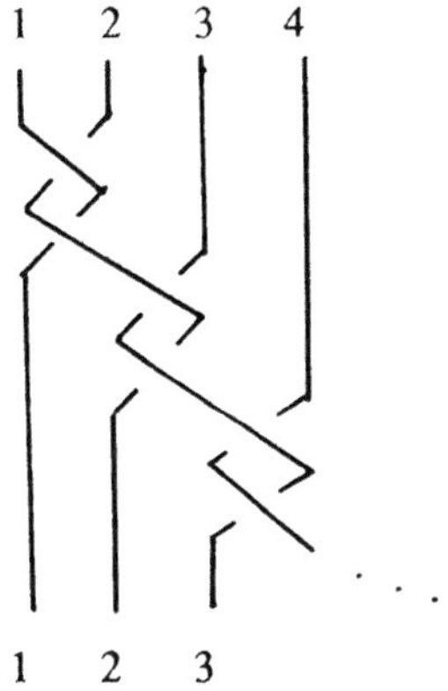

Figure 8. The braid $\sigma_1^2\sigma_2^2\sigma_3^2\ldots\sigma_n^2\ldots$

(ix) $\ldots\sigma_{2n}\sigma_{2n-1}^{-1}\ldots\sigma_3^{-1}\sigma_2\sigma_1^{-1}$ is a convergent braid which has a picture of the form given in Figure 9. It is not considered here as the string at ∞ has an infinite number of changes from underpassing to overpassing.

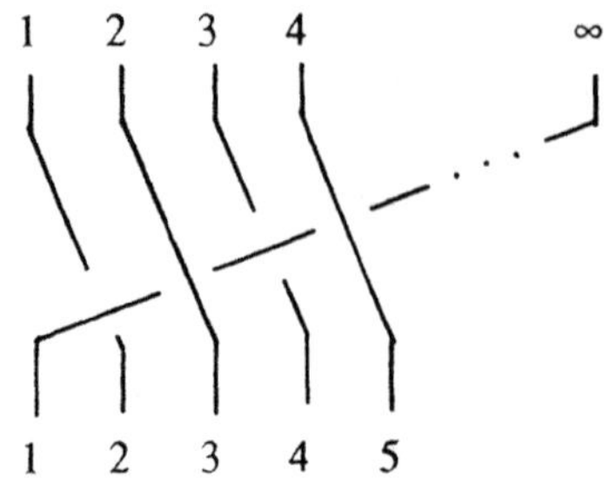

Figure 9. The braid $\ldots\sigma_{2n}\sigma_{2n-1}^{-1}\ldots\sigma_3^{-1}\sigma_2\sigma_1^{-1}$

3. Let ${}_{-\infty}P_{\infty}$ denote the subgroup of all pure braids on the strings $\ldots, -2, -1, 0, 1, 2, \ldots$ in the braid group ${}_{-\infty}B_{\infty}$. Let

$$A_{i,n} = \sigma_{n-1}\sigma_{n-2}\ldots\sigma_{i+1}\sigma_i^2\sigma_{i+1}^{-1}\ldots\sigma_{n-2}^{-1}\sigma_{n-1}^{-1}$$

for all $i < n$, where n is a fixed integer. Then the subgroup $<A_{i,n}; i < n> = V_n$ is a free group freely generated by the elements $A_{i,n}$ with $i < n$ for every n. Also the product

$$V^{(n)} = V_n.V_{n+1}.V_{n+2}\ldots = <A_{i,j}; i < j \geq n>$$

(only a finite number of elements in an infinite product being $\neq e$) is a normal subgroup of ${}_{-\infty}P_{\infty}$. These two results follow from standard properties of the elements $A_{i,j}$.

Now, as is well known, we can consider $A_{i,j}$ to be an automorphism $\bar{A}_{i,j}$ of the free group $F(\{\ldots, x_{-2}, x_{-1}, x_0, x_1, x_2, \ldots\})$ for every i and j. Define the automorphism Γ of this free group by

$$x_i\Gamma = x_{-i}^{-1} \quad \text{for all } i \;.$$

Then $\Gamma^2 = id$. Also

$$\Gamma.\bar{\sigma}_i.\Gamma = \bar{\sigma}_{-i-1}^{-1} \quad \text{and} \quad \Gamma.\bar{A}_{i,j}.\Gamma = \bar{A}_{-j,-i}^{-1}$$

for all i and j. Now define

$${}_nV = <A_{n,j}; j > n> \quad \text{and} \quad {}^{(n)}V = \ldots {}_{n-2}V.{}_{n-1}V.{}_nV$$

for all n. Then

$$\Gamma.{}_nV.\Gamma = V_{-n} \quad \text{and} \quad \Gamma.{}^{(n)}V.\Gamma = V^{(-n)}$$

for all n. So similar results hold for ${}_nV$ and ${}^{(n)}V$ as those proved for V_n and $V^{(n)}$. In particular we have ${}^{(n)}V$ is a normal subgroup of ${}_{-\infty}P_{\infty}$. Also the factor group ${}_{-\infty}P_{\infty}/({}^{(-n)}V.V^{(n)})$ is naturally isomorphic to the pure braid group

$${}_{-n+1}P_{n-1} = <A_{i,j};\ i > -n \ \text{ and } \ j < n>\,,$$

where here n is a fixed positive integer.

The group ${}_{-\infty}P_{\infty}$ is a topological group with $\{{}^{(-n)}V.V^{(n)}\}_{n\geq 1}$ being a set of fundamental neighbourhoods of the unit element. So every element of ${}_{-\infty}P_{\infty}$ can be represented as an element of

$$\prod_{n<0} {}_nV \,.\, \prod_{n\geq 0} V_n \;,$$

where only a finite number of elements in an infinite product are not equal to e. Hence every element in the normal subgroup ${}_{-\infty}P^c_\infty$ of pure braids (lying inside the group ${}_{-\infty}B^{cf}_\infty$) can be represented as an element of the doubly infinite product

$$\prod_{n<0} {}_nV \,.\, \prod_{n\geq 0} V_n \,. \tag{3.1}$$

As we shall now see not every element lying inside the double infinite product (3.1) belongs to ${}_{-\infty}P^c_\infty$.

(i) $\ldots A_{-n-1,0} \cdot A_{-n,0} \cdots A_{-2,0} \cdot A_{-1,0}$ does not belong to ${}_{-\infty}P^c_\infty$, since the 0-th string has an infinite number of variations from underpasses to overpasses. It has a picture of the form given in Figure 10.

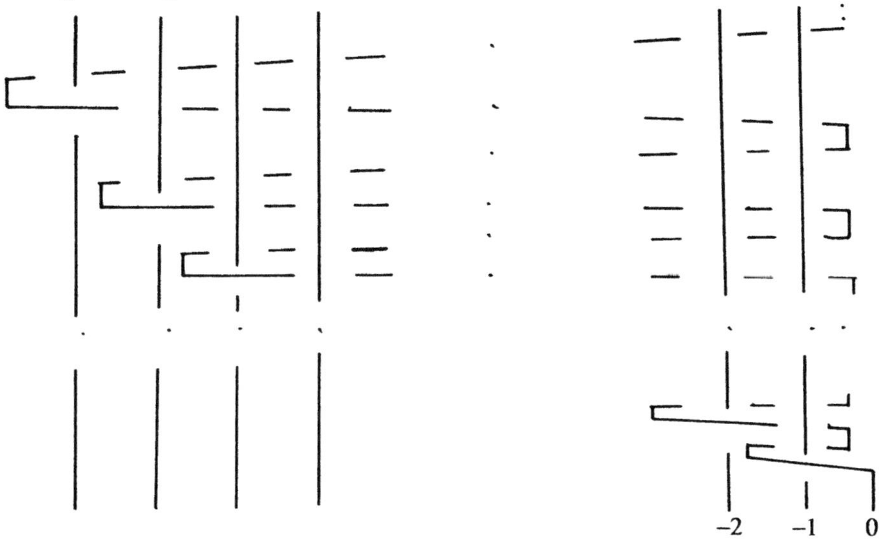

Figure 10. The braid $\ldots A_{-n-1,0} \cdot A_{-n,0} \cdots A_{-2,0} \cdot A_{-1,0} \cdot$

(ii) $A_{-1,1} \cdot A_{-2,2} \cdot A_{-3,3} \cdots A_{-n,n} \cdots$ does not belong to ${}_{-\infty}P^c_\infty$, since the 0-th string overpasses an infinite number of strings. It has a picture of the form given in Figure 11.

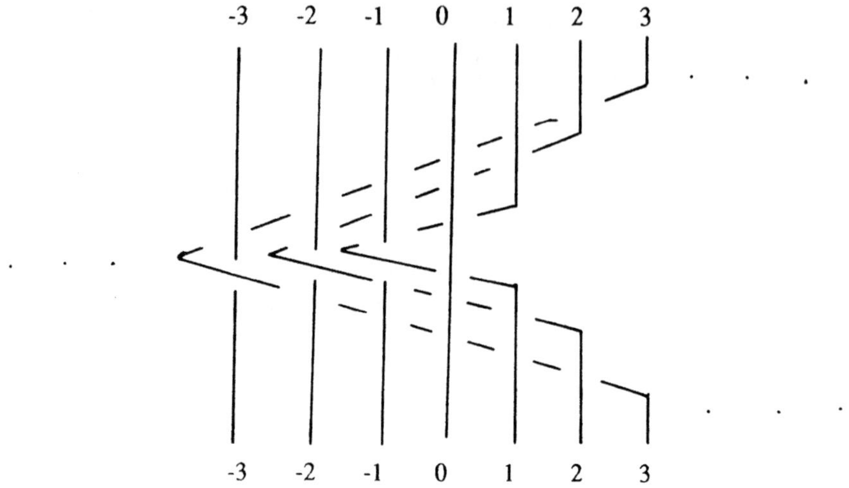

Figure 11. The braid $A_{-1,1} \cdot A_{-2,2} \cdot A_{-3,3} \ldots A_{-n,n} \ldots$.

Suppose that the element $\prod_{n<0} {}_n\mathrm{v} \cdot \prod_{n\geq 0} \mathrm{v}_n$ of (3.1) belongs to ${}_{-\infty}P^c_\infty$. Then we have that

$$\mathrm{v}_n \in \langle A_{i,n};\ i_n \leq i < n \rangle \text{ for } n \geq 0 \text{ while}$$

$$\mathrm{v}_n \notin \langle A_{i,n};\ n > i \text{ with } i \geq i_n + 1 \rangle \text{ and}$$

$${}_m\mathrm{v} \in \langle A_{m,j};\ j_m \geq j > m \rangle \text{ for } m < 0 \text{ while}$$

$${}_m\mathrm{v} \notin \langle A_{m,j};\ m < j \leq j_m - 1 \rangle \text{ for nontrivial elements.}$$

Hence we must have that

$$i_n \to \infty \text{ as } n \to \infty \text{ and } j_m \to -\infty \text{ as } m \to -\infty$$

in order for the above element to belong to ${}_{-\infty}P^c_\infty$.

4. There exists a natural group homomorphism of the free group $F(\{x_{-n-1}, \ldots, x_0, \ldots, x_{n+1}\})$ onto the free group $F(\{x_{-n}, \ldots, x_0, \ldots, x_n\})$, which is defined by mapping x_{-n-1} and x_{n+1} onto the unit element e and leaving the remaining free generators fixed. The corresponding inverse limit is denoted by

$$\underset{-\infty<i<\infty}{H} <x_i> .$$

It was first introduced and studied by G. Higman [4]. It has the following series of closed normal subgroups which are generated by

$$\underset{i\leq -n}{H} <x_i> * \underset{i\geq n}{H} <x_i> \quad \text{with} \quad n \geq 1$$

as a fundamental basis of neighbourhoods of the unit element. The following subgroup

$$\underset{-\infty<i<\infty}{T} <x_i> = \bigcap_{n\geq 1} (\underset{i<-n}{H} <x_i> * <x_{-n}> * \ldots * <x_0> * \ldots * <x_n> * \underset{i>n}{H} <x_i>)$$

was first considered by G. Higman [4] and was also studied by H.B. Griffiths [3], J.W. Morgan and I.A. Morrison [6].

This latter group has the subgroups $\underset{i\le 0}{T} <x_i>$ and $\underset{i\ge 0}{T} <x_i>$, which are naturally either isomorphic or anti-isomorphic (depending on the topological conventions used). This is also true for the subgroups $\underset{i\le 0}{H} <x_i>$ and $\underset{i\ge 0}{H} <x_i>$ of $\underset{-\infty<i<\infty}{H} <x_i>$. We have that

$$\underset{-\infty<i<\infty}{T} <x_i> = \underset{i<0}{T} <x_i> * \underset{i\ge 0}{T} <x_i> .$$

Let D be a closed disc in $\mathbf{R}^2$ with an infinite number of points $P_1, P_2, \ldots, P_i, \ldots$ removed so that $d(P_i, P_{i+1}) \to 0$ as $i \to \infty$. Since D is pathwise connected, the fundamental group of D is independent (upto isomorphism) of the choice of the base point P. We assume that D is of the form given in Figure 12, where $P_i = (-\frac{7}{8}(\frac{3}{4})^{i-1}, 0)$ for $i \ge 1$ and $P = (0,0)$

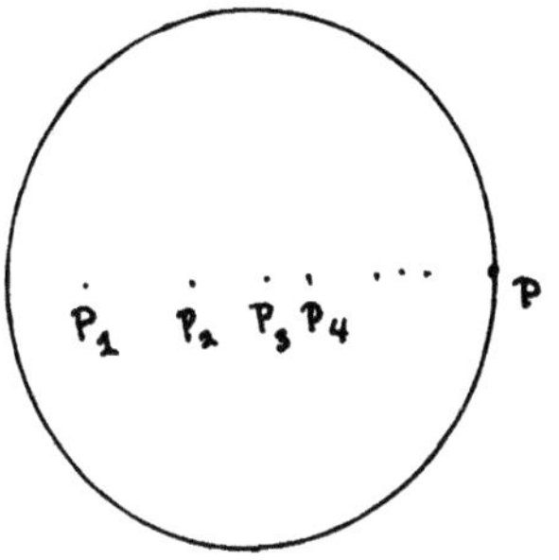

Figure 12. The disc D with an infinite number of points removed.

The space D is the union of spaces D_i of the form given in Figure 13, where the outer circle has diameter $(\frac{3}{4})^{i-1}$, while the inner circle has diameter $(\frac{3}{4})^i$ and P_i lies halfway in between the two circles.

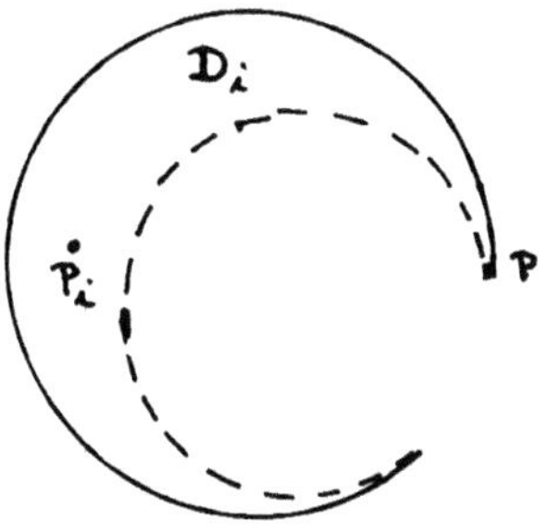

Figure 13. the space D_i.

So

$$D = \bigcup_{i\ge 1} D_i \quad \text{and} \quad D_i \cap D_j = \{P\} \quad \text{for } i \ne j .$$

It follows from H.B. Griffiths [3], J.W. Morgan and I.A. Morrison [6] that

$$\pi(D, P) \cong \underset{i\ge 1}{T} <x_i> ,$$

where x_i denotes a loop in D at P which encircles the missing point P_i only, that is, of the form given in Figure 14, where the radius of the small circle is $\frac{1}{3}\, d(P_i, P_{i+1})$ for all i

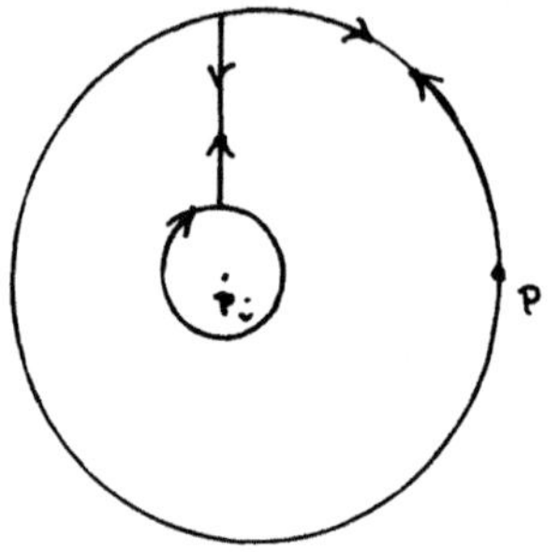

Figure 14. The loop x_i in D.

A similar argument to that given above together with the Theorem of Seifert and van Kampen shows that the fundamental group of a closed circular disc Y with missing points

$$P_0,\ P_{\pm 1},\ P_{\pm 2}, \ldots$$

and two further points of the form

$$\begin{matrix} P_{-\infty} & & P_{-3} & P_{-2} & P_{-1} & P_0 & P_1 & P_2 & P_3 & & P_{\infty} \\ \bullet & \cdots & \bullet & \bullet & \bullet & \bullet & \bullet & \bullet & \bullet & \cdots & \bullet \end{matrix} \quad \text{in } \mathbf{R}^2,$$

where $\lim\limits_{i\to-\infty} P_i \neq P_{-\infty}$ and $\lim\limits_{i\to\infty} P_i \neq P_{\infty}$ while $d(P_{-i}, P_{-i-1}) \to 0$ and $d(P_i, P_{i+1}) \to 0$ as $i \to \infty$, has fundamental group

$$<x_{-\infty}> * \underset{-\infty<i<\infty}{T} <x_i> * <x_{\infty}>$$

which we also denote by

$$\underset{-\infty\leq i\leq\infty}{T} <x_i> .$$

5. Let W_m denote the subgroup of all automorphisms of the free group

$${}_{-\infty}F_{\infty} = F(\{\ \ldots, x_{-n}, \ldots, x_{-1}, x_0, x_1, \ldots, x_n, \ldots\})$$

which leave all but a finite number of x_i fixed, including all those x_i with $|i| \leq m$, and which induce an automorphism of the free group $\prod\limits_{|i|>m}^{*} <x_i>$ for every $m \geq 1$. The subgroups

$$W_1 \supset W_2 \supset \ldots \supset W_m \supset \ldots$$

form a basis of the neighbourhoods of id in the group $A = \mathrm{Aut}^f({}_{-\infty}F_{\infty})$ of all those automorphisms of the free group ${}_{-\infty}F_{\infty}$ which leave all but a finite number of the x_i fixed. The corresponding completion $\hat{A}$ of A consists of all Cauchy sequences $\{\alpha^{(\mu)}\}$ of automorphisms so that given m there exists a positive integer $N = N(m)$ with

$$W_m \circ \alpha^{(\mu)} \circ W_m = W_m \circ \alpha^{(\nu)} \circ W_m$$

for all $\mu, \nu \geq N$.

$\{\alpha^{(\mu)}\}$ is said to be equivalent to $\{\beta^{(\mu)}\}$ if and only if given a positive integer m there exists a positive integer $N = N(m)$ so that

$$W_m \circ \alpha^{(\mu)} \circ W_m = W_m \circ \beta^{(\mu)} \circ W_m$$

for all $\mu \geq N$.

The *convergent product*, when it is unambiguously defined, is taken to be

$$\{\alpha^{(\mu)}\}\ \{\beta^{(\mu)}\} = \{\alpha^{(\mu)} \circ \beta^{(\mu)}\}.$$

There is always the *composition* of the mappings $\{\alpha^{(\mu)}\}$ and $\{\beta^{(\mu)}\}$, provided it is defined.

It is clear that every element α of $\hat{A}$ can be expressed as an ordered composition of mappings of the form

$$\alpha = \prod^{o}_{-\infty<j<\infty} \alpha_{i_j} = \lim_{n\to\infty} \Big(\prod^{o}_{-n<j<n} \alpha_{i_j} \Big),$$

where α_{i_j} belongs to $W_{|i_j|}$ for all j and $|i_j| > m$ for all j such that

$$|j| \geq \text{ some positive integer } \quad N = N(m).$$

Further it is clear that every element α of $\hat{A}$ gives a homomorphism of the free group ${}_{-\infty}F_{\infty}$ into $\underset{-\infty<i<\infty}{H} <x_i>$. If we know that α is continuous on the free group ${}_{-\infty}F_{\infty}$, then this homomorphism extends in a natural way to an endomorphism on $\underset{-\infty<i<\infty}{H} <x_i>$. If

$$\alpha^{-1} = \prod^{o}_{\infty>j>-\infty} \alpha_{i_j}^{-1}$$

is also continuous, then α can be considered to be a topological automorphism of $\underset{-\infty<i<\infty}{H} <x_i>$.

We have the well known group isomorphism of the braid group

$$f : {}_{-\infty}B_{\infty} \to {}_{-\infty}\bar{B}_{\infty} \subset A,$$

where $(U_m)f \subset W_m$ for all m, and so f is continuous. Hence we have

$$\hat{f} : {}_{-\infty}\hat{B}_{\infty} \to {}_{-\infty}\hat{\bar{B}}_{\infty} \subset \hat{A}$$

is a continuous partial isomorphism into. If

$$\prod_{-\infty<j<\infty} \sigma_{i_j}^{\varepsilon_j} \quad \text{belongs to} \quad {}_{-\infty}\hat{B}_{\infty},$$

then its image under $\hat{f}$ can be taken to be

$$\prod_{-\infty<j<\infty} \bar{\sigma}_{i_j}^{\varepsilon_j}.$$

Further if $\prod_{-\infty<j<\infty} \bar{\sigma}_{i_j}^{\varepsilon_j}$ represents an element of ${}_{-\infty}B^{c}_{\infty}$, then we have that $\prod_{-\infty<j<\infty} \bar{\sigma}_{i_j}^{\varepsilon_j}$ can *sometimes* be taken to be a continuous isomorphism of $\underset{-\infty\leq i\leq\infty}{T} <x_i>$ into $\underset{-\infty<i<\infty}{T} <x_i>$. This can be seen by means of the following process. Let σ be an element of ${}_{-\infty}B^{c}_{\infty}$. Then *sometimes* we have another way of looking at $\sigma\hat{f} = \bar{\sigma}$. Take a loop in the upper punctured plane of the tops of the strings of σ and push the loop down to the lower punctured plane via a cylinder with σ removed. As we shall see, it is not always possible to reach the lower punctured plane. If it turns out to be possible, then this *pushing down process* gives the required continuous isomorphism in the case when σ has strings at $-\infty$ and ∞. There are of course a number of other cases to consider – depending on how many strings occur in σ.

We now investigate as to when the pushing down process is possible for a fixed braid σ of ${}_{-\infty}B^c_\infty$. Clearly it is always true that

$$x_j\bar{\sigma} \quad \text{belongs to} \quad \underset{-\infty<i<\infty}{T} <x_i>$$

for every j. However it is not always true that $\bar{\sigma}$ maps an arbitrary infinite product in $\underset{-\infty<i<\infty}{T} <x_i>$ into $\underset{-\infty<i<\infty}{T} <x_i>$. A finite product is no problem. For example

$$x_i\,(\ldots\bar{\sigma}_n^{-2}\ldots\bar{\sigma}_2^{-2}\,\bar{\sigma}_1^{-2}) = (x_{i+1}^{-1}\ldots).x_i.(x_{i+1}^{-1}\ldots)^{-1}$$

for every i and so

$$(\prod_{i\geq 1} x_i)\,(\ldots\bar{\sigma}_n^{-2}\ldots\bar{\sigma}_2^{-2}\,\bar{\sigma}_1^{-2}) \neq (\prod_{i\geq 1} x_i)\,.$$

This shows that the pushing down process is not possible for this braid. For if the pushing down process is possible for σ, then

$$x_i\bar{\sigma} = A_i\,.\,x_{i\phi}\,.\,A_i^{-1}\ ,$$

where every A_i is an element of $\underset{-\infty<i<\infty}{T} <x_i>$ and

$$(\prod_{-\infty\leq i\leq\infty} x_i)\,\bar{\sigma} = \prod_{-\infty<i<\infty} x_i\,. \tag{5.1}$$

Finally we suppose that σ is a sane braid which is the braid product of a finite number of sane convergent braids, that is,

$$\sigma = \sigma^{(1)}\,\sigma^{(2)}\,\ldots\,\sigma^{(k)}$$

and such that the pushing down process is possible in each $\sigma^{(i)}$. Then in the semigroup of all continuous endomorphisms of $\underset{-\infty\leq i\leq\infty}{T} <x_i>$ we can take

$$\overline{\sigma^{(1)}}\,.\,\overline{\sigma^{(2)}}\,.\,\ldots\,.\,\overline{\sigma^{(k)}}$$

to be the endomorphism $\bar{\sigma}$ corresponding to σ and also the pushing down process is possible in σ.

6. We now give some miscellaneous results concerning braids. First of all we have the following four fundamental sane braids which have strings at $\pm\infty$

$$\sigma_{-\infty,i} = \sigma^{1}_{-\infty,i} = \ldots\,\sigma_k\,\sigma_{k+1}\ldots\sigma_i\ ,$$
$$\sigma^{-1}_{-\infty,i} = \ldots\,\sigma_k^{-1}\,\sigma_{k+1}^{-1}\ldots\sigma_i^{-1}\ ,$$

$$\sigma_{\infty,i} = \sigma^{1}_{\infty,i} = \ldots\,\sigma_k\,\sigma_{k-1}\ldots\sigma_i\ ,$$
$$\sigma^{-1}_{\infty,i} = \ldots\,\sigma_k^{-1}\,\sigma_{k-1}^{-1}\ldots\sigma_i^{-1}\ .$$

Every element of ${}_{-\infty}B^c_\infty$ can be expressed in one of the following forms

$$\sigma^{(1)}.\sigma^{\varepsilon}_{\pm\infty,i}\,.\,\sigma^{(2)}.\sigma^{\mu}_{\mp\infty,j}\,.\,\sigma^{(3)},\quad \sigma^{(1)}\,.\,\sigma^{\varepsilon}_{-\infty,i}\,.\,\sigma^{(2)},\quad \sigma^{(1)}\,.\,\sigma^{\varepsilon}_{\infty,j}\,.\,\sigma^{(2)},\quad \sigma^{(1)}$$

where $\sigma^{(1)}$, $\sigma^{(2)}$, $\sigma^{(3)}$ are sane braids without strings at $-\infty$ or at ∞ and ε, $\mu=\pm 1$. A sane braid has a string at $\pm\infty$ if and only if the corresponding "permutation" is an element of the completion ${}_{-\infty}S^c_\infty$ of the usual permutation group ${}_{-\infty}S_\infty$ of all finite permutations of the integers but does not belong to ${}_{-\infty}S^{cf}_{\infty}$. The topology is similar to the one used in ${}_{-\infty}B_{\infty}$. By ${}_{-\infty}S^{cf}_{\infty}$ we denote the group of those "permutations" which are in fact permutations of the integers, that is, are onto mappings. Every element of ${}_{-\infty}S^{cf}_{\infty}$ can be expressed uniquely (upto order) as a product of disjoint cycles. Some of these cycles can be infinite, in which case they will be doubly infinite of the form

$$(\ldots a\, j\, i\, b \ldots) \ .$$

Now let

$$(\ldots a\, j\, i\, b \ldots)$$

be *any* cycle – finite or infinite of the above form.

(i) If there exists a smallest integer i in this cycle, then we write

$$(\ldots a\, j\, i\, b \ldots) = (i\ j)\ (\ldots a\, j\, b \ldots) \ .$$

Repeated application of this procedure will give a representation of this cycle as a product (possibly infinite to the right) of transpositions.

(ii) If there exists a largest integer j in the above given cycle, then we write

$$(\ldots a\, j\, i\, b \ldots) = (\ldots a\, i\, b \ldots)\,(i\ j) \ .$$

Repeated application of this procedure will give a representation of this cycle as a product (possibly infinite to the left) of transpositions.

(iii) If neither of the above conditions hold, then one proceeds via integers of smallest modulus and thus gets a situation similar to that described in case (i).

One obtains a corresponding sane convergent braid on replacing

$$(i\ j) \ \text{ by } \ \sigma_{i,j} \ \text{ when } \ i < j \ ,$$

where $\sigma_{i,j} = \sigma_{j-1}\ \sigma_{i,j-1}\ \sigma_{j-1}$ and $\sigma_{i,i+1} = \sigma_i$.

There exists a group homomorphism of ${}_{-\infty}B^{cf}_{\infty}$ onto ${}_{-\infty}S^{cf}_{\infty}$ which has the pure braid group ${}_{-\infty}P^{c}_{\infty}$ as kernel. Hence one has the following fact:

Every sane braid is the braid product of at most eight sane convergent braids.

We end this section by giving a large class of sane braids in which the pushing down procedure is possible. If σ is a sane braid, then the *graph* of σ is obtained from the picture of σ by joining up the gaps left by the underpassing line segments whenever they occur as given in Figure 15

Figure 15. Going from picture of σ to the graph of σ.

The *initial descendant* of the j-th string of a sane braid σ without strings at $\pm\infty$ is defined as follows.

(a) Suppose j is a nonnegative integer. Then the initial descendant of the j-th string of σ is the smallest integer k obtained by starting at P_j and transversing down the graph of σ till Q_k is obtained.

(b) Suppose that j is a negative integer. Then the initial descendant of the j-th string of σ is the largest integer k obtained by starting at P_j and transversing down the graph of σ till Q_k is reached.

The graphs of the sane braids

$$\ldots \sigma_n^{-2} \ldots \sigma_2^{-2} \sigma_1^{-2} \quad \text{and} \quad \ldots \sigma_n^{2} \ldots \sigma_2^{2} \sigma_1^{2}$$

are both equal to the graph given in Figure 16.

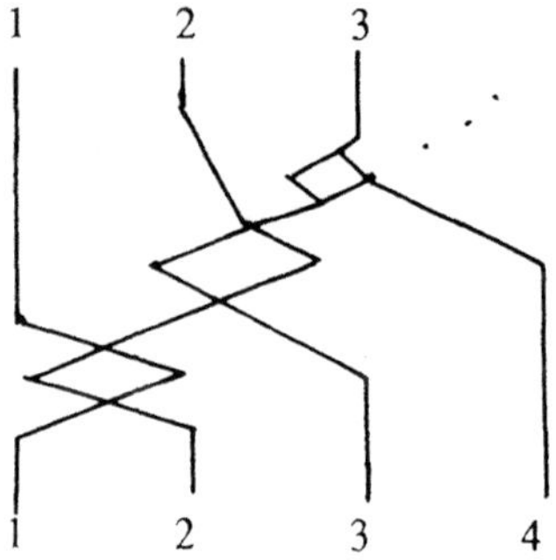

Figure 16. The graph of two different sane braids.

The initial descendant of the j-th string of both these braids is 1 for every positive integer j. The pushing down procedure is not possible in both these cases. The first case was dealt with in the previous section. The second example follows since

$$(x_1\, x_3 \ldots x_{2n-1} \ldots)\ (\ldots \sigma_i^{-2} \ldots \sigma_2^{-2}\, \sigma_1^{-2})$$

contains an infinite number of distinct occurences of $x_1^{\pm 1}$.

The pushing down procedure is possible in the following sane braids:

$$\sigma_{-\infty,i}^{1}\ ,\ \sigma_{-\infty,i}^{-1}\ ,\ \sigma_{\infty,i}^{1}\ ,\ \sigma_{\infty,i}^{-1}\ ,$$

any sane braid without strings at $\pm\infty$ which has the property that the initial descendant of the j-th string tends to $-\infty(+\infty)$ as j tends to $-\infty(+\infty$ respectively) whenever there are an infinite number of strings to the left (right respectively)

and any meaningful finite product of the above braids.

7. We form the link $L(\sigma)$ associated with the sane braid σ, which is obtained from σ by identifying P_i with Q_i for every integer i. Further we join P_∞ to $\lim_{i\to\infty} P_i$ by a line of the form

$$\lim_{i\to\infty} P_i \ \text{———————}\ P_\infty$$

which lies in the upper plane of σ. We do the same for $P_{-\infty}$ and $\lim_{i\to\infty} P_{-i}$, namely, we have the line

$$P_{-\infty} \;\text{———————}\; \lim_{i\to\infty} P_{-i}$$

The group of the link $L(\sigma)$ is $G(L(\sigma))$, which is the fundamental group $\pi(C_{\mathbf{R}^3}(L(\sigma)))$. *If the pushing down procedure is possible in* σ, *then the group of the link* $L(\sigma)$ *is isomorphic to the factor group of* $\underset{-\infty<i<\infty}{T} <x_i>$ *modulo the closed normal subgroup generated by the relations*

$$x_i^{-1} \,.\, (x_i\,\bar{\sigma}) \qquad \textit{for all integers } i \;.$$

Note that the relations

$$x_\infty = x_\infty \bar{\sigma} \quad \text{and} \quad x_{-\infty} = x_{-\infty} \bar{\sigma}$$

can both be omitted, since both $x_{-\infty}\,\bar{\sigma}$ and $x_\infty\,\bar{\sigma}$ belong to $\underset{-\infty<i<\infty}{T} <x_i>$.

We outline a proof based on that given in S. Moran [5] Chapter 6 for ordinary braids. A proof based on that given in J. Birman [2] Theorem 2.2 would seem to require stronger assumptions.

Let *cyl* denote a solid cylinder in $\mathbf{R}^3$ which encloses the braid σ. By l we denote the straight line path joining the base point Q (in the lower plane) to the base point P (in the upper plane). Now let $C^*_{cyl}(\sigma)$ denote the space obtained from $C_{cyl}(\sigma)$ by removing the following lines from the upper plane

$$P_{-\infty} \;\text{————}\; \lim_{i\to\infty} P_{-i} \qquad\qquad \lim_{i\to\infty} P_i \;\text{————}\; P_\infty$$

and lines corresponding to these in the lower plane. Now in $C^*_{cyl}(\sigma)$ we have that

$$l^{-1}\, x_i\, l \qquad \text{is homotopic to} \qquad x_i\,\bar{\sigma}$$

for all i, by the pushing down procedure. Now let T be the torus obtained from *cyl* by adopting the obvious identification. Then there exists the natural continuous mapping

$$f:\; C^*_{cyl}(\sigma) \to C_T(L(\sigma)) \;.$$

This gives the group homomorphism

$$f_\pi :\; \pi(C^*_{cyl}(\sigma), Q) \to \pi(C_T(L(\sigma)), P)$$

which is into. Here the former group is isomorphic to $\underset{-\infty<i<\infty}{T} <x_i>$. Note that infinite products of elements of $\underset{-\infty<i<\infty}{T} <x_i>$ are not always given explicitly on a picture, since the lines such as the one joining P_∞ to $\lim_{i\to\infty} P_i$ gets in the way. Thus we have that the group

$$\pi(C_T(L(\sigma)), P)$$

is generated by l and $\underset{-\infty<i<\infty}{T} <x_i>$ and has the relations

$$l^{-1}\,(\lim_{n\to\infty} x^{(n)})\, l \;=\; \lim_{n\to\infty}\,(x^{(n)}\,\bar{\sigma})$$

for every $\lim_{n\to\infty} x^{(n)}$ in $\underset{-\infty<i<\infty}{T} <x_i>$. Now define the group G to be the group that is generated by l and $\underset{-\infty<i<\infty}{T} <x_i>$ and has the defining relations

$$l^{-1}\,(\lim_{n\to\infty} x^{(n)})\, l \;=\; \lim_{n\to\infty}\,(x^{(n)}\,\bar{\sigma})$$

for every $\lim_{n\to\infty} x^{(n)}$ in $\underset{-\infty<i<\infty}{T} <x_i>$. Now it is not difficult to see that every element of G is conjugate to an element of the form

$$w(\ldots, x_{-i}, \ldots, x_{-1}, x_0, x_1, \ldots, x_i, \ldots) \,.\, l^{-m} ,$$

where m is an integer, or its inverse. Next we map the group G homomorphically onto the group

$$\pi(C_T(L(\sigma)), P) \,.$$

This preserves the above given facts. We have that some element of the form $w \,.\, l^{-m}$ is equal to the unit element if and only if

$$w = e \qquad \text{and} \qquad m = 0$$

as can be seen in $C_T(L(\sigma))$. Hence

$$\pi(C_T(L(\sigma)), P) \equiv G \,.$$

The proof can now be completed by using the Theorem of Seifert and van Kampen.

8. We now work out the groups of links associated with some sane braids.

(i) $$x_k \ldots \bar{\sigma}_n \ldots \bar{\sigma}_{i+1}\, \bar{\sigma}_i = \begin{cases} x_i\, x_{k+1}\, x_i^{-1} & \text{for } i \le k < \infty \\ x_k & \text{for } i > k \,. \end{cases}$$

Hence the group $G(L(\ldots \sigma_n \ldots \sigma_{i+1}\, \sigma_i))$ is isomorphic to $\underset{k \ge i}{T} \langle x_i \rangle$ modulo the closed normal subgroup generated by $x_k^{-1}\, x_i\, x_{k+1}\, x_i^{-1}$ for all $k \ge i$, where we consider the braid to be an element of ${}_iB^c_\infty$. This factor group is isomorphic to the multiplicative group $U(Z)/R(Z)$. Here $U(Z)$ is the unrestricted direct product

$$\prod_{1 \le i < \infty}^{UX} \langle t_i \rangle$$

of copies of the infinite cyclic group and $R(Z)$ is the restricted direct product of the same groups. Note that the loop

$$\prod_{1 \le i < \infty} x_i$$

is represented by the coset

$$\left(\prod_{1 \le i < \infty} t_i^i \right) R(Z) \,.$$

(ii) $$x_k \ldots \bar{\sigma}_n^{-1} \ldots \bar{\sigma}_{i+1}^{-1}\, \bar{\sigma}_i^{-1} = \begin{cases} x_{k+1} & \text{for } i \le k < \infty \\ x_k & \text{for } i > k \,. \end{cases}$$

Consider this sane braid to be an element of ${}_iB^c_\infty$. Then the group of the corresponding knot is isomorphic to $\underset{k \ge i}{T} \langle x_i \rangle$ modulo the closed normal subgroup determined by

$$x_i = x_{i+1}, \quad x_{i+1} = x_{i+2}, \ldots \,.$$

This is the multiplicative group $U(Z)/R(Z)$ which is given in (i) above.

(iii) The groups of the knots corresponding to the sane braids $\sigma_{-\infty,i}$ and $\sigma_{-\infty,i}^{-1}$ (considered as elements of ${}_{-\infty}B^c_{i+1}$) also have groups isomorphic to the multiplicative group $U(Z)/R(Z)$ which is given in (i) above.

(iv) Consider $\sigma_1^2\, \sigma_2^2 \ldots \sigma_n^2 \ldots$ as an element of ${}_1P^c_\infty$. Now

$$x_k\, \bar{\sigma}_1^{-2} \bar{\sigma}_2^{-2} \ldots \bar{\sigma}_n^{-2} \ldots = \begin{cases} x_{k-1}\, u_k\, x_{k-1}^{-1} & \text{for } k \ge 2 \\ u_1 & \text{for } k = 1 \,, \end{cases}$$

where $x_k\bar{\sigma}_k^{2}\bar{\sigma}_{k+1}^{2}\ldots = u_k = (x_k u_{k+1}) \,.\, x_k \,.\, (x_k u_{k+1})^{-1}$ for $k \geq 1$. The group of the link is isomorphic to $\underset{i\geq 1}{T} <x_i>$ modulo the closed normal subgroup generated by the relations

$$x_k \;=\; x_{k-1} \,.\, u_k \,.\, x_{k-1}^{-1} \quad \text{for all } k \geq 2 \text{ and } x_1 \;=\; u_1 \,.$$

This gives that $u_k \;=\; x_{k-1}^{-1} x_k x_{k-1}$ and so we have that

$$(x_k,\, x_{k+1}) \;=\; x_{k+1}^{-1} \,.\, (x_{k-1},\, x_k) \,.\, x_{k+1}$$

for all $k \geq 2$. Repeated application of this relation gives that

$$(x_k,\, x_{k+1}) = (x_{k+1}^{-1}\; x_k^{-1}\; \ldots x_3^{-1}) \,.\, (x_1,\, x_2) \,.\, (\,\ldots)^{-1} \tag{8.1}$$

for all $k \geq 2$. We also have that

$$u_1 = (x_1 u_2) \,.\, x_1 \,.\, (x_1 u_2)^{-1}\,, \quad u_1 = x_1, \quad u_2 = x_1^{-1} x_2 x_1 \,.$$

This gives that

$$(x_1,\, x_2) \;=\; 1 \,.$$

So the relation (8.1) gives that the group of this link is isomorphic to the group $\underset{i\geq 0}{T} <x_i>$ modulo the closed normal subgroup generated by

$$(x_k,\, x_{k+1}) \qquad \text{for all} \qquad k \geq 1 \,.$$

(v) Consider $\ldots \sigma_n \ldots \sigma_2 \sigma_1 \,.\, \sigma_1^2 \sigma_2^2 \ldots \sigma_n^2 \ldots$ as an element of ${}_1B_\infty^c$. Now by (i) and (iv) above we have that the group of the corresponding knot has "defining relations"

$$x_i \;=\; (u_1 x_i) \,.\, u_{i+1} \,.\, (u_1 x_i)^{-1} \qquad \text{for all} \qquad i \geq 1 \tag{8.2}$$

where

$$u_k \;=\; (x_k u_{k+1}) \,.\, x_k \,.\, (x_k u_{k+1})^{-1} \qquad \text{for all} \qquad k \geq 1 \,. \tag{8.3}$$

These relations for $i = k = 1$ give that $u_1 = u_2$ and a proof by induction on i shows that $u_i = u_1$ for all i. So the group of this knot has topological generators

$$x_1,\, x_2,\, \ldots,\, x_n,\, \ldots$$

and topological defining relations

$$u_i \;=\; u_1 \qquad \text{for all } i \tag{8.4}$$

where u_k is defined by equation (8.3). This group is the Topologist's free product of a countably infinite number of copies of B_3 with one subgroup being amalgamated. The resulting "amalgamated" subgroup is isomorphic to the group $U(Z)/R(Z)$ which is given in (i) above.

(vi) The group of the knot corresponding to the sane braid

$$\ldots \sigma_n^{-1} \ldots \sigma_2^{-1}\, \sigma_1^{-1} \,.\, \sigma_1^2\, \sigma_2^2 \ldots \sigma_n^2 \ldots$$

in ${}_1B_\infty^c$ is isomorphic to the multiplicative group $U(Z)/R(Z)$ which is given in (i) above.

(vii) The sane braid

$$\ldots \sigma_{-n} \ldots \sigma_{-2}\, \sigma_{-1} \quad . \quad \ldots \sigma_n \ldots \sigma_2\, \sigma_1\, \sigma_0 \quad (n \geq 1)$$

has a picture of the form given in Figure 17 as an element of ${}_{-\infty}B^{c}_{\infty}$.

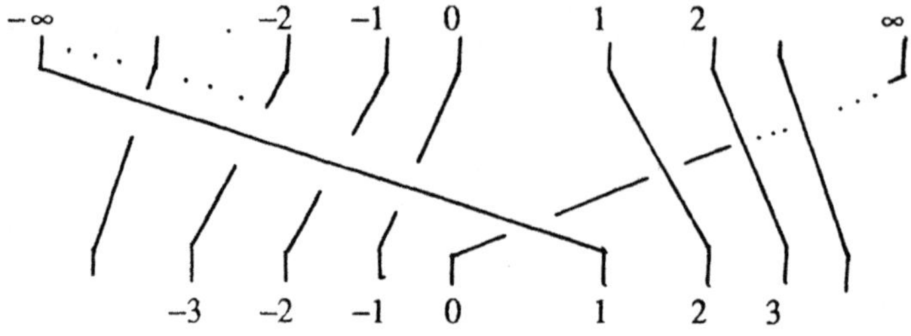

Figure 17. The braid $\ldots\sigma_{-n}\ldots\sigma_{-2}\,\sigma_{-1}\ldots\sigma_{n}\ldots\sigma_{2}\,\sigma_{1}\,\sigma_{0}\quad(n\geq 1)$

The group of the corresponding knot has topological generators

$$\ldots, x_{-k}, \ldots, x_{-1}, x_0, x_1, \ldots, x_k, \ldots$$

and topological defining relations

$$x_{-k}=x_{-k-1}\ \text{ for all }\ k\geq 0\ \text{ and }\ x_0^{-1}x_kx_0=x_{k+1}\ \text{ for all }\ k\geq 1\,.$$

Hence this group is isomorphic to the free product $<x_1>*(U(Z)/R(Z))$.

9. We conclude this paper by enumerating some properties of the groups of these links and pointing out some open problems.

(i) If τ is a sane braid without strings at $\pm\infty$ and the pushing down process is possible both in τ and τ^{-1}, then

$$G(L(\tau))\cong G(L(\tau^{-1}))\ .$$

(ii) If σ and τ are sane braids so that $\sigma\tau$ is defined, τ has no strings at $\pm\infty$ and the pushing down process is possible both in σ and in $\tau^{\pm 1}$, then

$$G(L(\sigma\tau))\cong G(L(\tau\sigma))\ .$$

(iii) The knots and links constructed in this paper are not closed subsets of $\boldsymbol{R}^3$ – they are relatively compact.

(iv) If σ is a sane braid in which the pushing down process is possible, then the group of the link $L(\sigma)$ modulo the closure of its commutator subgroup is either isomorphic to the unrestricted direct product of countably infinite number of infinite cyclic groups or contains $U(Z)/R(Z)$ as a subgroup. In the first case $L(\sigma)$ is a link with a countably infinite number of finite components and σ does not have strings at $\pm\infty$. In the second case $L(\sigma)$ has a component of infinite length – this is so for instance if σ has strings at $\pm\infty$.

(v) If σ is a sane braid with strings both at $-\infty$ and ∞ such that $L(\sigma)$ is a knot, then $L(\sigma)$ is not pathwise connected.

(vi) There is a dual theory to that of sane braids. One can go over to such braids by replacing every sane braid by its formal inverse. Example (i) of Section 2 is an example of such a braid.

(vii) We have left open the questions as to the relationships between the varying

equalities, products, topologies

defined throughout the paper. However we note the following:

(a) the convergent braids

$$\ldots\sigma_n\ldots\sigma_2\sigma_1 \quad \text{and} \quad \ldots\sigma_{n+1}^{-n}\,\sigma_n^{n}\ldots\sigma_3^{-2}\sigma_2^{2}\,.\,\sigma_2^{-1}\sigma_1$$

are equivalent (as defined in Section 0) although one is sane while the other is not;

(b) the convergent product (as defined in Section 0) of the convergent braid $\ldots\sigma_n\ldots\sigma_2\sigma_1$ with itself (if it exists) would have a string at ∞ and also at $\infty+1$ (or more appropriately at ω and at $\omega+1$) – this is different from the braid product (if one were to define it even for more general braids as is implicitly done in the proof of the main result in Section **7** and (e) below by introducing extra lines at the points of infinity);

(c) in some ways it would be natural to explore the theory of braids with strings at $\omega+1$ and so on, but we shall not do so here;

(d) if

$$\sigma_\nu\ldots\sigma_2\sigma_1\sigma_1^2\sigma_2^2\ldots\sigma_\nu^2=\sigma^{(\nu)} \quad \text{and} \quad \sigma_1^3\sigma_2^3\ldots\sigma_\nu^3=\tau^{(\nu)},$$

then $L(\sigma^{(\nu)})$ and $L(\tau^{(\nu)})$ are string isotopic knots for every positive integer ν, but $\lim\limits_{\nu\to\infty}\sigma^{(\nu)}$ gives a sane braid while $\lim\limits_{\nu\to\infty}\tau^{(\nu)}$ does not give a sane braid (see (v) of Section 8 and (v) of Section 2);

(e) suppose that the sane braid

$$\sigma=\sigma^{(1)}\,.\,\sigma^{(2)}\,.\;\ldots\;.\,\sigma^{(k)}$$

is the braid product of the sane convergent braids $\sigma^{(1)}, \sigma^{(2)}, \ldots, \sigma^{(k)}$, where

$$\sigma^{(i)}=\lim_{n_i}\,(\sigma^{(i)}\,(n_i))$$

is the appropriate limit of finite braids for $i=1, 2, \ldots, k$. Then

$$\sigma=\lim_{n_k}\,(\ldots(\lim_{n_2}\,(\lim_{n_1}\,(\sigma^{(1)}(n_1)\,.\,\sigma^{(2)}(n_2)\,.\;\ldots\;.\,\sigma^{(k)}(n_k))))\ldots)$$

and

$$L(\sigma)=\lim_{n_k}\,(\ldots(\lim_{n_2}\,(\lim_{n_1}\,(L(\sigma^{(1)}(n_1)\,.\,\sigma^{(2)}(n_2)\,.\;\ldots\;.\,\sigma^{(k)}(n_k)))))\ldots).$$

References

[1] *The Collected Works of Emil Artin*, (1965). Edited by Serge Lang and John T. Tate. Addison-Wesley.

[2] Birman, Joan S. (1974), *Braids, Links and Mapping Class Groups*. Princeton.

[3] Griffiths, H.B. (1956). Infinite products of semigroups and local connectivity. *Proceedings London Mathematical Society*, **6**, 455-485.

[4] Higman, Graham (1952), Unrestricted free products of groups and varieties of topological groups. *Journal London Mathematical Society*, **27**, 73-81.

[5] Moran, Siegfried (1983), *The Mathematical Theory of Knots and Braids. An Introduction*. North-Holland.

[6] Morgan, John W. and Morrison, Ian A. (1986), A van Kampen theorem for weak joins. *Proceedings London Mathematical Society*, **53**, 562-576.

Acknowledgements
I am grateful to the Pure Mathematics Departments of the Universities of Bielefeld, Bochum, Exeter and Erzurum for opportunities to talk about the above described theory of infinite braids and thus clarify my thoughts on this matter. Hugh Morton kindly helped me to eliminate a mistake. Thanks are also due to Mrs Julie Snook for so ably typing a somewhat abstruse manuscript.

INVARIANTS OF LINKS AND 3-MANIFOLDS FROM SKEIN THEORY AND FROM QUANTUM GROUPS.

H. R. MORTON
Department of Pure Mathematics
University of Liverpool
PO Box 147
Liverpool L69 3BX
England.

ABSTRACT. Starting with Kauffman's bracket polynomial the techniques of linear skein theory are used to present and package a family of polynomial invariants for a framed link. An equivalent family of invariants is derived from representations of the quantum group $SU(2)_q$. Specialisation of the variable q leads to invariants of a 3-manifold defined by surgery on a framed link, in terms of the invariants of the link. A similar programme is outlined relating the invariants constructed from the Homfly polynomial to those derived from the quantum groups $SU(k)_q$.

Introduction.

In this series of talks I shall start by discussing the knot invariants and algebra related to Kauffman's bracket polynomial, and the construction of 3-manifold invariants from them. The whole area can alternatively be viewed in terms of representations of the quantum group $SU(2)_q$; I shall exhibit descriptions which have a convenient interpretation in either light, and also give the means for translating between them. My presentation here is based on the bracket polynomial, and has much in common with the work of Lickorish, [16], and Blanchet, Habegger, Masbaum and Vogel, [2].

A direct approach on the quantum group route is given in my paper with Strickland, [23], which draws directly on the early work of Kirillov and Reshetikhin, [13]. A more general basis for the use of quantum group representations in constructing knot invariants is given in the work of Reshetikhin and Turaev, [31]. Detailed descriptions of representations for $SU(2)_q$ can be found in Kirby and Melvin, [12]; while these are based on specialisations of $SU(2)_q$ in which the parameter q is a root of unity they do present careful and explicit details which enable the less complicated case of generic q to be handled as well.

The reason for their treatment is to give an account of the invariants of a 3-manifold M which depend on the choice of a root of unity, and a quantum group (in this case $SU(2)_q$), in terms of the invariants of any framed link which determines the 3-manifold M by the process of surgery on the link. These 3-manifold invariants were first constructed in this way by Reshetikhin and Turaev, [32]; their existence and general properties were proposed originally by Witten, [43], based on interpretations of constructions from theoretical physics. Other accounts are given in [24], [16] and [2]. Those in [16] and [2] are based entirely on the bracket polynomial, while the account in [24] uses the quantum group representations at generic q as a means of establishing properties of the knot invariants, and then makes constructions based on the evaluations of these at a given root of unity, without having to consider the more complicated representation theory which arises at the root of unity.

My presentation of the 3-manifold invariants uses the techniques appropriate to the bracket polynomial. I shall restate the point that, however the knot invariants are con-

M.E. Bozhüyük (ed.), Topics in Knot Theory, 107–155.

structed, whether by quantum group representations or by bracket polynomials, there is a common halfway stage reached to which each of the constructions brings its own insights. The final attack on the question of manifold invariants can then be made from this point, no matter how it has been reached, although the representation theory provides invaluable guidance at this stage in, for example, setting up and choosing a suitable basis for a naturally occurring finite dimensional vector space.

I believe that a similar two-stage process is appropriate in constructing 3-manifold invariants from other quantum groups. Such a construction is done by slightly different means, for example, by Turaev and Wenzl, [37], and a general framework is given by Walker, [38], in the spirit of Segal's modular functors. It is possible to make a nice comparison of the knot invariants defined from the quantum groups $SU(k)_q$, for different k, with knot invariants based on the Homfly polynomial, [29], [41], [19]. This permits an analogous two-stage process, allowing the definition of 3-manifold invariants in terms of the knot invariants for generic q, with a root of unity substituted for q; the representation theory to be used in the first stage only requires the study of generic q, when the representations mirror directly those of the corresponding classical group. In the final section I shall give a description of the $SU(k)_q$ knot invariants from the point of view of Homfly polynomials, in a similar framework to the earlier talks, which can be thought of as dealing with the case $k = 2$. More details will be found in [19]; this gives a preparation of the common ground which could be used for the production of manifold invariants by specialising q to be a root of unity.

Readers of earlier versions of this paper should note some minor amendments in section 6, where the substitutions $v = s^{-k}, x = s^{-1/k}$ replace those used previously.

1. Knot invariants derived from Kauffman's bracket.

1.1 The bracket invariant.

In 1986, Kauffman showed how to construct an element $< D > \in \mathbf{Z}[A^{\pm 1}]$ for every plane diagram D of a knot or link in $\mathbf{R}^3$, which is determined (up to a constant) by two properties. These are

(1) $$< D_+ > = A < D_0 > + A^{-1} < D_\infty >,$$

or more pictorially

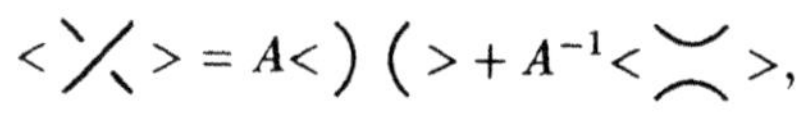

where D_+, D_0 and D_∞ are three link diagrams which only differ as shown.

(2) $$< D \sqcup O > = \delta < D >,$$

where $\delta = -A^2 - A^{-2}$, and $D \sqcup O$ is a diagram containing one component O which has no self-crossings, or crossings with the rest of the diagram D.

Example. Properties (1) and (2) allow the simplification of  as

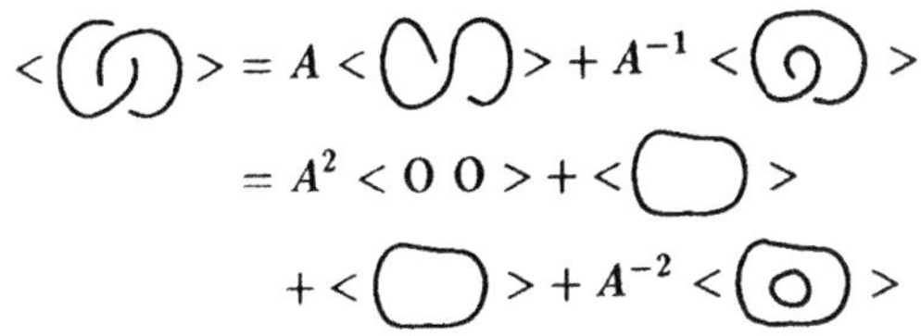

$$= (A^2 + A^{-2})\,\delta < O > + 2 < O >$$
$$= -(A^4 + A^{-4}) < O > .$$

In a similar way, $< D >$ can be written in terms of $< O >$ for any D; in Kauffman's original work the Laurent polynomial $< D >$ was normalised by taking $< O > = 1$, but now it is more often chosen to include use of the empty diagram ϕ, with the condition that $< \phi > = 1$, and consequently $< O > = \delta < \phi > = \delta$.

The reason for using properties (1) and (2) is given by Kauffman's theorem, which can be readily established.

THEOREM 1.1 (Kauffman) . *When a diagram D is altered by one of the Reidemeister moves R_{II} or R_{III} the value of $< D >$ is unchanged.* □

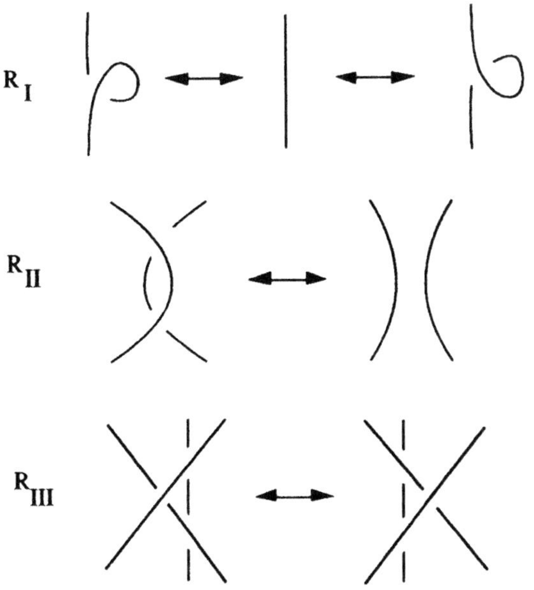

Reidemeister's moves

Reidemeister's moves R_I, R_{II} and R_{III} alter one diagram to another which represents a different view of the same knotted curve in space, up to a natural equivalence of closed curves in space corresponding to physical manipulations of pieces of rope. The classical theorem of Reidemeister states that any two diagrams D_1 and D_2 of two curves which are equivalent in space can be transformed from one to the other by a sequence of Reidemeister's moves, (allowing diagrams to be distorted between moves as shown).

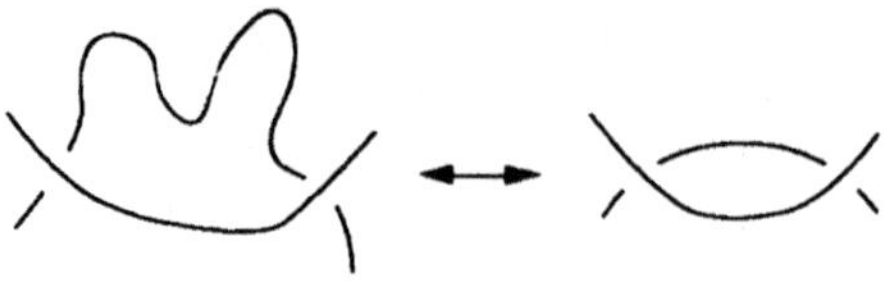

Thus Kauffman's bracket *almost* defines an invariant of a curve C in space, by calculating $< D >$ for a diagram D of the curve. The element $< D > \in \Lambda$ would indeed depend only on the curve C if it were to be unaltered by all of the three Reidemeister moves. Now R_{II} and R_{III} have no effect on $< D >$. However the bracket, $< D >$, *is* altered when D is changed by a move of type R_I. All the same, this change is quite limited, and consists of multiplication by a fixed scalar $\lambda^{\pm 1}$, depending on whether a left-handed or right-handed curl is removed.

This can be summarised as

(3) 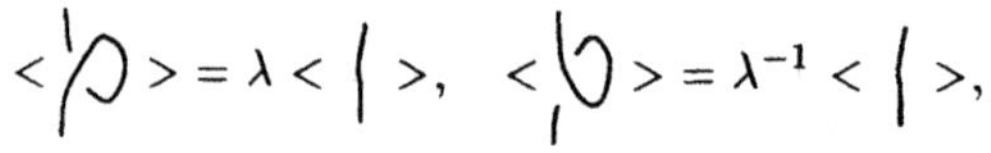

where properties (1) and (2) show readily that $\lambda = -A^3$.

Kauffman's theorem is proved in [8]. It leads immediately, using property (3), to an invariant of oriented curves in space, which can be seen as follows.

In an oriented diagram each crossing c can be given a sign $\varepsilon(c) = \pm 1$, defined as shown,

 $\varepsilon = +1$,  $\varepsilon = -1$.

Now define the *writhe* $w(D)$ of the oriented diagram D to be $w(D) = \sum \varepsilon(c)$, the sum of the signs of the crossings in D. Since $w(D)$ is unaltered by Reidemeister moves R_{II} and R_{III}, and changes by ± 1 under move R_I, the function

$$\lambda^{-w(D)} < D >$$

is unaltered by all Reidemeister moves, and hence gives an invariant of an oriented curve C in space in terms of any choice of diagram D representing C. Kauffman showed that this invariant could be identified with Jones' invariant, introduced in 1984, which has been the foundation for much recent work in relating knot theory with other topics.

1.2 Linear skein theory for the Kauffman bracket.

In this section I shall develop the notation and ideas of linear skein theory in using diagrams of various sorts to define certain linear spaces, or more accurately Λ-modules, with the properties (1) and (2) of the bracket polynomial closely in mind. The general methods were first used by Conway in dealing with versions of the Alexander polynomial.

Notation. Let F be a planar surface, for example $\mathbf{R}^2$ itself, or an annulus $S^1 \times I \subset \mathbf{R}^2$, or a rectangular disc. When F has a boundary we also specify a finite, possibly empty, set of points on its boundary. A *diagram* in F consists of any number of closed curves, together with arcs joining the specified boundary points of F. As in the standard case of knot diagrams, the curves and arcs have a finite number of crossing points where two strands cross. At a crossing the strands are distinguished in the conventional way as an over-crossing and an under-crossing, so that the diagram can be interpreted as a view of some curves lying within $F \times I$.

Write Λ for the ring $\mathbf{Z}[A^{\pm 1}]$, and $\mathcal{D}(F)$ for the set of Λ-linear combinations of diagrams in F.

For example, when $F = \mathbf{R}^2$, $\mathcal{D}(F)$ consists of linear combinations of knot (and link) diagrams, such as $AK_1 - (A + 2A^{-1})K_2$ for the diagrams K_1 and K_2 shown.

K_1 K_2

Notation. When F is a rectangular disc with m points specifed on the top edge, and n points on the bottom edge, denote F by R_n^m, and call a diagram in F an (m,n)-*tangle*.

An example of a $(4,2)$-tangle is shown below.

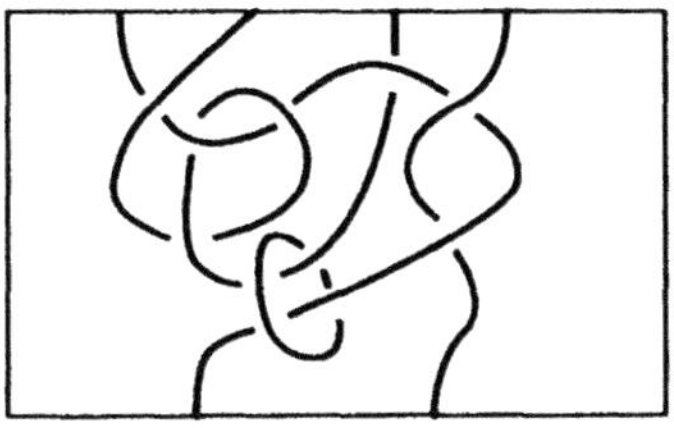

The linear combination of $(2,2)$-tangles $\sigma - AI - A^{-1}H$ is an element of $\mathcal{D}(R_2^2)$ for the tangles $\sigma =$  , $I =$  and $H =$  .

Definition. The *linear skein* $\mathcal{S}(F)$ of a planar surface F, with a distinguished finite, (possibly empty), subset of boundary points, is the quotient of $\mathcal{D}(F)$ by the linear relations

(1) $$D_+ = AD_0 + A^{-1}D_\infty$$

(2) $$D \sqcup O = \delta D \quad (= -(A^2 + A^{-2})D)$$

where D_+, D_0 and D_∞ are any three diagrams in F which differ only as in the bracket relation (1), and $D \sqcup O$ consists of a diagram D together with a disjoint simple closed curve O which is *null-homotopic in* F.

Thus condition (2) allows us to replace

 but not  by $\delta \times$ 

in the linear skein of the annulus, $\mathcal{S}(S^1 \times I)$.

THEOREM 1.2. *As a Λ-module, $\mathcal{S}(F)$ is spanned by diagrams with no crossings and no null-homotopic closed curves.*

Proof: By induction on the number of crossings and null-homotopic curves. Relation (1) in the definition of $\mathcal{S}(F)$ allows us to replace a diagram by a linear combination of two diagrams with fewer crossings, while relation (2) allows the removal of null-homotopic closed curves. □

COROLLARY 1.3. *The linear skein $\mathcal{S}(\mathbf{R}^2)$ is spanned as a Λ-module by the empty diagram ϕ, (or, if the empty diagram is excluded, by the simple unknot diagram O).* □

Remark. For any diagram D in $\mathbf{R}^2$ we can write $D = < D > \phi$ in $\mathcal{S}(\mathbf{R}^2)$; this provides an isomorphism $\mathcal{S}(\mathbf{R}^2) \cong \Lambda$, induced by mapping D to $< D >$.

THEOREM 1.4. *Two diagrams in F which differ by a Reidemeister move within F of type R_{II} or R_{III} represent the same element of $\mathcal{S}(F)$.*

Proof: Relations (1) and (2) in $\mathcal{S}(F)$ are exactly what is used in the proof of Kauffman's theorem. □

1.3 SKEIN MAPS.

Conway's framework, as described by Lickorish [15], for relating skeins of different surfaces can be helpfully used here to provide a range of linear and multilinear maps between skeins.

The central idea is to place one planar surface F inside another F', and include some fixed 'wiring', W, in the region between F and F', consisting of one or more closed curves and arcs, arranged so that the boundary points of the arcs consist exactly of the distinguished boundary points of F and F'.

Definition. A *wiring* W of F into F' means a choice of inclusion of F in F', and a fixed diagram of curves and arcs in $F' - F$ whose endpoints consist of all the distinguished points on the boundaries of F and F'.

Any diagram D inserted in the surface F is then extended by W to give a diagram $W(D)$ in F'.

Examples. (1) The rectangle R_n^n can be wired into the annulus $S^1 \times I$ as shown. For a tangle T the extended diagram $W(T)$ in the annulus, or more usually in $\mathbf{R}^2$, is called the *closure* of T, and will be denoted by $\widehat{T}$.

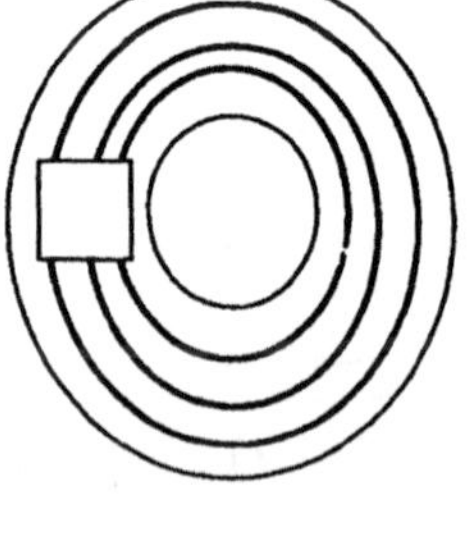

(2) The annulus itself can be wired into $\mathbf{R}^2$ by simple inclusion, without any extra curves.

(3) The *plat closure* of a $(2m, 2m)$-tangle is the diagram in $\mathbf{R}^2$ induced by the wiring shown.

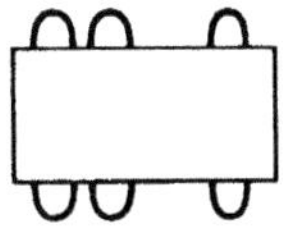

(4) A partial closure of an (n, n)-tangle T is the $(n-1, n-1)$-tangle $W(T)$ induced by the wiring of R_n^n into R_{n-1}^{n-1} shown below.

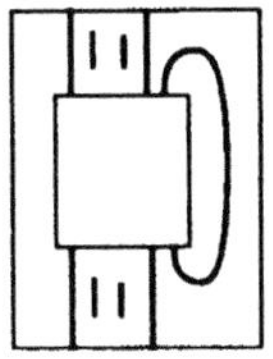

Any wiring W of F into F' determines a linear map

$$\mathcal{D}(W) : \mathcal{D}(F) \to \mathcal{D}(F')$$

by $D \mapsto W(D)$. It is clear that this induces a linear map between the skeins $\mathcal{S}(F)$ and $\mathcal{S}(F')$.

THEOREM 1.5. *A wiring W of F into F' induces a linear map*

$$\mathcal{S}(W) : \mathcal{S}(F) \to \mathcal{S}(F'),$$

defined on a diagram D in F by $D \mapsto W(D)$.

Proof: It is enough to observe that if diagrams in F satisfy skein relations (1) or (2) then they continue to do so when extended by W to diagrams in F', so the relations in $\mathcal{S}(F)$ are respected by the map. □

It is clear from theorem 1.4 that the wiring W can be altered by Reidemeister moves R_{II} or R_{III} in $F' - F$ without changing the map $\mathcal{S}(W)$.

1.4 MULTILINEAR EXTENSIONS.

The wiring construction can be used to wire several surfaces at once, $F_1, \ldots, F_k$ say, into F'. Any such wiring will induce a map

$$\mathcal{S}(W) : \mathcal{S}(F_1) \times \ldots \times \mathcal{S}(F_k) \to \mathcal{S}(F')$$

which is multilinear.

For example, we can very simply wire the rectangles R_n^m and R_p^n into R_p^m, one above the other, inducing a bilinear product

$$\mathcal{S}(R_n^m) \times \mathcal{S}(R_p^n) \to \mathcal{S}(R_p^m).$$

In the case $m = n = p$ this diagram-based product determines a multiplication which turns $\mathcal{S}(R_n^n)$ into an *algebra* over Λ.

Notation. Write $TL_n = \mathcal{S}(R_n^n)$ for this algebra, which is isomorphic to the n-th Temperley-Lieb algebra.

Theorem 1.2 shows that TL_n is spanned by diagrams in R_n^n with no closed curves, and no crossings.

When $n = 3$ there are just five such diagrams,

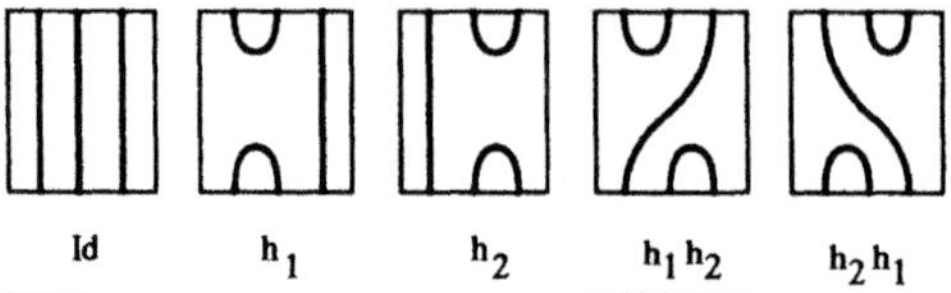

Note that $h_1^2 =$  $= \delta h_1$ and $h_1 h_2 h_1 =$ 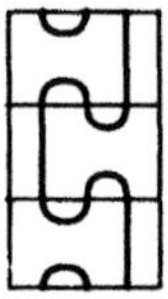 $= h_1$.

For general n, TL_n is spanned by $\binom{2n}{n}/(n+1)$ such diagrams; the number of diagrams is the n-th Catalan number.

Kauffman proved in [8] that TL_n can be presented as an algebra with generators $h_1, \ldots, h_{n-1}$, similar to h_1 and h_2 above, and only the obvious relations, namely

$$\begin{aligned} h_i h_j &= h_j h_i, \quad | \, i - j \, | > 1, \\ h_i^2 &= \delta h_i, \\ h_i h_{i \pm 1} h_i &= h_i. \end{aligned}$$

He was thus able to identify this algebra with the Temperley-Lieb algebra, which appears from a totally different viewpoint in Jones' original work.

1.5 The braid groups.

An n-string braid is a diagram in R_n^n consisting only of n arcs, which all run monotonically from bottom to top. Two n-braids are composed by placing one below the other. Braids, up to Reidemeister moves R_{II} and R_{III}, form Artin's n-string braid group, B_n, described by him in [1].

PROPOSITION 1.6. *There is a multiplicative homomorphism $B_n \to TL_n$ determined by representing $\beta \in B_n$ by a diagram in R_n^n and reading the diagram as an element of the skein TL_n.*

Proof: Diagrams which differ only by moves R_{II} and R_{III} represent the same element in the skein, so the map is well-defined. It is clearly a homomorphism, since composition is defined in the same way in each case. □

The image of B_n under this homomorphism spans TL_n, since each of the generators h_i of TL_n satisfies the relation $\sigma_i = A\,\mathrm{Id} + A^{-1} h_i$, where σ_i is the elementary braid

$\sigma_i =$ 

,

and thus $h_i = A\sigma_i - A^2\,\mathrm{Id}$. The presentation of TL_n can be rewritten in terms of the generators σ_i. The relations then include the relations in B_n together with the additional relations

$$(\sigma_i - A)(\sigma_i + A^{-3}) = 0,$$

or in other words $(\sigma_i + A^{-3})h_i = 0$.

1.6 CALCULATIONAL METHODS.

It is possible to make use of the algebra TL_n in calculating the bracket invariant of a link L which has been presented as a closed braid $\hat{\beta}$ on n strings, simply by combining the map $B_n \to TL_n$ with the linear map $TL_n \to \Lambda = \mathcal{S}(\mathbf{R}^2)$ induced by the closure wiring on R^n_n. We must thus write the braid β as a linear combination $\beta = \sum \lambda_g T_g$ of the $\binom{2n}{n}/(n+1)$ spanning elements $\{T_g\}$ of TL_n, with $\lambda_g \in \Lambda$. It is then enough to know the bracket invariant $< \hat{T}_g >$ of the closure of each T_g, to get

$$< L > = < \hat{\beta} > = \sum \lambda_g < \hat{T}_g >.$$

The expression of β in terms of $\{T_g\}$ can be built up from knowledge of β as a word in the elementary braids σ_i, by knowing simply how to write each product $T_g\sigma_i$, as defined in section 1.4, in terms of the basis $\{T_g\}$ of TL_n.

The amount of calculation required does not grow rapidly with the number of crossings, for braids on a fixed number of strings. Such calculations still give one of the quickest ways of handling invariants of quite complicated links; see Morton and Short, [21, 22], for further analysis and comments. In principle the bracket invariant of any knot can be found in this way, as every knot can be presented as a closed n-braid for some n, although calculations become rapidly more impracticable with increasing n.

For a simple related illustration, note first that TL_2 is spanned by just two elements, 1 ($=$ Id) and h ($= h_1$), with the 2-braid σ given by $\sigma = A + A^{-1}h$. The diagram illustrated, with r and k half-twists respectively in the two boxes, arises by wiring two copies of R^2_2 into the plane with the wiring W shown, and then inserting σ^r and σ^k into the copies of R^2_2.

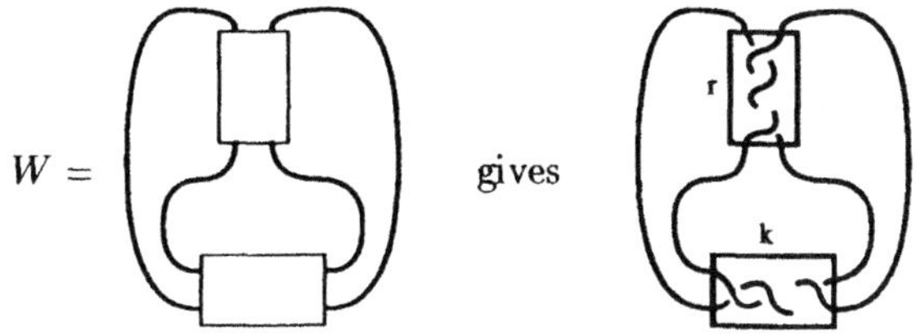

The induced bilinear map

$$\mathcal{S}(W) : TL_2 \times TL_2 \to \Lambda$$

then evaluates the bracket invariant of the complete diagram, when applied to (σ^r, σ^k). We can write $\sigma^r = (A + A^{-1}h)^r = P_r + Q_r h \in TL_2$ in terms of the basis elements 1 and h, and similarly σ^k. Combine this information with the calculation of $\mathcal{S}(W)$ on pairs of basis elements, to complete the calculation. It is easy to see that $\mathcal{S}(W)(1,1) = \mathcal{S}(W)(h,h) = \delta$ while $\mathcal{S}(W)(1,h) = \mathcal{S}(W)(h,1) = \delta^2$, so that the required invariant can be written

$$(P_r \quad Q_r)\begin{pmatrix} \delta & \delta^2 \\ \delta^2 & \delta \end{pmatrix}\begin{pmatrix} P_k \\ Q_k \end{pmatrix}.$$

In calculating $\sigma^r \in TL_2$ it can be more efficient to use a different basis of TL_2 which reflects better its algebraic structure. In each TL_n there is one element which will be of further algebraic use. This is related to one of the two non-zero homomorphisms from TL_n to Λ. It is clear from the presentation of TL_n that there is a Λ-linear homomorphism $\varphi : TL_n \to \Lambda$, defined by $\varphi(1) = 1$, $\varphi(h_i) = 0$. In terms of braids this corresponds to $\varphi(\sigma_i) = A$, $\varphi(1) = 1$. (The other homomorphism, ψ, is defined by $\psi(\sigma_i) = -A^{-3}$.)

In the next section I shall exhibit an element $f_n \in TL_n$ with the property that $Tf_n = f_nT = \varphi(T)f_n$ for every $T \in TL_n$. Before doing this, I shall look in further detail at the skein of the annulus.

1.7 The skein of the annulus.

Notation. Write $\mathcal{B} = \mathcal{S}(S^1 \times I)$ for the skein of the annulus.

The linear map $\mathcal{B} \to \mathcal{S}(\mathbf{R}^2) \cong \Lambda$ induced by the inclusion as in example (2) above will sometimes be denoted simply by $v \mapsto \langle v \rangle$ as it is induced on a diagram in the annulus by taking its bracket invariant when regarded as a diagram in the plane.

We can wire two copies of the annulus into the annulus itself by running one copy parallel to the other without adding extra wiring. This defines a bilinear product $\mathcal{B} \times \mathcal{B} \to \mathcal{B}$, under which $\mathcal{B}$ becomes an algebra over Λ.

For example, the element of $\mathcal{B}$ represented by

 is the product of  and 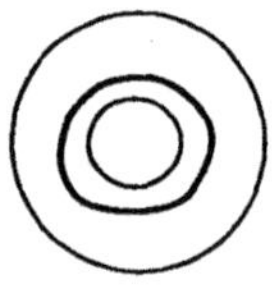.

Write α = 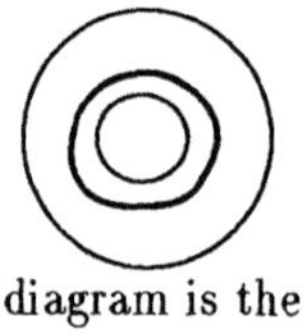 as an element of $\mathcal{B}$. Then k parallel curves represent α^k,

while the empty diagram is the unit element, 1, of the algebra $\mathcal{B}$.

By theorem 1.2, $\mathcal{B}$ is spanned by diagrams with no crossings and no null-homotopic curves. Any such diagram is either empty, or consists of k parallel curves around the annulus, for some k, so that $\mathcal{B} \cong \Lambda[\alpha]$, the ring of polynomials in α.

For example, use of the skein relations shows that the diagram

is equal to $(-1 - A^{-4}) + (1 - A^4)\alpha^2$ in $\mathcal{B}$.

Proposition 1.7. *The evaluation map $\langle \cdot \rangle : \mathcal{B} \to \Lambda \cong \mathcal{S}(\mathbf{R}^2)$ is a ring homomorphism.*

Proof: This follows at once from the structure of $\mathcal{B}$ since $< \alpha^k > = \delta^k$ by skein relation (2). Even without this knowledge it is enough to observe that the two parallel copies of the annulus containing diagrams to be multiplied in $\mathcal{B}$ can be moved apart without change, using R_{II} and R_{III}, before evaluating each separately. □

2. Satellite knots.

Suppose that we want to use the bracket invariant to compare two knots C_1 and C_2. Let us draw diagrams of each knot and calculate its bracket invariant. If the knots are equivalent, and the diagrams used have the same writhe, then we will get the same answer in each case. Hence different answers, from diagrams with the same writhe, guarantee that the given knots are different.

We might, however, get the same answer from two knots which we suspect to be different. It is still possible that we may be able to show that the knots are different by a less direct use of the bracket invariant. First, 'decorate' the two knots in the same way, to give two more complicated knots K_1 and K_2. Make sure that if the decoration is done in the same way, and the two knots C_1 and C_2 are equivalent, then the decorated knots are equivalent. Then use the bracket invariant again to compare K_1 and K_2; if these give different answers then C_1 and C_2 must be different.

Such a project might be doomed to failure. If, for example, the bracket invariant of K_i could be calculated in terms of the bracket invariant of C_i and the decoration, as is the case for the classical Alexander polynomial, then two knots with the same bracket invariant would, after being decorated in the same way, still give two knots with the same invariant.

Happily, there is a chance of using the bracket invariant in this way. One of the early discoveries [21] about the recent knot invariants was the existence of pairs of knots with the same invariant which can be distinguished by calculating the invariant of the knots resulting from suitable decoration.

2.1 Construction of satellites.

I shall now describe how to decorate a knot. Starting with a given knot C we draw a diagram of it. This selects a 'parallel' curve to C, determined by keeping just to one side of C in the diagram. Altering the diagram by R_{II} or R_{III} does not change this 'diagrammatic' parallel curve when thought of as a curve in space relative to C, while R_I introduces a full twist of the parallel around C.

Definition. A *framed knot* is a curve C in $\mathbf{R}^3$, with a choice of a neighbouring parallel curve; a *framed link* has a choice of parallel for each component of the link.

In much of what follows we shall be dealing with framed knots and links. I shall normally assume that any diagram of a framed link is drawn so that the chosen parallels agree with the diagrammatic parallels. Suitable insertion of curls in the diagram allows the diagrammatic parallels to be adjusted so that this is the case.

The study of framed knots and links is almost equivalent to the study of diagrams of the knots and links up to the moves R_{II} and R_{III}. As noted, the diagrammatic parallels are unaltered by R_{II} and R_{III} ; conversely we can pass between diagrams with the same parallel curves by using R_{II} and R_{III} if we are also allowed to move curls from one side of the string to the other, as shown.

See Kauffman [10] for further comments. In the applications given here this last move will be permissible, so I shall assume that any statements about framed links can be interpreted in terms of diagrams up to moves R_{II} and R_{III}, and vice versa.

To continue then with the construction, we shall assume that we have a diagram of C, or equivalently a framing of C. Now select a diagram P in the annulus. We decorate C with P as follows. Place the annulus with one edge following C and one following its parallel, and copy P into this annulus. The image of P forms a new diagram, which is the knot C decorated by P. Changing the exact positioning of the copy of P as it is placed to lie around C will alter this new diagram, but only by moves R_{II} and R_{III}. Write $C * P$ for this new diagram, defined up to R_{II} and R_{III}. For example, when C is the trefoil with framing as shown,

$C =$ 

and $P =$  then $C * P =$  .

Alteration of the diagram of C itself by R_{II} or R_{III} will alter $C * P$ only by a sequence of moves R_{II} or R_{III} respectively, and so $C * P$, as a framed knot, depends only on C as a framed knot and on P. Altering the framing of C, i.e. altering its diagram by a move R_I, will in general alter $C * P$ substantially; for this reason a framing of C must be specified in some way.

From a more 3-dimensional viewpoint, the decoration P can be viewed as lying in a solid torus, which is then embedded in $\mathbf{R}^3$ as a neighbourhood of the curve C. The resulting image of P is called a *satellite* of C, while C is known as its *companion*. Again, some specification, amounting to a decision on framing, is needed to describe exactly how the solid torus is to be embedded.

2.2 The total bracket invariant.

Our immediate study can be seen as the study of C by means of the bracket invariant of its various satellites, as we change the decoration pattern P. As with the wiring construction, we can show that the process of decorating a fixed diagram C by a pattern P in the annulus induces a linear map $\mathcal{B} \to \Lambda \cong \mathcal{S}(\mathbf{R}^2)$.

THEOREM 2.1. *Let C be a knot diagram. Then there is a linear map $J_C : \mathcal{B} \to \Lambda \cong \mathcal{S}(\mathbf{R}^2)$ induced by mapping a diagram P in the annulus to the diagram $C * P$.*

Proof: When diagrams in the annulus satisfy skein relations (1) or (2) then the diagrams which result from decorating C will also satisfy the same skein relation. The map J_C is thus well-defined on the skein $\mathcal{B}$. □

As in the case of wiring diagrams, there is an extension of this result where C is replaced by a link diagram L with k components. Each component can be decorated independently, giving a multilinear map

$$J_L : \mathcal{B} \times \ldots \times \mathcal{B} \to \Lambda,$$

from k copies of $\mathcal{B}$. It is clear that if L is changed by R_{II} or R_{III} then the map J_L is unaltered; indeed J_L is an invariant of the framed link L, its *total bracket invariant*.

We can make a further generalisation on this construction to the case where D is a diagram with k closed components in a surface F. By decorating each component of D, following its diagrammatic parallel, with a linear combination of diagrams in the annulus, we induce a multilinear map

$$J_D : \mathcal{B} \times \ldots \times \mathcal{B} \to \mathcal{S}(F).$$

When a diagram, L say, arises by decoration of another diagram we can use such a map, taking F itself as an annulus, to write the total invariant of the diagram L as the composite of simpler maps.

For example, suppose that L is a link of $k+1$ components which can be drawn with one of the components, L_{k+1} say, as a simple closed curve. Then, after suitable adjustment by moves R_{II} and R_{III}, the remaining components can be arranged to form a diagram $D = \widehat{T}$ in an annulus, as shown, so that the link L itself is arranged as the Hopf diagram H, with one component decorated by D.

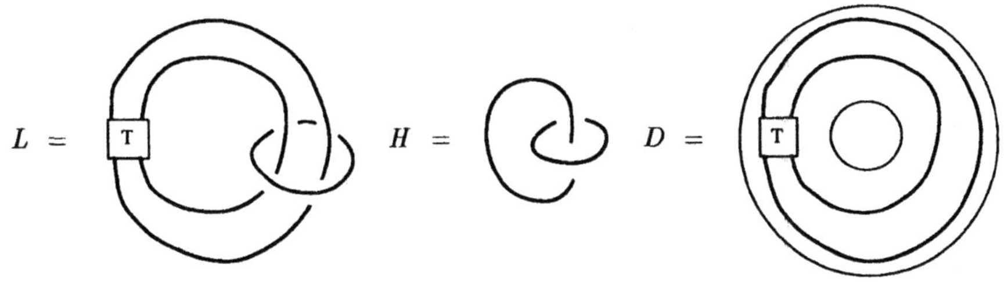

THEOREM 2.2. *The invariant J_L is the composite*

$$\mathcal{B} \times \ldots \times \mathcal{B} \xrightarrow{J_D \times 1} \mathcal{B} \times \mathcal{B} \xrightarrow{J_H} \Lambda.$$

Proof: Decorate each component of L by diagrams $P_1, \ldots, P_{k+1}$. The decorations $P_1, \ldots, P_k$ determine a diagram in the annulus which is just the decoration of D. The final diagram is the Hopf diagram H with this complicated diagram, representing $J_D(P_1, \ldots, P_k)$ in $\mathcal{B}$, decorating one component, while the other component is decorated by P_{k+1}. Then $J_L(P_1, \ldots, P_{k+1}) = J_H(J_D(P_1, \ldots, P_k), P_{k+1})$. The result follows by linearity. □

2.3 The satellite formula.

We may also use this framework to calculate the total invariant J_K of a knot $K = C * P$ which is a satellite of C constructed by decorating the framed knot C by a diagram P in the annulus. Assuming that P has one component we may decorate P by any diagram Q in the annulus, to get a diagram $P * Q$ also in the annulus. It is easy to see that, up to R_{II} and R_{III}, the diagrams $C * (P * Q)$ and $K * Q = (C * P) * Q$ are the same. It is then immediate that the invariant $J_K : \mathcal{B} \to \Lambda$ is the composite

$$\mathcal{B} \xrightarrow{J_P} \mathcal{B} \xrightarrow{J_C} \Lambda.$$

This equation, and its counterpart for links and patterns with more than one component, will be termed the *satellite formula*. In this simple case we may also write it as

$$J_{C*P} = J_C \circ J_P.$$

The satellite formula shows that, unlike the bracket polynomial alone, we know the total invariant J of a satellite once we know J for the companion and for the annulus diagram P used in constructing the satellite. (Where C or P have more than one component, the corresponding multilinear maps should be used, and composed appropriately, depending on the component of the companion which is decorated.)

The total bracket invariant J_C contains all the information about bracket invariants of satellites of the knot C. It is known once its values $J_C(\alpha^k)$ on the basis $\{\alpha^k\}$ of $\mathcal{B}$ are known. To determine the bracket invariant of the satellite when C is decorated by a pattern P it is enough to write out $P = a_0 + a_1\alpha + \ldots + a_r\alpha^r$ in $\mathcal{B}$ and calculate the bracket invariant of C decorated by α^k, for $0 \le k \le r$. Then

$$J_C(P) = \sum_{k=0}^{r} a_k J_C(\alpha^k).$$

Now $J_C(1) = 1$ and $J_C(\alpha) = \langle C \rangle$, since decoration of C by α just gives C again. However, as remarked earlier, there is in general no way to determine $J_C(\alpha^k)$ from $J_C(\alpha)$, when $k \ge 2$.

For the unknot and the Hopf link, and also for other torus knots and links, the map J_L is known, but not for any other knots. There are examples known, though, of inequivalent knots C_1 and C_2 for which $J_{C_1} = J_{C_2}$; these examples include all mutant pairs of knots, such as the famous pair of Conway and Kinoshita-Teresaka, [25].

The relation given above for the invariant J of a satellite knot is equivalent to the 'satellite formula' of [23] which relates the total invariant of a given satellite to those of the companion, the Hopf link H and the 'pattern link', namely the satellite of H constructed from the same annulus diagram P as the given satellite. The pattern link consists of P and one extra component, which can be compared to the axis of a closed braid, and which gives the means for recovering P as a diagram in the annulus from the pattern link in S^3.

To get the appropriate reinterpretation of [23] it is simply necessary to identify $\mathcal{B}$ with the representation ring $\mathcal{R}$ of the quantum group $SU(2)_q$.

In section 4 I shall give a brief account of the translation between the two viewpoints, but the important features of either approach are the existence of the multilinear invariant J_L for a framed link L, and its natural behaviour on satellites.

2.4 Framing change and the total invariant.

To complete this stage in the understanding of the invariant J_L for a framed link L we must discuss the behaviour of J when the framing of L is altered.

To see more clearly what happens I shall look at the case when L has one component. Suppose that L' has the same diagram as L, except for the addition of a single right-handed curl, so that the underlying knots are equivalent, but the framing has been altered by a single twist. If we use the simple decoration by α then $J_{L'}(\alpha) = \langle L' \rangle = \lambda \langle L \rangle = \lambda J_L(\alpha)$, where $\lambda = -A^3$. However, $J_{L'}(\beta)$ is not in general a simple multiple of $J_L(\beta)$. For example, we can calculate $J_{L'}(\alpha^2)$ in terms of J_L by using the diagram shown

to decorate L. This diagram represents $A^8\alpha^2 - (A^8 - 1)$ in $\mathcal{B}$.

In general the change of framing can be expressed in terms of the map $\mathcal{F} : \mathcal{B} \to \mathcal{B}$ induced by decorating the diagram T in the annulus.

$T =$ 

THEOREM 2.3. *Let L' be a knot given from L by adding one right-hand twist to the framing. Then $J_{L'} = J_L \circ \mathcal{F}$, where $\mathcal{F} = J_T$, induced by the diagram shown above in the annulus.*

Proof: The diagram L' is just $L * T$, and so the result follows from the satellite formula. □

The map $\mathcal{F}$ has an inverse, induced similarly by the left-hand curl.

When the framing on a link of several components is altered, the total invariant J, as a multilinear map on $\mathcal{B}$, is changed by applying a suitable power of the automorphism $\mathcal{F}$ to each copy of $\mathcal{B}$, depending on the change of framing to be made on the corresponding link component.

To describe the effect of framing change it is enough to determine the map $\mathcal{F}$, or equivalently to find $\mathcal{F}(\alpha^k)$ for each k. As noted above, it is not true that $\mathcal{F}(\alpha^k)$ is a multiple of α^k when $k > 1$, although it is easy to see that, as a polynomial in α, it must have degree at most k, and indeed that its degree is exactly k. In the 3-dimensional view, $\mathcal{F}$ arises when the solid torus formed by thickening the annulus is mapped to itself by cutting along a meridian disc and regluing after a full twist.

To handle the invariant J most readily, including its behaviour under framing change, it is natural to try to change the basis of $\mathcal{B}$ from $\{\alpha^k\}$ to one consisting of eigenvectors w_i of $\mathcal{F}$, if this is possible. Then $J_{L'}(w_i) = \lambda_i J_L(w_i)$, where λ_i is the eigenvalue of w_i, and the value of $J_{L'}(\beta)$ can be found readily in terms of J_L by writing β in terms of the basis w_i.

2.5 The Temperley-Lieb algebra.

I shall now use the Temperley-Lieb algebra to help construct enough eigenvectors of $\mathcal{F}$ to form a basis of $\mathcal{B}$. Some of the properties of these eigenvectors are most readily appreciated in the alternative view of $\mathcal{B}$ as the representation ring of $SU(2)_q$ in which the eigenvectors appear naturally as the irreducible representations. For this reason I shall index the eigenvectors as $w_1, \ldots, w_i, \ldots$, where w_i, which is a monic polynomial in α of degree $i-1$, will correspond to the irreducible representation of dimension i, in conflict with the notation used by Lickorish [16], who indexes by the degree of the polynomial. In the corresponding construction in [2], Blanchet et al focus heavily on the eigenvector property, without using the Temperley-Lieb algebra at all.

Using the closure wiring referred to earlier to map (n,n)-tangles into annulus diagrams I shall construct elements of $\mathcal{B}$ from the closure of elements in TL_n; in particular, the closure of the element f_n mentioned at the end of section 1.6 is a multiple of the desired eigenvector w_{n+1}.

It is easy to see the effect of $\mathcal{F}$ on any element of $\mathcal{B}$ which is in the closure of TL_n, in terms of the multiplication in TL_n. For suppose that X is an (n,n)-tangle. Then the closure of the tangle $Q_n X$, where Q_n is the right-handed curl on n parallel strings, as shown,

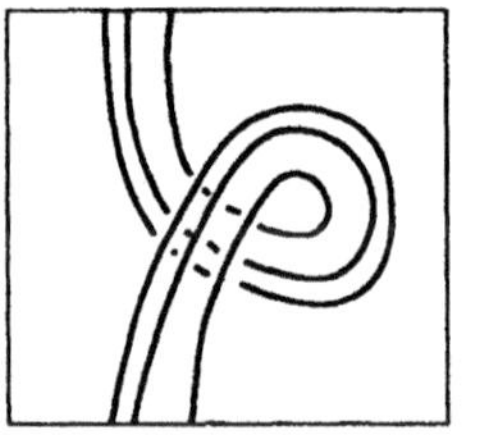

will represent $\mathcal{F}(\widehat{X})$ as an element of $\mathcal{B}$. Write $\varphi_A : TL_n \to \Lambda$ for the linear homomorphism defined by $\varphi_A(\sigma_i) = A$ for each i. In what follows I shall define the elements $f_n \in TL_n$ with the property that $Tf_n = \varphi_A(T) f_n$ for all $T \in TL_n$. It is then immediate that the closure of f_n is an eigenvector of $\mathcal{F}$ since we can write $Q_n f_n = \varphi_A(Q_n) f_n$. Take $X = f_n$; its closure $\widehat{f}_n$ is then an eigenvector with eigenvalue $\varphi_A(Q_n)$.

Now by removing n right-hand curls, one from each component, we can write Q_n as a multiple of the right-hand full-twisted braid Δ_n^2, as an element of TL_n, so we have $Q_n = (-1)^n A^{3n} \Delta_n^2$. Since Δ_n^2 is a braid it is easy to calculate $\varphi_A(\Delta_n^2)$ in terms of the crossings in the braid, as $\varphi_A(\sigma_i) = A$ for each i. Now after removal of the n curls from Q_n there remain $n^2 - n$ crossings in the braid Δ_n^2, all in the same sense, so we have $\varphi_A(\Delta_n^2) = A^{n^2-n}$ and thus the eigenvalue for $\widehat{f}_n$ is $(-1)^n A^{n^2+2n}$.

Define elements $w_i \in \mathcal{B} \cong \Lambda[\alpha]$ by the relations

$$w_1 = 1, \quad w_2 = \alpha,$$
$$w_{i+1} = \alpha w_i - w_{i-1},\ i > 1.$$

Each w_k is clearly a monic polynomial of degree $k-1$, and can be recognised as the Chebyshev polynomial of the second kind, resulting from writing $\sin k\theta / \sin\theta$ as a polynomial in $\alpha = 2\cos\theta$, (cf. Lickorish [16]).

The final result in this section is to establish that $\varphi_A(f_n)w_{n+1} = \widehat{f}_n$ so that each w_i is an eigenvector of $\mathcal{F}$ with eigenvalue $\lambda_i = (-1)^{i-1}A^{i^2-1}$, provided that $\varphi_A(f_n) \neq 0$.

While it appears more appealing to divide $\widehat{f}_n$ by $\varphi_A(f_n)$ in order to map exactly to w_{n+1} this can only be done by extending the ring Λ to allow suitable denominators. At the present stage this need cause no problems, but later developments which require substitution of the variable A in Λ then become more difficult as there is a chance that $\varphi_A(f_n)$ may become zero. Lickorish in fact uses carefully controlled denominators to define an element denoted by $f^{(n)}$ whose closure is exactly w_{n+1}. However, the definition of f_n without the factor, as given here, is also quite natural.

2.6 Positive permutation braids.

I shall construct the element $f_n \in TL_n$ by means of positive permutation braids. These have been used in [22] as a convenient basis for the Hecke algebra, and are discussed more fully in [4]. In the algebraic context the construction of f_n given here is a special case of a construction of Jones in the Hecke algebra [7]; this method has also been noted more recently by Kauffman [10].

Definition. For each permutation $\pi \in S_n$ there is an n-braid $w_\pi(\sigma_1, \ldots, \sigma_{n-1})$, called a *positive permutation braid*. It is uniquely determined by the following properties.

(1) String i joins the point numbered i at the bottom of the braid to the point numbered $\pi(i)$ at the top, $i = 1, \ldots, n$.

(2) At any crossing, string i always crosses *over* string j if $i < j$.

We may view the strings in the braid as lying in layers, with string 1 above string 2, and so on, so that each string can be moved independently of the others. This ensures the uniqueness of w_π, which can be drawn, if required, so that pairs of strings cross at most once. In this form, condition (2) is equivalent to asking that each crossing be positive, when all strings are oriented from bottom to top.

Let us now consider an algebra $\mathcal{A}$ in which the n-string braid group B_n is represented. In what follows, we shall be primarily interested in the Temperley-Lieb algebra, TL_n, but the arguments will work as well in a more general setting. I shall continue to write σ_i for the element of $\mathcal{A}$ which represents the elementary braid σ_i. We may then define an element $E_n(\sigma_1, \ldots, \sigma_{n-1})$ in the algebra $\mathcal{A}$ as the sum

$$E_n(\sigma_1, \ldots, \sigma_{n-1}) = \sum_{\pi \in S_n} w_\pi(\sigma_1, \ldots, \sigma_{n-1}).$$

Thus $E_3 = 1 + \sigma_1 + \sigma_2 + \sigma_1\sigma_2 + \sigma_2\sigma_1 + \sigma_1\sigma_2\sigma_1$, noting that the last braid in the sum, corresponding to the permutation (13), could equally well be written as $\sigma_2\sigma_1\sigma_2$. It is a convenient property of the permutation braids that it is only necessary to remember them by their permutation of the strings, without having to specify each braid as a word in $\{\sigma_i\}$.

THEOREM 2.4. *For each i we can factorise E_n in the given algebra $\mathcal{A}$ as $E_n = E_n^{(i)}(\sigma_i + 1)$.*

Proof: Given i, we can pair the permutations as follows. For each permutation π consider its composite $\pi' = \pi \circ (i\ i{+}1)$ with the transposition $(i\ i{+}1)$. Exactly one of the pair preserves the order of i and $i+1$. Suppose that it is π, so that $\pi(i) < \pi(i+1)$. Then the braid $w_\pi \sigma_i$ satisfies property (2) above, and so is itself a positive permutation braid. Since its permutation is π' we have $w_\pi \sigma_i = w_{\pi'}$. Then

$$\begin{aligned} E_n &= \sum_{\pi(i)<\pi(i+1)} w_\pi + \sum_{\pi'(i)>\pi'(i+1)} w_{\pi'} \\ &= \sum_{\pi(i)<\pi(i+1)} w_\pi + \sum_{\pi(i)<\pi(i+1)} w_\pi \sigma_i \\ &= E_n^{(i)}(\sigma_i + 1), \end{aligned}$$

where $E_n^{(i)} = \sum_{\pi(i)<\pi(i+1)} w_\pi$. □

If λ is a scalar, then we may substitute $\lambda\sigma_i$ for σ_i and rewrite the element $w_\pi(\lambda\sigma_1, \ldots, \lambda\sigma_{n-1})$ as $\lambda^{l(\pi)} w_\pi(\sigma_1, \ldots, \sigma_{n-1})$ in $\mathcal{A}$, where $l(\pi)$ is the writhe of the braid w_π. This is the same as the length of w_π when written as a monomial in positive powers of the elementary braids σ_i. It is equal to the number of reversals of the permutation π, i.e. the number of pairs $i < j$ for which $\pi(i) > \pi(j)$.

Suppose now that all the elementary braids satisfy the quadratic equation

$$(\sigma_i - a)(\sigma_i - b) = 0$$

in the algebra $\mathcal{A}$. Substitute $\lambda\sigma_i$ for σ_i in E_n, with $\lambda = -a^{-1}$ or $\lambda = -b^{-1}$, to define

$$a_n = E_n(-a^{-1}\sigma_1, \ldots, -a^{-1}\sigma_{n-1}), \quad b_n = E_n(-b^{-1}\sigma_1, \ldots, -b^{-1}\sigma_{n-1}).$$

THEOREM 2.5. *Suppose that the algebra $\mathcal{A}$ is spanned by braids, that $(\sigma_i - a)(\sigma_i - b) = 0$ in $\mathcal{A}$ and that φ_a and φ_b are linear homomorphisms from $\mathcal{A}$ to the scalars defined by $\varphi_a(\sigma_i) = a$, $\varphi_b(\sigma_i) = b$. Then every $T \in \mathcal{A}$ satisfies*

$$a_n T = \varphi_b(T) a_n = T a_n, \quad b_n T = \varphi_a(T) b_n = T b_n.$$

Proof: To establish the left-hand equality in each case it is enough to show that $a_n \sigma_i = \varphi_b(\sigma_i) a_n = b a_n$ for each i, and similarly that $b_n \sigma_i = a b_n$. We can factorise a_n using the theorem above, as

$$a_n = E_n^{(i)}(-a^{-1}\sigma_1, \ldots, -a^{-1}\sigma_{n-1}) \times (-a^{-1}\sigma_i + 1) = Q_n(\sigma_i - a), \text{ say},$$

giving $a_n(\sigma_i - b) = Q_n(\sigma_i - a)(\sigma_i - b) = 0$, so that $a_n \sigma_i = b a_n$. Similarly $b_n(\sigma_i - a) = 0$.

The factorisation of E_n as $(\sigma_i + 1)E_n^{'(i)}$ is also possible, proving the right-hand equalities $T a_n = \varphi_b(T) a_n$ and $T b_n = \varphi_a(T) a_n$. □

Remark. When $\mathcal{A}$ is the group algebra of the symmetric group, $\mathbf{Z}[S_n]$, and each σ_i is represented as a transposition, the quadratic equation is $\sigma_i^2 - 1 = (\sigma_i - 1)(\sigma_i + 1) = 0$. The elements a_n and b_n are then the classical symmetriser and skew-symmetriser,

$$b_n = \sum_{\pi \in S_n} \pi, \quad a_n = \sum_{\pi \in S_n} \varepsilon(\pi)\pi .$$

The Temperley-Lieb algebra TL_n is generated by the n-braids $\sigma_1,\ldots,\sigma_{n-1}$ which satisfy the relation $\sigma_i = A+A^{-1}h_i$ with $h_i^2 = \delta h_i$ $(= -(A^2+a^{-2})h_i)$. Then $\sigma_i h_i = -A^{-3}h_i$, so that $(\sigma_i + A^{-3})(\sigma_i - A) = 0$.

Definition. In TL_n we define an element f_n by

$$f_n = E_n(A^3\sigma_1,\ldots,A^3\sigma_{n-1}) \quad (= \sum_{\pi\in S_n} A^{3l(\pi)} w_\pi(\sigma_1,\ldots,\sigma_{n-1})).$$

COROLLARY TO THEOREM 2.5. *Every $T \in TL_n$ satisfies the multiplicative property*

$$f_n T = T f_n = \varphi_A(T) f_n,$$

where $\varphi_A : TL_n \to \Lambda$ is the linear homomorphism defined by $\varphi_A(\sigma_i) = A$ for each i.

Proof: We can apply the theorem to TL_n with $a = -A^{-3}$ and $b = A$. Then $f_n = a_n$ and the result follows. □

Remarks. The element $b_n \in TL_n$ is identically zero for $n \geq 3$.

The general algebra to which the theorem applies is some quotient of the Hecke algebra. Jones [7] notes the elements a_n and b_n for the Hecke algebra when $a = q, b = -1$; any other case can be rewritten in this way if σ_i is replaced throughout by a suitable multiple.

2.7 THE ALTERNATIVE BASIS FOR THE SKEIN OF THE ANNULUS.

Having established the definition and multiplicative property of f_n in TL_n I now want to relate the closure of f_n in $\mathcal{B}$ to the element w_{n+1}, defined inductively above by $w_{n+1} = \alpha w_n - w_{n-1}$, with $w_1 = 1$ and $w_2 = \alpha$.

THEOREM 2.6. *In $\mathcal{B}$, the skein of the annulus, we have $\widehat{f}_n = \varphi_A(f_n) w_{n+1}$ for all $n \geq 1$.*

Proof: We have $f_1 = 1$ in TL_1 as a braid on one string, so $\widehat{f}_1 = \alpha = w_2$. Indeed, we could consider $f_0 = \phi$ in TL_0, which gives $\widehat{f}_0 = \phi = w_1$, noting that the empty diagram in the annulus represents the identity element $w_1 = 1$ in the algebra $\mathcal{B}$. The rest of the proof is by induction on n, and depends on establishing the appropriate relation between $\widehat{f}_n$, $\widehat{f}_{n-1}$ and $\widehat{f}_{n-2}$. This in turn depends on rewriting some of the permutation braids w_π which appear in the sum E_n.

Corresponding to the inclusion $i : S_{n-1} \subset S_n$ in which $\pi' \in S_{n-1}$ is extended to $\pi' \in S_n$ by $\pi'(n) = n$ there is an inclusion of the braid group B_{n-1} in B_n by adjoining an n-th straight string. The same procedure defines an inclusion $i : TL_{n-1} \subset TL_n$; this can even be seen to come from a simple wiring of R_{n-1}^{n-1} into R_n^n which adjoins the extra string. The element $i(E_{n-1})$ is then $\sum_{\pi(n)=n} w_\pi$. Because of the extra string the closure of $i(T)$, for any $T \in TL_{n-1}$, can be written as $\alpha\widehat{T}$ in $\mathcal{B}$.

Define braids $\gamma_r \in B_n$, $r = 0,\ldots,n-1$, by $\gamma_0 = 1$, $\gamma_r = \sigma_{n-1}\sigma_{n-2}\ldots\sigma_{n-r}$. In γ_r the string ending at position n crosses exactly r others, while no other strings cross each other. The braids $w_\pi\gamma_r$ with $\pi(n) = n$ are then positive permutation braids for all such

π and for all $r = 0, \ldots, n-1$. All permutations of strings arise exactly once on this list, so all positive permutation braids are counted exactly once as

$$E_n = \sum_{\pi \in S_n} w_\pi = \left(\sum_{\pi(n)=n} w_\pi \right) \left(\sum_{r=0}^{n-1} \gamma_r \right).$$

Replace σ_i by $A^3\sigma_i$ to get

$$(*)_n \qquad f_n = i(f_{n-1}) \left(\sum_{r=0}^{n-1} A^{3r}\gamma_r \right).$$

We can calculate $\varphi_A(f_n)$ inductively, using $(*)_n$, since $\varphi_A(\gamma_r) = A^r$. For we have

$$\varphi_A(f_n) = \varphi_A(f_{n-1}) \left(\sum_{r=0}^{n-1} A^{4r} \right) = [n]_q \varphi_A(f_{n-1}),$$

where $[n]_q = 1 + q + \cdots + q^{n-1}$ $(= n$ when $q = 1)$ and $q = A^4$. Consequently,

$$\varphi_A(f_n) = [n]_q[n-1]_q \ldots [1]_q = [n]_q!.$$

To complete the proof of the theorem it will be enough to establish the relation

$$\widehat{f}_n = [n]_q \alpha \widehat{f}_{n-1} - [n]_q[n-1]_q \widehat{f}_{n-2},$$

as the right hand side is then, by the induction hypothesis, $\varphi_A(f_n)(\alpha w_n - w_{n-1}) = \varphi_A(f_n)w_{n+1}$.

We now use $(*)_n$ to find the closure $\widehat{f}_n$. For any elements T_1 and T_2 in TL_n the products T_1T_2 and T_2T_1 have the same closure in $\mathcal{B}$. We can then replace f_n by the product $P_n = \left(\sum_{r=0}^{n-1} A^{3r}\gamma_r \right) i(f_{n-1})$. Now $\sigma_j\, i(f_{n-1}) = A\, i(f_{n-1})$, for $j < n-1$, by the multiplicative property of f_{n-1}, so $\gamma_r i(f_{n-1}) = A^{r-1}\sigma_{n-1} i(f_{n-1})$, for $r > 0$. Then

$$\begin{aligned} P_n &= i(f_{n-1}) + \left(\sum_{r=1}^{n-1} A^{4r-1} \right) \sigma_{n-1} i(f_{n-1}) \\ &= \left(\sum_{r=0}^{n-1} A^{4r} \right) i(f_{n-1}) + \left(\sum_{r=1}^{n-1} A^{4r-2} \right) h_{n-1} i(f_{n-1}) \\ &= [n]_q i(f_{n-1}) + A^2[n-1]_q h_{n-1} i(f_{n-1}), \end{aligned}$$

since $\sigma_{n-1} = A + A^{-1}h_{n-1}$ in TL_n. Hence $\widehat{f}_n = \widehat{P}_n = [n]_q \alpha \widehat{f}_{n-1} + A^2[n-1]_q \widehat{Q}_n$, where $Q_n = h_{n-1} i(f_{n-1}) \in TL_n$.

We complete the proof by showing that $\widehat{Q}_n = -A^{-2}[n]_q \widehat{f}_{n-2}$.

By $(*)_{n-1}$ we have

$$Q_n = h_{n-1} i(f_{n-1}) = h_{n-1} i(f_{n-2}) \left(\sum_{r=0}^{n-2} A^{3r}\gamma'_r \right),$$

where $\gamma'_r = \sigma_{n-2} \ldots \sigma_{n-r-1}$.

Then Q_n has the same closure as $\left(\sum_{r=0}^{n-2} A^{3r}\gamma'_r\right) h_{n-1}i(f_{n-2}) = R_n$, say. Now σ_j commutes with h_{n-1} and $\sigma_j i(f_{n-2}) = A\, i(f_{n-2})$, for $j < n-2$, so as above we get

$$\begin{aligned} R_n &= h_{n-1}i(f_{n-2}) + \left(\sum_{r=1}^{n-2} A^{4r-1}\right) \sigma_{n-2}h_{n-1}i(f_{n-2}) \\ &= \left(\sum_{r=0}^{n-2} A^{4r}\right) h_{n-1}i(f_{n-2}) + \left(\sum_{r=1}^{n-2} A^{4r-2}\right) h_{n-2}h_{n-1}i(f_{n-2}) \\ &= [n-1]_q h_{n-1}i(f_{n-2}) + A^2[n-2]_q h_{n-2}h_{n-1}i(f_{n-2}). \end{aligned}$$

Now for $T \in TL_{n-2}$ the closures of the elements $h_{n-1}\, i(T)$ and $h_{n-2}h_{n-1}\, i(T)$ in TL_n are respectively $\delta\widehat{T}$ and $\widehat{T}$, as seen below.

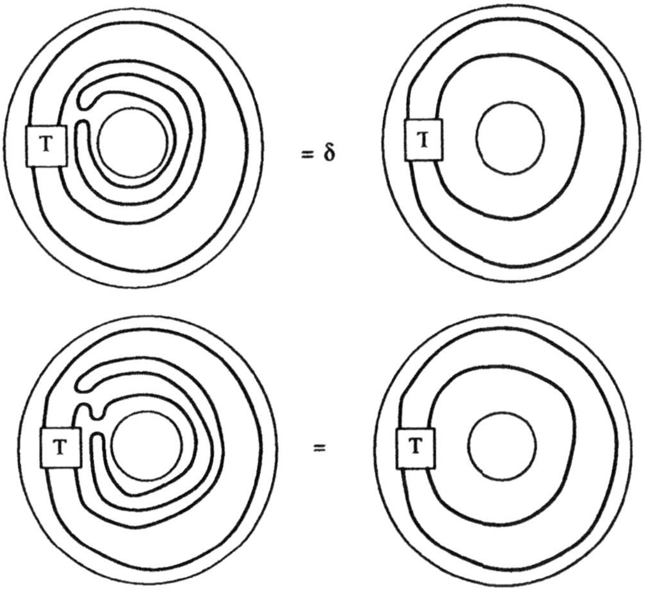

Thus

$$\begin{aligned} \widehat{Q}_n = \widehat{R}_n &= ([n-1]_q\delta + A^2[n-2]_q)\widehat{f}_{n-2} \\ &= -A^{-2}\left((1+A^4)[n-1]_q - A^4[n-2]_q\right)\widehat{f}_{n-2} \\ &= -A^{-2}(A^4[n-1]_q + 1)\widehat{f}_{n-2} = -A^{-2}[n]_q\widehat{f}_{n-2}. \end{aligned}$$

This completes the last step in the proof. □

As remarked earlier, this result establishes that the elements w_i are eigenvectors of the twist-induced map $\mathcal{F}$, with eigenvalue $\lambda_i = (-1)^{i-1}A^{i^2-1}$.

3. Invariants of 3-manifolds.

3.1 Surgery on framed links.

A description of closed orientable 3-manifolds has been known for some time in terms of surgery on framed links in S^3.

Given a framed link L in S^3, the technique of surgery produces a manifold $M(L)$ by removing a solid torus neighbourhood V_i of each link component L_i from S^3, leaving the 'exterior' of L, a compact 3-manifold whose boundary consists of k tori. The closed manifold $M(L)$ is built up from this piece and k solid tori, by gluing each solid torus to one of the boundary components. On the boundary of each solid torus there is a distinguished family of closed curves, the meridians, which bound discs in the solid torus. To specify $M(L)$ we must say which curves on the boundary of the exterior of L are to be matched with the meridians by the gluing.

We use the framing of L to determine this match. The framing of the component L_i specifies a choice of curves parallel to L_i which determines a distinguished family of curves on the corresponding boundary component of the exterior of L; the surgery is defined by matching these curves with the meridians.

We may think of the link L as giving us a view in S^3 of a large part of the manifold $M(L)$, namely the exterior of L. All that remains unseen are the added solid tori, and the picture provides a good indirect knowledge of these as well. Of course there can be other views of the same 3-manifold, based on a different link L' say, in other words we may find links L and L' for which $M(L) \cong M(L')$.

The study of 3-manifolds by means of framed links is greatly simplified by the results of Kirby [11] and Fenn and Rourke [5].

THEOREM 3.1 (Kirby, Fenn-Rourke).

(1) *Every closed oriented 3-manifold arises as $M(L)$ for some framed link L.*

(2) *There is an orientation preserving homeomorphism $M(L) \cong M(L')$ if and only if L and L' are related by a finite sequence of Kirby moves.*

Kirby moves are of two types, shown below.

L ⟼ $\varphi_+(L)$

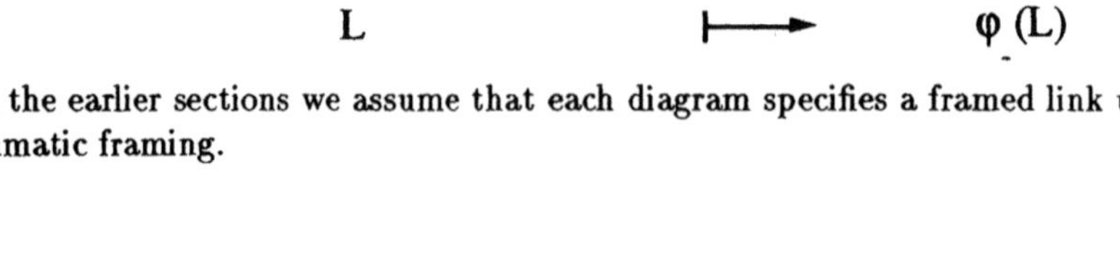

L ⟼ $\varphi_-(L)$

As in the earlier sections we assume that each diagram specifies a framed link using the diagrammatic framing.

These moves have been used by Reshetikhin and Turaev [32], and subsequently several other authors, as a means of approaching the family of 3-manifold invariants described by Witten [W]. The central idea is to look for an element $\Omega \in \mathcal{B}$ with the property that the value $J_L(\Omega,\ldots,\Omega) \in \Lambda$, possibly normalised in some way, is unaltered when L is changed by Kirby moves. If such an Ω were to exist, then $J_L(\Omega,\ldots,\Omega)$ would depend only on the manifold $M(L)$, and so would give an element of Λ which is an invariant of $M(L)$. Unfortunately this does not prove to be possible without some modification, even allowing Ω to be a formal power series in α rather than a polynomial.

The modification which works is to decide initially on a 'level' l, or equivalently to select a $4r$-th root of unity, with $r = l+2$, which is to be substituted for the variable A in Λ. Having decided on l, it is *then* possible to choose $\Omega \in \mathcal{B}$, (depending on l), so that the complex number given by substituting a $4r$-th root of unity in $J_L(\Omega,\ldots,\Omega)$ is, after suitable normalisation, unaltered by the Kirby moves, and is thus an invariant of $M(L)$.

In keeping with Segal's view of Witten's invariants it is worth noting that if a link invariant J_L is eventually to be evaluated at a $4r$-th root of unity then elements of $\mathcal{B}$ whose difference lies in an ideal $\mathcal{I}_r$ independent of L will always determine the same value. The element Ω can then be considered as an element of $\mathcal{B}/\mathcal{I}_r$, which is a finite-dimensional algebra over Λ. In section 3.2 I shall write $\mathcal{B}_r$ for this algebra after substituting a chosen $4r$-th root of unity for A in the coefficient ring Λ. In this context it is possible to describe further modifications to deal in a similar way with the more general constructions of manifolds by 'rational' surgery on a link, in which a different family of curves is used to match with the meridians when the solid tori are glued to the link exterior, [24].

3.2 Evaluations of the total invariant at roots of unity.

I shall start by discussing the evaluation of J_L on the ideal in $\mathcal{B}$ generated by one of the elements w_r. For any given component of the diagram of L it is possible, using moves R_{II} and R_{III}, to draw it in an annulus as the closure of some $(1,1)$-tangle T so that the chosen component, L_1 say, is the single arc in T while the remaining components $L_2,\ldots,L_k$ lie entirely in T. This diagram $\widehat{T}$ in the annulus induces a multilinear map $J_{\widehat{T}} : \mathcal{B}\times\ldots\times\mathcal{B} \to \mathcal{B}$ and J_L is the composite of this with the evaluation map $<\cdot>: \mathcal{B} \to \Lambda$.

Theorem 3.2. *For any n, and any $\beta_2,\ldots,\beta_k \in \mathcal{B}$ we have*

$$J_{\widehat{T}}(w_{n+1},\beta_2,\ldots,\beta_k) = \lambda w_{n+1}, \text{ for some } \lambda \in \Lambda.$$

Proof: From the tangle T construct an (n,n)-tangle $T^{(n)}$ with n parallel arcs in place of the single arc. Decorate the $k-1$ closed curves by $\beta_2,\ldots,\beta_k$ to give an element, G say, in TL_n. The closure of f_nG will then be $J_{\widehat{T}}(\widehat{f}_n,\beta_2,\ldots,\beta_k) \in \mathcal{B}$. Now the multiplicative property of f_n allows us to write $f_nG = \lambda f_n$, where $\lambda = \varphi_A(G) \in \Lambda$. Thus $J_{\widehat{T}}(\widehat{f}_n,\beta_2,\ldots,\beta_k) = \lambda \widehat{f}_n$, and the result follows since $\widehat{f}_n$ is a multiple of w_{n+1}. □

Theorem 3.3. *Let L be a link diagram. Then*

(a) $J_L(w_r \times \mathcal{B} \times \ldots \times \mathcal{B}) \subset \Lambda < w_r >$, the ideal generated by $< w_r >$ in Λ, and

(b) $J_L(\mathcal{I}_r \times \mathcal{B} \times \ldots \times \mathcal{B}) \subset \Lambda < w_r >$, where $\mathcal{I}_r \subset \mathcal{B}$ is the ideal of $\mathcal{B}$ generated by w_r.

Proof: Part (a) is an immediate corollary of the previous theorem, with $r = n+1$, on drawing L to lie appropriately in the annulus. The result holds when any of the components is decorated by w_r.

To prove part (b) it is enough to deal with the element $w_r\beta \in \mathcal{I}_r$ for any $\beta \in \mathcal{B}$. We can use the multiplication in $\mathcal{B}$ to write

$$J_L(w_r\beta, \beta_2, \ldots, \beta_k) = J_{L'}(w_r, \beta, \beta_2, \ldots, \beta_k),$$

where L' is the link with two parallel components in place of the first component of L but otherwise identical to L. The result now follows from (a) applied to L'. □

The evaluation map $< \cdot > : \mathcal{B} \to \Lambda$ is a ring homomorphism, and hence

$$< w_{n+1} > = < \alpha >< w_n > - < w_{n-1} >.$$

Starting from $< w_1 > = 1$ and $< w_2 > = < \alpha > = \delta = -(A^2 + A^{-2})$ it follows readily that $< w_r > = (-1)^{r-1}\dfrac{A^{2r} - A^{-2r}}{A^2 - A^{-2}} \in \Lambda$. Then $< w_r > = 0$ when $A^{4r} = 1,\ A^4 \neq 0$.

Notation. Write $\Lambda_r \subset \mathbf{C}$ for the image of Λ when A is mapped to a primitive $4r$-th root of unity, e.g. $A = e^{\pi i/2r}$.

Equivalently, take Λ_r to be the quotient of Λ by the ideal generated by Euler's polynomial $\varphi_{4r}(A)$.

Write also $\mathcal{B}_r$ for the finite-dimensional Λ_r-module $(\mathcal{B}/\mathcal{I}_r)\otimes\Lambda_r$, where, as noted above, the coefficient ring has been changed from Λ to Λ_r by substitution for A. Theorem 3.3 can then be reformulated.

THEOREM 3.4. *For any link diagram L the invariant $J_L : \mathcal{B} \times \ldots \times \mathcal{B} \to \Lambda$ induces a multilinear map $J_L^{(r)} : \mathcal{B}_r \times \ldots \times \mathcal{B}_r \to \Lambda_r$.*

Proof: The value of $J_L^{(r)}$ can be calculated by choosing decorations $(\beta_1, \ldots, \beta_k)$ in $\mathcal{B} \times \ldots \times \mathcal{B}$, and substituting the chosen $4r$-th root of unity for A in $J_L(\beta_1, \ldots, \beta_k)$. The previous theorem shows that this number in Λ_r is unchanged when an element of the ideal $\mathcal{I}_r$ is added to any β_i, by multilinearity of J_L, and so the result depends only on the elements represented by β_i in $\mathcal{B}_r$. □

3.3 STRUCTURE OF THE ALGEBRAS $\mathcal{B}$ AND $\mathcal{B}_r$.

The product $w_j w_k$ of two basis elements in $\mathcal{B}$ can be written as a sum $\sum_i n_{ijk} w_i$, with structure constants $n_{ijk} \in \Lambda$. It can be established inductively that $n_{ijk} \in \mathbf{N}$, and that $n_{1jk} = \delta_{jk}$; in fact

$$n_{ijk} = \begin{cases} 1, & \text{if } i+j+k = 1 \bmod 2 \text{ and } |j-k| < i < j+k, \\ 0, & \text{otherwise.} \end{cases}$$

This is more obvious once we are able to identify $\mathcal{B}$ with the representation ring of $SU(2)$ and w_i with the irreducible representations.

Then n_{ijk} is the coefficient of w_1 in the product $w_i w_j w_k$. Since $\mathcal{B}$ is commutative, n_{ijk} is unchanged by permutation of i, j and k.

The algebra $\mathcal{B}_r$ has a basis $w_1, \ldots, w_{r-1}$, or properly speaking the images of these elements. Each $w_i \in \mathcal{B}$ represents some integer linear combination of $w_1, \ldots, w_{r-1}$ in $\mathcal{B}_r$, and we can write $w_j w_k = \sum_{i=1}^{r-1} m_{ijk} w_i$ in $\mathcal{B}_r$, for some integers m_{ijk}. It can be shown [24] that m_{ijk} is also symmetric in i, j and k when $j, k < r$ and only takes the values 0 or 1.

3.4 THE 3-MANIFOLD INVARIANTS.

Let us now compare the invariants of two links related by a positive Kirby move. Suppose that the two links are as shown in the Kirby move diagram, and that the second link $\varphi_+(L)$ has k components, corresponding to the first k components of the link L. Regard the closure $\widehat{T}$ as a diagram in the annulus, determining $J_{\widehat{T}} : \mathcal{B} \times \ldots \times \mathcal{B} \to \mathcal{B}$. Choose any decoration $\beta_1, \ldots, \beta_k$ of $\widehat{T}$ and write $X = J_{\widehat{T}}(\beta_1, \ldots, \beta_k) \in \mathcal{B}$. The satellite formula shows that $J_L(\beta_1, \ldots, \beta_k, \Omega) = J_M(X, \Omega)$, where M is the link shown below.

$M =$  $K =$ 

Since M itself is two parallel copies of the diagram K, which in turn is the unknot with a positive curl, we can write $J_M(X, \Omega) = J_K(X\Omega) = \langle F(X\Omega) \rangle$. We want to compare this with the invariant after the Kirby move, namely $J_{\varphi_+(L)}(\beta_1, \ldots, \beta_k) = \langle X \rangle$.

THEOREM 3.5. *Given r there exists $\Omega \in \mathcal{B}$ and $c_\pm \neq 0 \in \Lambda_r$ depending only on r such that, for any choice of L and decorations $\beta_1, \ldots, \beta_k$,*

$$J_L(\beta_1, \ldots, \beta_k, \Omega) = c_\pm J_{\varphi_\pm(L)}(\beta_1, \ldots, \beta_k)$$

when evaluated in Λ_r.

Proof: Choose $\Omega = \sum_{k=1}^{r-1} a_k w_k$ with $a_k = \langle w_k \rangle$. By the calculations above, it is enough to find c_+ so that $J_M(X, \Omega) = c_+ \langle X \rangle$ in Λ_r for all $X \in \mathcal{B}$. Since we are evaluating in Λ_r it is enough to check for X in a spanning set of $\mathcal{B}_r$, e.g. $X = w_j$, $j = 1, \ldots, r-1$.

Now $J_M(w_j, \Omega) = J_K(w_j\Omega)$. Again it is enough to work with $w_j\Omega$ as an element of $\mathcal{B}_r$, since we are only concerned with the evaluation in Λ_r, so that in Λ_r we have

$$\begin{aligned} J_K(w_j\Omega) &= J_K\Big(\sum_{k=1}^{r-1} a_k w_j w_k\Big) \\ &= J_K\Big(\sum_{k=1}^{r-1}\sum_{i=1}^{r-1} m_{ijk} a_k w_i\Big) \\ &= \Big\langle \sum_{k=1}^{r-1}\sum_{i=1}^{r-1} m_{ijk} a_k \lambda_i w_i \Big\rangle \\ &= \sum_{i=1}^{r-1} \lambda_i \langle w_i \rangle \sum_{k=1}^{r-1} m_{ijk} a_k \,. \end{aligned}$$

On the other hand, $< w_i >< w_j > = < w_i w_j > = \sum_{k=1}^{r-1} m_{kij} < w_k > = \sum_{k=1}^{r-1} m_{ijk} a_k$ by symmetry of the coefficients m_{ijk}. Thus

$$\begin{aligned} J_M(w_j, \Omega) &= \sum_{i=1}^{r-1} \lambda_i < w_i >< w_i >< w_j > \\ &= c_+ < w_j >, \end{aligned}$$

where $c_+ = \sum_{i=1}^{r-1} \lambda_i < w_i >^2$.

The assignment $c_- = \sum_{i=1}^{r-1} \lambda_i^{-1} < w_i >^2$ will handle the negative Kirby move similarly, as the only difference is in the use of F^{-1} in place of F to deal with the left-handed curl. Noting that $c_\pm$ are complex conjugates in Λ_r since $|A| = 1$ we can write $c_\pm = \rho c^{\pm 1}$ in polar form, with $\rho > 0$ and $|c| = 1$. It is possible to calculate c, ρ in terms of the root of unity A, and check also that $\rho \neq 0$. □

Assignment of Ω to each component then gives an element of Λ_r which is invariant under the Kirby moves, except for the appearances of $c_\pm$. It is not difficult to introduce a normalising factor to correct for this, as follows.

To a framed *oriented* link $L = L_1 \cup L_2 \cup \ldots \cup L_k$ we can associate a quadratic form with $k \times k$ matrix (l_{ij}) where

$$l_{ij} = \mathrm{lk}(L_i, L_j),\ i \neq j, \quad l_{ii} = \text{framing on } L_i.$$

Write $\mathrm{sig}(L)$ for the signature of this form.

Then $\mathrm{sig}(L)$ is independent of the choice of orientation of L, and

$$\mathrm{sig}\, \varphi_\pm(L) = \mathrm{sig}\, L \mp 1.$$

COROLLARY 3.6. *When $M(L)$ is given by surgery on the framed link L with k components the complex number*

$$\rho^{-k} c^{-\mathrm{sig}\, L} J_L(\Omega, \ldots, \Omega),$$

evaluated at the given root of unity, is an invariant of the 3-manifold $M(L)$.

Proof: It is enough to show that the number is unaltered by a Kirby move on L. Consider the case of the positive Kirby move, giving $\varphi_+(L)$ with $k-1$ components. Then

$$\begin{aligned} \rho^{-k} c^{-\mathrm{sig}\, L} J_L(\Omega, \ldots, \Omega) &= c_+ \rho^{-k} c^{-\mathrm{sig}\, L} J_{\varphi_+(L)}(\Omega, \ldots, \Omega) \\ &= \rho^{-(k-1)} c^{-(\mathrm{sig}\, L - 1)} J_{\varphi_+(L)}(\Omega, \ldots, \Omega) \\ &= \rho^{-(k-1)} c^{-\mathrm{sig}\, \varphi_+(L)} J_{\varphi_+(L)}(\Omega, \ldots, \Omega), \end{aligned}$$

which is the corresponding number for $\varphi_+(L)$. The negative Kirby move works similarly, with c_- in place of c_+ covered by the alteration in signature. □

Remarks. There has only been a limited amount of calculation of these invariants. A recent tabulation of known evaluations is given in [27]. Kirby and Melvin have been able to give a closed formula for the invariants for Lens spaces as r varies, and also show how the value for $r = 2, 3, 4$ or 6 can be related in general to known topological invariants. Strickland has also developed programs to compute for Lens spaces, using knowledge of J for torus knots. The difficulty in general comes in calculating $J_L(\Omega)$ for larger values of r, as this requires knowledge of $J_L(w_k)$, at least in Λ_r, for all $k < r$. This in turn is equivalent to knowing $J_L(\alpha^j)$ for $j < r-1$, in other words, the bracket invariant of the j-fold parallels of L. As a computational exercise this rapidly becomes impractical with increasing j, even when L has a braid presentation on as few as 3 strings.

4. The quantum group approach.

In this section I shall discuss the alternative view of the invariants J_L of a framed link which was pioneered by Reshetikhin [29] and Turaev [34].

The starting point here is a quantum group $\mathcal{G}_q$, most conveniently one which is associated to a classical Lie group G; in the present context it is enough to consider $G = SU(2)$. The quantum group is an algebra over a ring Λ which includes a parameter q. Many of the constructions involve polynomials in $q^{\pm\frac{1}{4}}$ at the worst, and with care the ring can be regarded as $\mathbf{Z}[q^{\pm\frac{1}{4}}]$.

Finite-dimensional representations of the quantum group $\mathcal{G}$ (i.e. $\mathcal{G}$-modules) play a central role in the definition of link invariarts. The most impoitant property of $\mathcal{G}$ is that it is a Hopf algebra, in other words it admits a comultiplication $\Delta : \mathcal{G} \to \mathcal{G} \times \mathcal{G}$ which has a sufficiently natural interaction with the algebra multiplication to allow tensor products of $\mathcal{G}$-modules to be themselves regarded as $\mathcal{G}$-modules.

The map Δ is not symmetric, in the sense that $\tau \circ \Delta \neq \Delta$ where $\tau : \mathcal{G} \otimes \mathcal{G} \to \mathcal{G} \otimes \mathcal{G}$ is induced by $\tau(g \otimes h) = h \otimes g$. Consequently, when V and W are two $\mathcal{G}$-modules, the simple interchange map $\tau : V \otimes W \to W \otimes V$ need not be an isomorphism of $\mathcal{G}$-modules, since Δ is used in the definition of $V \otimes W$ as a $\mathcal{G}$-module. There is, however, as part of the definition of a quantum group, an element R in a suitable extension of $\mathcal{G} \otimes \mathcal{G}$ which relates $\tau \circ \Delta$ and Δ. From this 'universal R-matrix' R there arises a $\mathcal{G}$-module isomorphism $R_{VW} : V \otimes W \to W \otimes V$ for all modules V and W, which is not the simple interchange map; thus in general $R_{VW}^{-1} \neq R_{WV}$.

4.1 Construction of link invariants.

The aim is to start with any (m,n)-tangle T and choose a 'colouring' of its components by finite-dimensional $\mathcal{G}$-modules, in other words, select a $\mathcal{G}$-module for each component. Then try to represent coloured tangles by $\mathcal{G}$-module homomorphisms in such a way that when the strings at the bottom of the tangle T have been coloured by modules $V_1, \ldots, V_n$ and the strings at the top by $W_1, \ldots, W_m$ then the coloured tangle is represented by a module homomorphism $V_1 \otimes \ldots \otimes V_n \to W_1 \otimes \ldots \otimes W_m$, while the composite of two consistently coloured tangles placed one above the other is represented by the composite of the two homomorphisms.

Every tangle can be built up as the composite of a number of elementary tangles which are either a simple crossing  or  or a local maximum  or minimum

 , alongside a number of parallel straight strings. Once it is decided how to assign a homomorphism to each of these elementary tangles, with colouring, the homomorphism for the whole tangle will be determined as a composite. To show that the homomorphism defined in this way for a coloured tangle is independent of how the tangle is drawn, up to say moves R_{II} and R_{III}, it is sufficient to show that certain combinations of the elementary tangles determine the same homomorphism.

To make the assignments for the elementary coloured tangles we require homomorphisms $V \otimes W \to W \otimes V$ for each of the $(2,2)$-tangles  and  , for which we use R_{VW} and R^{-1}_{WV} respectively. The identity $(1,1)$-tangle,  , is represented by 1_V; when placed alongside other elementary tangles a number of parallel straight strings are represented by taking the tensor product with the appropriate identity homomorphism.

When a tangle has no points at the top or bottom, the appropriate $\mathcal{G}$-module to use as domain or target is the trivial module, in other words the coefficient ring Λ. Thus the local minimum $(2,0)$-tangle, $U =$  , requires a homomorphism $\Lambda \to V \otimes V$, while the local maximum tangle, $V =$  , requires a homomorphism $V \otimes V \to \Lambda$. Turaev observed that only a small number of checks on these are needed to ensure invariance of the homomorphism when the strings of the tangle are moved. These are shown pictorially below, and should be read as the equality of the composites of the homomorphisms determined when the tangle is coloured arbitrarily, and regarded as the product of elementary tangles.

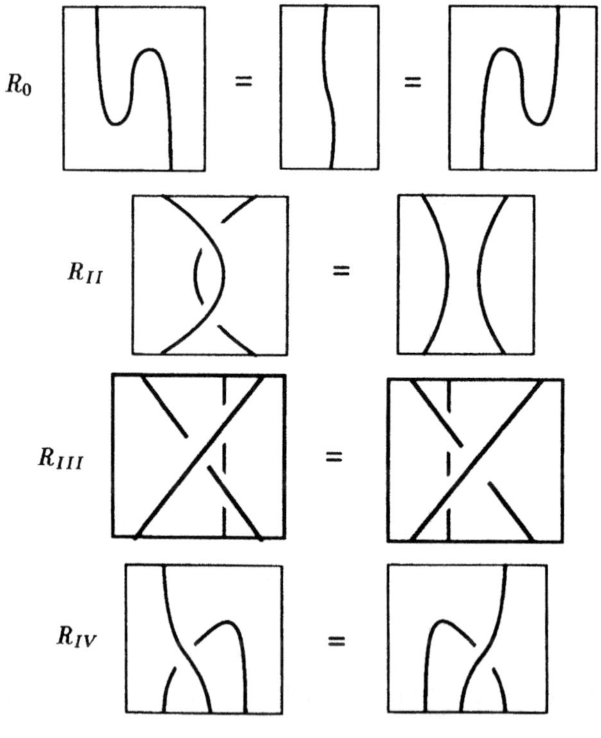

A little care is needed in defining the homomorphisms to represent the local maximum and minimum coloured by the general module V. Reshetikhin and Turaev [31] give details in a wider context; for irreducible V and the quantum group $SU(2)_q$ there is an almost canonical choice, and having made this choice to satisfy R_0 the other relations are guaranteed by the nature of the universal R-matrix. The consequence of the definition is that a link diagram L, regarded as a $(0,0)$-tangle, determines a homomorphism $\Lambda \to \Lambda$ for each assignment of modules $V_1, \ldots, V_k$ to its components. This homomorphism is simply multiplication by some scalar $J(L; V_1, \ldots, V_k)$ which depends only on L up to moves R_{II} and R_{III} and so gives an invariant of the framed link L.

Whatever definition of the homomorphisms representing  and  is used, a little care can be taken to ensure that

(1) $J(L)$ is multilinear on sums of modules,

(2) when one component of L, say the first, is coloured with the tensor product $V \otimes W$ then

$$J(L; V \otimes W, V_2, \ldots, V_k) = J(L'; V, W, V_2, \ldots, V_k),$$

where the link L' has two parallel components in place of the first component of L, coloured with V and W separately.

A fuller account is given in [23], in which condition (1) is forced by working primarily with irreducible representations, and then (2) has to be proved. In [31] the definitions guarantee property (2) immediately, while (1) then needs a little proof. Rosso [33] has shown that in the general case, where $\mathcal{G}$ is regarded as an algebra over the field of rational functions in an indeterminate $q^{\pm\frac{1}{4}}$, finite dimensional $\mathcal{G}$-modules are completely reducible, (i.e. isomorphic to a direct sum of irreducible modules). In this generic case write $\mathcal{R}$ for the representation ring of $\mathcal{G}$, as an algebra over Λ. An element of $\mathcal{R}$ is then a finite Λ-linear combination of finite dimensional irreducible $\mathcal{G}$-modules, and every $\mathcal{G}$-module can be written in $\mathcal{R}$ as a positive-integer combination of irreducible modules. Tensor product of modules makes $\mathcal{R}$ into a ring.

4.2 The total quantum invariant.

The multilinear property (1) of $J(L)$ means that it can be extended to give a multilinear map $J(L) : \mathcal{R} \times \ldots \times \mathcal{R} \to \Lambda$. By definition, $\mathcal{R}$ has a basis consisting of irreducible representations of $\mathcal{G}$; in this case with $\mathcal{G} = SU(2)_q$ we know that $\mathcal{R}$ is isomorphic to the representation ring of $SU(2)$ having one irreducible module W_i in each dimension $i \geq 1$. Details of these modules and the corresponding R-matrices are given in [13]; an account following the universal R-matrix prescription of Drinfeld is given in [12].

The generic case, where the parameter q is treated as an indeterminate, has the advantage that the representation ring is isomorphic to the representation of the corresponding classical Lie group, and so its structure is understood. Construction of link invariants can also be done when $\mathcal{G}$ is replaced by a finite dimensional algebra, and the coefficient ring is altered by specialising q to a root of unity. In this case the representation theory becomes more complicated, as modules are not always completely reducible, so that a direct interpretation of the link invariant as a function on the representation ring is no longer possible, and more detailed work is needed to handle the invariant comfortably, as in [32] and [12].

Returning to the generic case, it is straightforward to use properties (1) and (2) for $J(L)$, and knowledge of the ring $\mathcal{R}$, to identify $\mathcal{R}$ with the ring $\mathcal{B}$ and $J(L)$ with the total bracket invariant J_L.

THEOREM 4.1. *The Λ-linear map $h : \mathcal{R} \to \mathcal{B}$ defined by $h(W_i) = w_i$ is a ring isomorphism, where $\Lambda = \mathbf{Z}[A^{\pm 1}]$, $A^4 = q$. For a framed link L the invariants $J(L)$ and J_L can be identified by*

$$J(L; V_1, \ldots, V_k) = J_L(h(V_1), \ldots, h(V_k)).$$

Proof: It is a classical result that the representation ring of $SU(2)$ is a polynomial ring generated by the fundamental 2-dimensional irreducible representation, so that $\mathcal{R}$ is the polynomial ring generated by W_2. Hence there is an isomorphism from $\mathcal{R}$ to $\mathcal{B}$ carrying W_2 to $\alpha = w_2 \in \mathcal{B}$. To establish that this is the map h it is enough to show that the elements W_i satisfy the recurrence relation $W_{n+1} = W_2 W_n - W_{n-1}$ in $\mathcal{R}$. Now it is readily established from the representation theory of $SU(2)$ that the tensor product $W_2 \otimes W_n$ decomposes as the direct sum of irreducibles $W_{n-1} \oplus W_{n+1}$ so that in $\mathcal{R}$ we have $W_2 W_n = W_{n-1} + W_{n+1}$.

Using the fact that $\mathcal{R}$ is spanned by the powers of W_2 we may evaluate the invariant $J(L)$ by evaluating it simply on modules $V_j = W_2^j$, for varying j. When the invariant $J(L)$ is evaluated at W_2^j on one component of L we may use property (2) to replace this by the link L' with j components in place of the one component, each coloured by W_2. In this way comparison of $J(L)$ and J_L reduces to showing that for each link L

$$J(L; W_2, \ldots, W_2) = J_L(w_2, \ldots, w_2).$$

Now $J_L(w_2, \ldots, w_2) = < L >$ so it remains to identify $J(L; W_2, \ldots, W_2)$ with the bracket polynomial of L. It is enough to show that the three linear maps from $W_2 \otimes W_2$ to itself representing the diagrams

$\sigma =$ , Id $=$  and $H =$ 

satisfy the relation $\sigma = A\,\mathrm{Id} + A^{-1}H$, and that the invariant of the simple unknot, as a $(0,0)$-tangle, is $\delta = -A^2 - A^{-2}$.

When all strings are coloured by W_2 the $(2,2)$-tangles σ, Id and H are each represented by an endomorphism of the module $W_2 \otimes W_2$. These endomorphisms are $R_{W_2 W_2}$, $1_{W_2 \otimes W_2}$ and the composite of the local minimum and local maximum maps for W_2 respectively. It is possible, given the detailed information from the quantum group, to calculate these maps explicitly and confirm that they satisfy the linear relation corresponding to the equation $\sigma = A\,\mathrm{Id} + A^{-1}H$. We can also confirm from the explicit maps that the composite of the local maximum and local minimum maps when coloured with W_2 represents the simple unknotted circle by the map from Λ to Λ which is multiplication by $\delta = -A^{-2} - A^2$. Consequently the linear map $\mathcal{D}(\mathbf{R}^2) \to \Lambda$ defined on the diagram L by $J(L; W_2, \ldots, W_2)$ respects the defining relations for $\mathcal{S}(\mathbf{R}^2)$ and hence factors through $\mathcal{S}(\mathbf{R}^2)$. Thus, applied to the diagram L, we have

$$J(L; W_2, \ldots, W_2) = < L > J(\phi;) = < L >,$$

since $L = < L > \phi$ in $\mathcal{S}(\mathbf{R}^2)$. □

Remark. It is in fact more accurate to take the isomorphism determined by $W_2 \mapsto -w_2$, and the identification of A with $-e^{-h/4}$, where $q = e^h$. The quantum group homomorphism $R_{W_2 W_2}$ is then given directly by Drinfeld's universal R-matrix for $SU(2)_q$.

We may thus use either the bracket invariant approach or the quantum group approach to determine the same multilinear invariant $J(L)$ in terms of $\mathcal{B}$, the skein of the annulus, or equally of $\mathcal{R}$, the representation ring of $SU(2)$. In this second guise some of the properties of the invariant which we have already discussed appear quite naturally, in particular that w_i is an eigenvector of the map $\mathcal{F} : \mathcal{B} \to \mathcal{B}$. The framing change in the quantum view requires the insertion of a curl on the component of a link, to which some element of $\mathcal{R}$ has been attached. Suppose that this element is one of the irreducibles, W_i. We may draw the diagram after the framing change so that the extra curl is viewed as a $(1,1)$-tangle coloured with W_i, inserted at some point in the original diagram. This $(1,1)$-tangle is represented by a module homomorphism from W_i to W_i. Since W_i is irreducible, such a map must, by Schur's lemma, be a scalar multiple, λ_i say, of the identity. Hence the curl can be removed at the expense of multiplying $J(L)$ by λ_i without any other change.

Having made the identification of the two descriptions for the generic link invariant it is possible to move on to discuss the 3-manifold invariant, as in the previous section, via the quotient ring $\mathcal{B}_r$ (or $\mathcal{R}_r$) without having to consider the actual representations of $SU(2)_q$ at the root of unity.

4.3 The Temperley-Lieb algebra again.

One further link between the two viewpoints arises when we apply the quantum group viewpoint to tangles in which every component is coloured with the fundamental representation $W_2 = V$, say. Each $(m.n)$-tangle then determines a linear map from $V^{\otimes n}$ to $V^{\otimes m}$, which is a $\mathcal{G}$-module homomorphism, while composition of tangles induces composition of maps. Because the skein relations are satisfied when W_2 is used on all strings there is an induced map from the skein $\mathcal{S}(R_n^n) = TL_n$ to the linear endomorphisms of $V^{\otimes n}$. This gives a representation, which is in fact faithful, of the Temperley-Lieb algebra TL_n as an algebra of $2^n \times 2^n$ matrices, with coefficients in Λ. Since $V^{\otimes n}$ is a $\mathcal{G}$-module, and the tangles are all represented by module endomorphisms, we can see further that TL_n is represented as a subalgebra of all $\mathcal{G}$-module endomorphisms of $V^{\otimes n}$. Indeed, if the coefficient ring Λ is extended to include sufficient denominators then the image of TL_n can be shown to be the algebra of all $\mathcal{G}$-module endomorphisms of $V^{\otimes n}$.

There is just one submodule of $V^{\otimes n}$ which is isomorphic to the irreducible W_{n+1}. Projection to this submodule determines a $\mathcal{G}$-module endomorphism of $V^{\otimes n}$, and hence an element of TL_n. This element of TL_n is in fact the element f_n discussed earlier, divided by $\varphi_A(f_n)$. The multiplicative property of f_n is seen in this context from the fact that $V^{\otimes k}$ with $k < n$ has no summands isomorphic to W_{n+1}, so that the composition of the projection with the map representing any (n,k)-tangle, $k < n$, must be zero. Now each generator h_i of TL_n is the composite of an $(n, n-2)$-tangle with an $(n-2, n)$-tangle, so that the projection when composed with any of these must be zero. This leads to the equation $f_n h_i = 0$, and thus to the multiplicative property, given that φ_A can also be recognised by the property that $\varphi_A(h_i) = 0$.

The representation of TL_n on $V^{\otimes n}$ can be quickly recovered from the two maps representing the local maximum and minimum. These can be chosen to have matrices

$$(0 \quad A \quad -A^{-1} \quad 0) \text{ and } (0 \quad -A \quad A^{-1} \quad 0)^T,$$

representing the linear maps Max $: V \otimes V \to \Lambda$ and Min $: \Lambda \to V \otimes V$ respectively, where V has a basis v_1, v_2 and the basis elements of $V \otimes V$ are written in the order $v_1 \otimes v_1$, $v_1 \otimes v_2$, $v_2 \otimes v_1$, $v_2 \otimes v_2$. These maps satisfy the condition R_0 and can be combined as Max.Min to represent H. The matrix representing σ is then given by $\sigma = A + A^{-1}H$, while the value of δ can be checked by calculating the product Min.Max.

This representation of TL_n can be used as a means of calculating explicitly the bracket polynomial of the closure of any (n,n)-tangle. It also provides a representation of the braid group B_n on $V^{\otimes n}$ in which the generators σ_i satisfy a quadratic relation, and so have only two eigenvalues. This representation preserves each $\mathcal{G}$-submodule of $V^{\otimes n}$ which consists of the sum of all submodules isomorphic to a given irreducible W_i, and hence it breaks up into a number of lower dimensional representations of B_n and indeed of TL_n. Details of this are discussed in Reshetikhin's papers [29]. Other representations of the braid group arise in a similar way, with higher degree minimal polynomial for σ_i, using (n,n)-tangles coloured by one of the other irreducible modules W_j in place of W_2.

5. A geometric view of the invariants.

In defining the 'generic' type of link invariant J_L, taking values in a ring Λ containing an indeterminate $A = q^{\frac{1}{4}}$, I have described two different approaches which arrive at essentially the same end result. The interpretations of the parameter space $\mathcal{B} = \mathcal{R}$ in terms of 'decorations' or 'colourings' which can be applied to the link components highlight different properties, depending on whether the view as quantum group representations or as bracket invariants of satellites is uppermost in the mind.

Either of these views constitutes a first stage for the invariants. The second stage arises when they are used to build invariants of general 3-manifolds, typically in terms of evaluations of the generic invariants, where the indeterminate is replaced by a specified root of unity. The account given so far has made use of some features which are special to $SU(2)_q$, or equally to the bracket invariant, but there is much which will work readily in a wider context. In the final section I shall give a brief account of the generic stage in constructing invariants, using the quantum groups $SU(k)_q$ on one hand, and linear skein theory based on the Homfly polynomial on the other. Similar work relates Kauffman's Dubrovnik polynomial with the quantum groups of the B, C and D series, coming from the orthogonal and symplectic groups, [42]. The corresponding second stage can be pursued, with a little care, following the general lines of section 3.

In the remainder of this section I look first at the generic invariant from the 3-dimensional point of view, and then note how this and the second stage invariants fit in to the framework of Witten.

5.1 The generic invariant and modular functors.

Both approaches, from linear skein theory and from the representation theory of $SU(2)_q$, lead to a framed link invariant $J_L : \mathcal{B}^k \to \Lambda$, and a satellite formula relating J_K for a satellite K to J_C for its companion C and $J_P : \mathcal{B}^k \to \mathcal{B}$ for the pattern P, viewed as a diagram in the annulus.

There are two alternative views of the pattern

(1) as a k-component diagram in the annulus, and

(2) as a $k+1$-component link P' consisting of P together with one distinguished unknotted component which determines the annulus.

View (1) determines a multilinear map $J_P : \mathcal{B}^k \to \mathcal{B}$, while view (2) gives a map $J_{P'} : \mathcal{B}^{k+1} \to \Lambda$. These can be related by regarding P' as a satellite of the Hopf link H using the pattern P, so that $J_{P'} = J_H \circ (J_P \times 1)$ as maps from $\mathcal{B}^k \times \mathcal{B}$ to Λ. The Hopf link invariant $J_H : \mathcal{B}^2 \to \Lambda$ thus provides a bilinear form which plays a central role in comparing the two views.

The remaining feature of the generic invariant is the linear automorphism $\mathcal{F} : \mathcal{B} \to \mathcal{B}$ describing the framing change, and the basis of $\mathcal{B}$ consisting of its eigenvectors.

When we move to a more 3-dimensional view one characteristic feature is the behaviour of the invariants when pieces of 3-manifold with boundary are glued together. In Witten's framework, once the choice of a quantum group $\mathcal{G}$ and a level k have been made there should then be determined a 'modular functor' from the category of cobordisms of surfaces to the category of complex vector spaces and linear maps.

Definition. We say that the boundary of a 3-manifold has been *marked* if for each boundary component of genus g there is an explicit choice of homeomorphism from a standard copy of the surface of genus g to that boundary component. We refer to the homeomorphism as a *marking*.

A full description of the required ingredients is given for example in [38]. The central idea is that every oriented 3-manifold M with boundary ∂M can be regarded as a cobordism when its boundary is marked and is partitioned into two parts, each consisting of a union of closed surfaces. In categorical terms, the objects of the category are unions of oriented surfaces, and the morphisms are oriented 3-manifolds with marked partitioned boundary, so that M with boundary $\partial M^- \cup \partial M^+$ is regarded as a morphism from the incoming boundary ∂M^- to the outgoing boundary ∂M^+. Morphisms (cobordisms) are composed by gluing the outgoing boundary of one manifold to the incoming boundary of the other, using the marking of each component to determine the gluing.

A *modular functor* is a functor from this category to the category of vector spaces and linear maps. It associates a vector space to each surface of genus g, and the tensor product of such spaces to a disjoint union of surfaces. The marked cobordism M provides a linear map from the space for ∂M^- to the space for ∂M^+. This map is assumed to be unchanged when the marking of a component is altered by isotopy. The functorial property ensures that composition of cobordisms translates into composition of linear maps. The marking of a boundary component may be altered by composing the original cobordism with another of the form surface $\times I$, in which different choices of marking are made at the two ends. Such cobordisms determine an automorphism of the vector space for each self-homeomorphism of the standard surface of genus g, and thus an action of the mapping class group of the surface on the vector space corresponding to the surface. In this way it is easy to take account of the effect of gluing two manifolds together in different ways, when the gluing is altered by a homeomorphism of one boundary component. The vector space corresponding to the empty surface ϕ is the trivial, 1-dimensional, vector space, and so a closed manifold M, which is a cobordism from ϕ to ϕ, yields a linear map from $\mathbf{C}$ to $\mathbf{C}$. Such a map is just multiplication by some scalar $\lambda(M)$, which is the invariant of the manifold, given the choices of quantum group and level required to set up the functor.

The generic invariants J_L which we have discussed earlier can be thought of as a similar more substantial cobordism invariant, but on a limited class of cobordisms. In this context, the framed link L determines its exterior, the manifold defined by removing a neighbourhood of each component from S^3. This leaves a 3-manifold with k boundary components, with the marking of each by $S^1 \times S^1$ determined by taking one factor to the curves chosen as parallels and the other factor to the meridian curves which bound discs in the deleted neighbourhood of the link component. If we think of the exterior as a cobordism from the union of k tori to the empty surface we can regard J_L as giving a partial cobordism functor in which a torus corresponds to the linear space $\mathcal{B}$. The multilinear map $J_L : \mathcal{B}^k \to \Lambda$ can be seen as a linear map on the tensor product $\mathcal{B}^{\otimes k}$ so that the cobordism from the union of k tori to the empty surface determines the linear map $J_L : \mathcal{B} \otimes \ldots \otimes \mathcal{B} = \mathcal{B}^{\otimes k} \to \Lambda$ between the linear spaces corresponding to the boundary.

The linear map $J_P : \mathcal{B}^{\otimes k} \to \mathcal{B}$ determined by a pattern P fits into this setting when the exterior of the pattern P in the solid torus given by thickening the annulus is viewed as a cobordism from the boundary of the neighbourhood of P to the boundary of the solid torus. In this setting the satellite formula shows that the composition of cobordisms making up the exterior of the satellite by gluing the exterior of the companion to the exterior of P translates exactly to the appropriate composition of linear maps, provided that the marking of the boundary of the solid torus is suitably chosen. Further comments on this point of view are made in [23].

We could try to base a limited cobordism functor on these definitions, with the restriction that the only boundary components allowed should be unions of tori. We do not however have enough freedom to do this; the most serious problem is that we are in general unable to change the linear map appropriately when we change the assignment of boundary components on a link exterior from incoming to outgoing. The case of the pattern link P' is a special case in which the component to be switched is unknotted in S^3; in this case the marking to be used on the outgoing component differs from that of the incoming components by switching the factors in $S^1 \times S^1$. The two maps $J_{P'} : \mathcal{B}^{\otimes k+1} \to \Lambda$ and $J_P : \mathcal{B}^{\otimes k} \to \mathcal{B}$ exhibit the sort of change that we would like to use generally when switching components from incoming to outgoing. They are related by the invariant $J_H : \mathcal{B} \otimes \mathcal{B} \to \Lambda$ of the Hopf link. This represents the exterior of H, in which both components are incoming. The same 3-manifold is homeomorphic to the product $(S^1 \times S^1) \times I$, represented by the identity $\mathcal{B} \to \mathcal{B}$ when viewed as a cobordism with one incoming and one outgoing component. The expected procedure for altering the map when a component is switched from incoming to outgoing would be to change a copy of $\mathcal{B}$ in the domain of the map to a copy of its dual $\mathcal{B}^*$ in the target, and then use the bilinear form J_H to identify $\mathcal{B}^*$ with $\mathcal{B}$. This would at least agree with the case of a pattern link P' and its unknotted component.

The problem with doing this in general is that $\mathcal{B}$ is infinite dimensional, so that J_H does not provide a good identification. The other missing ingredient is the ability to alter the marking of a boundary component, so as to allow freedom to glue boundaries together in different ways. The change of framing, which corresponds to certain changes of marking, can indeed be represented by use of the automorphism $\mathcal{F}$ on the vector space $\mathcal{B}$, but there is no immediate analogue available to account for the other homeomorphisms in the mapping class group of the torus.

5.2 The finite-dimensional invariants.

Both of these problems disappear when we fix the level l, and thus $r = l - 2$, and pass to the corresponding quotient ring $\mathcal{B}_r$ in place of $\mathcal{B}$ as the linear space to use for each boundary torus. The exterior of a link L can now be represented by the map $J_L^{(r)}$, regarded either as a multilinear map from $(\mathcal{B}_r)^k$ to $\Lambda_r \subset \mathbf{C}$ or equivalently as a linear map on the tensor product $(\mathcal{B}_r)^{\otimes k}$. This map is determined by the full polynomial invariant J_L after replacement of the variable A by a $4r$-th root of unity.

The complex vector space $\mathcal{B}_r$ is finite dimensional, and can be readily identified with its dual, using the non-degenerate bilinear form $J_H^{(r)}$. This permits link exteriors to be used in defining cobordism invariants, where any selection of boundary components may be taken as the incoming part of the boundary. With these as basic ingredients, a coherent assignment of linear maps can be made to cover the case of compact 3-manifolds with torus boundary components, up to a power of the number c (depending on r) mentioned in section 3. For example, the trivial knot, whose exterior is a solid torus, determines the invariant $<\cdot>: \mathcal{B}_r \to \Lambda$ when regarded as a cobordism from the torus to the empty set. As a cobordism from the empty set to the torus, it gives the element $w_1 \in \mathcal{B}_r$, regarded as a map from Λ to $\mathcal{B}_r$. In this setting, the torus is marked in such a way that composing this cobordism with a link exterior has the effect of gluing the solid torus to the boundary of the neighbourhood of one component of the link L so as to replace the neighbourhood exactly. The new cobordism is just the exterior of the link given by deleting the chosen component of L, and its invariant is given by decorating that component of L by w_1, i.e. by the empty decoration, as expected.

To perform surgery on the link exterior we must reglue the solid torus in a different way, or equivalently we must choose a different marking of its boundary torus, switching the two factors $S^1 \times S^1$. When working with $\mathcal{B}_r$ it is possible to represent the full mapping class group of the torus on $\mathcal{B}_r$, (up to a power of c), and in particular to represent the switching homeomorphism. The image of w_1 under the switch is $\rho^{-1}\Omega$, and so the solid torus glued in to one boundary component of a link exterior by surgery is a cobordism which is represented by the map $\Lambda_r \to \mathcal{B}_r$ which takes 1 to $\rho^{-1}\Omega$. The cobordism invariant of the new manifold is then given from that of the manifold before gluing by evaluation at $\rho^{-1}\Omega$ on the appropriate component. So we anticipate in this view that we might get an invariant of the manifold given by surgery on a framed link L by regarding the manifold as a composite of cobordisms, starting with k solid tori, and attaching them to the exterior of L. The resulting invariant would then be $J_L^{(r)}(\rho^{-1}\Omega, \ldots, \rho^{-1}\Omega)$ up to a power of c, which is indeed the form of the invariant discussed in section 3.

The invariant of a manifold constructed by general Dehn surgery from a framed link L, where solid tori are glued in to the link exterior using other markings of the boundaries, can similarly be found by evaluation of $J_L^{(r)}$ on suitably chosen elements of $\mathcal{B}_r$, depending on the nature of the marking for each individual boundary component. The determination of these elements is a matter of finding the image of w_1 under the automorphism of $\mathcal{B}_r$ corresponding to the self-homeomorphism of the torus which alters the chosen marking to the marking determined by the framing of L. They can be found once the action of the mapping class group of the torus on $\mathcal{B}_r$ has been established. The powers of c mentioned as an indeterminacy can be handled as in [24], or they can be incorporated into the cobordism invariant by regarding the marked 3-manifolds as also carrying a framing, adjustments to which account for multiplication by powers of c.

It is possible to extend the invariant from a similar point of view to handle general cobordisms in which the boundary components need not be tori. An account of the linear space related to the surface of genus g can be given in terms of the skein of a planar surface with g holes, just as $\mathcal{B}_r$ is described in terms of the skein of the annulus. See for example the recent account by Lickorish [18], following work of Vogel, or an earlier account by Kohno from the quantum group viewpoint [14].

6. Unitary invariants and the Hecke algebras.

In this final section I shall give a brief indication of the similarities and modifications to the previous work which are needed in considering the invariants related to the Homfly polynomial [6] by satellite constructions, or equivalently to the unitary quantum groups $SU(k)_q$, for different values of k. There is a similar relation between the orthogonal/symplectic quantum groups and Kauffman's 2-variable invariant. Wenzl [41] gives an account of this in which the quantum group approach, and the appropriate algebra, is much to the fore. He continues, with Turaev [37], to develop this to the second stage when a root of unity is involved, so as to discuss 3-manifold invariants based on modifications to the quantum group. This in turn entails a separate study of the representation theory for the modified quantum group, rather than using the classical representation theory based on the generic case. Although I will not attempt to move to this stage for the general quantum group, it is possible to reach the 3-manifold invariants in a similar way to the discussions above by dealing with invariants defined on what is in effect a natural quotient ring of the representation ring of the quantum group being used, or equivalently of the corresponding classical group.

6.1 The Homfly polynomial.

The Homfly polynomial $P_L(v,z) \in \mathbf{Z}[v^{\pm 1}, z^{\pm 1}]$ was developed independently by several groups shortly after the discovery of the Jones polynomial [6, 28]. It is an invariant of an *oriented* link, characterised by the Homfly skein relation

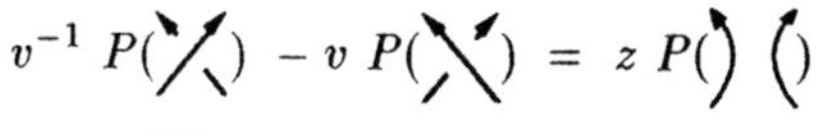

between oriented link diagrams differing only where shown. It is invariant under all three Reidemeister moves, and so $P_{L \amalg O} = \delta P_L$, where $\delta = (v^{-1} - v)/z$, and $L \amalg O$ consists of the diagram L together with a disjoint simple closed curve.

It provides a simultaneous generalisation of the Alexander polynomial and Jones' polynomial by

$$P_L(v,z) = \begin{cases} \Delta_K(t), & \text{the Alexander polynomial, when } v = 1, z = s - s^{-1}, t = s^2 \\ \nabla_K(z), & \text{Conway's version of the Alexander polynomial, when } v = 1, \\ V_K(t), & \text{the Jones polynomial, when } v = s^2 = t, z = s - s^{-1}. \end{cases}$$

In this original form P is normalised so that the unknot O has invariant 1; it is more convenient in work which relates to quantum groups to normalise so that the empty knot ϕ has invariant 1 and the unknot has invariant δ. I shall adopt this convention in the present work.

We may construct close relatives of the Homfly polynomial which are invariants of an oriented diagram D only up to R_{II} and R_{III} for any scalar λ by setting

$$X_D = \lambda^{w(D)} P_D(v,z),$$

where $w(D)$ is the writhe of the diagram D. Then X can be recognised by the properties

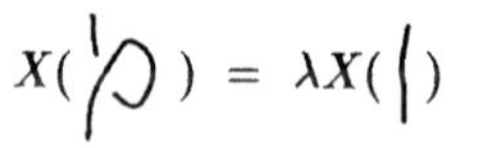

and the skein relation

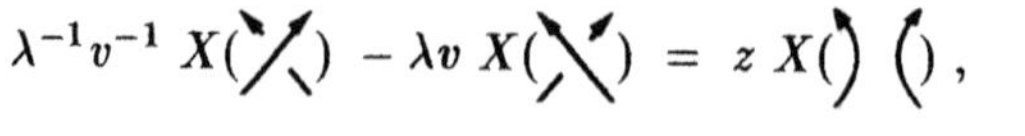

up to normalisation. In this way we can identify any invariant of oriented diagrams which satisfies a skein relation between  with such a variant of the Homfly polynomial, provided that it multiplies by a fixed scalar λ under R_I. The bracket polynomial, for example, arises with $\lambda = -A^3, z = A^{-2} - A^2$ and $v = A^{-4}$.

In general, when we write the relation as

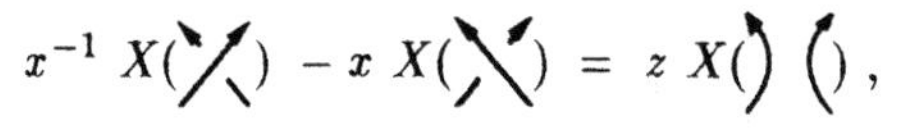

we have $X = \lambda^{w(D)} P_D(x\lambda^{-1}, z) = (xv^{-1})^{w(D)} P_D(v, z)$.

6.2 Skein theory.

We can use the Homfly skein relation to define skeins based on the Homfly polynomial, following the methods used in the first section for the bracket invariant. We shall consider diagrams, up to moves R_{II} and R_{III} in a planar surface F whose boundary contains a finite set of distinguished points. We insist that each boundary point is given an orientation either as an input or an output, and we consider *oriented* diagrams in F whose string orientation matches the orientation of the boundary points.

Definition. For a planar surface F the *Homfly skein* $\mathcal{S}_P(F)$ is the set of linear combinations of *oriented* diagrams in F subject to the relations

(1) 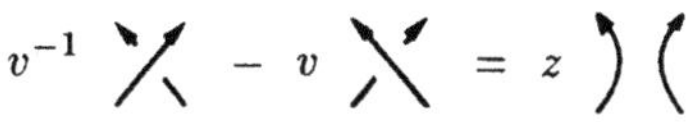

(2) 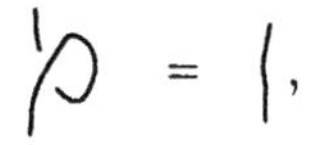

for diagrams which differ as shown.

The existence and uniqueness theorem for the Homfly polynomial shows that $\mathcal{S}_P(\mathbf{R}^2)$ is isomorphic to the scalars, and the diagram L represents the multiple $P_L(v,z) \times \phi$ of the empty diagram ϕ, given our convention that $P_\phi = 1$.

As an example, if we take F to be the rectangle R_n^n with n *inputs* at the bottom and n *outputs* at the top then the skein $\mathcal{S}_P(R_n^n)$, constructed from oriented (n,n)-tangles, forms an algebra with composition induced by putting rectangles one below the other, as for the Temperley-Lieb algebra. This algebra is spanned by $n!$ elements, represented by the positive permutation braids $w_\pi, \pi \in S_n$ discussed above. It is generated as an algebra by the elementary braids σ_i, oriented with all strings upwards, and it is known to be isomorphic to the n-th Hecke algebra H_n, as shown in [26].

A presentation for this algebra is given by generators σ_i satisfying the braid relations

$$\sigma_i\sigma_j = \sigma_j\sigma_i, \ |i-j| > 1,$$
$$\sigma_i\sigma_{i+1}\sigma_i = \sigma_{i+1}\sigma_i\sigma_{i+1}$$

and the skein relation $v^{-1}\sigma_i - v\sigma_i^{-1} = z$.

Variants on the skein definitions can be adopted, by use of a scalar λ as in the invariant X above, with $x = \lambda v$, from which we can define a variant skein $\mathcal{S}'_P(F)$ by the relations

(1′) 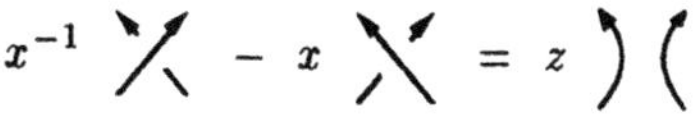

(2) 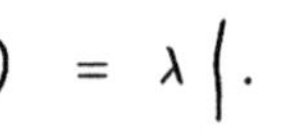

There is a linear isomorphism $\mathcal{S}_P(F) \to \mathcal{S}'_P(F)$ defined on each diagram D in F by $D \mapsto \lambda^{-w(D)}D$, where $w(D)$ is the writhe of the diagram. In the case when $F = \mathbf{R}^2$ the link diagram L, which represents $P_L(v,z) \times \phi$ in $\mathcal{S}_P(F)$, will represent $\lambda^{w(L)}P_L(v,z) \times \phi$ in $\mathcal{S}'_P(F)$.

One frequent choice for this variant is $x = 1$, and so $\lambda = v^{-1}$. When $F = R_n^n$ and $x = 1$ the isomorphism gives another presentation of the algebra H_n in terms of the elements c_i which are represented in $\mathcal{S}'(F)$ by the elementary braids. The isomorphism carries σ_i to vc_i and so the presentation is given by generators c_i satisfying the braid relations

$$c_ic_j = c_jc_i, \ |i-j| > 1,$$

$$c_ic_{i+1}c_i = c_{i+1}c_ic_{i+1}$$

and the skein relation $c_i - c_i^{-1} = z$ in $\mathcal{S}'(F)$. This presentation of H_n is used in [22] for calculating the Homfly polynomial; it has the advantage of involving only the variable z when representing braids in H_n as elements of $\mathcal{S}'(F)$.

In what follows, I shall write H_n for the algebra $\mathcal{S}_P(R_n^n)$ with generators σ_i and identify it with any of the variants $\mathcal{S}'_P(R_n^n)$ by means of the isomorphism.

Extend the ring of scalars to include rational functions of v and s, and set $z = s - s^{-1}$. The skein relation in H_n can then be written in the form

$$(\sigma_i - a)(\sigma_i - b) = 0,$$

with $a = -vs^{-1}$ and $b = vs$, or in the other generators, as $(c_i + s^{-1})(c_i - s) = 0$. We can then construct elements a_n and b_n in H_n by substituting either $sv^{-1}\sigma_i$ or $-v^{-1}s^{-1}\sigma_i$ for σ_i in the sum E_n of the positive permutation braids. As in our earlier work these elements have the property that

$$Ta_n = \varphi_b(T)a_n, \ Tb_n = \varphi_a(T)b_n$$

for any $T \in H_n$ where φ_a and φ_b are homomorphisms from H_n to the scalars defined by $\varphi_a(\sigma_i) = a$, $\varphi_b(\sigma_i) = b$. A similar calculation to that of section 2 shows that

$$\varphi_b(a_n) = [n]_c! \left(= \prod_{r=1}^{n}(1 + c + \ldots + c^{r-1}) \right) \quad \text{where } c = -b/a,$$

and similarly $\varphi_a(b_n) = [n]_d!$, where $d = -a/b = c^{-1}$.

6.3 The annulus.

As before we shall use the skein of the annulus, $\mathcal{S}_P(S^1 \times I) = \mathcal{C}$ say, which again forms an algebra in which composition is induced by placing two copies of the annulus one inside the other. A spanning set for this infinite dimensional skein is discussed by Turaev [35] under the name of the 'Conway module of the solid torus'. It is shown there that $\mathcal{C}$ is freely generated as an algebra by commuting elements $1, \alpha_1, \alpha_2, \ldots, \alpha_1^*, \alpha_2^*, \ldots$ where 1 is represented by the empty diagram, as in the skein $\mathcal{B}$, and α_1 is represented by the closure of the identity 1-string braid, like α in $\mathcal{B}$. The element α_1^* has the same diagram but with the orientation reversed. The element α_i is represented by the closure of the i-string braid $\sigma_{i-1}\sigma_{i-2}\ldots\sigma_1$; reversing the string orientation gives α_i^*. A diagram representing $\alpha_3\alpha_2^*$ is shown below.

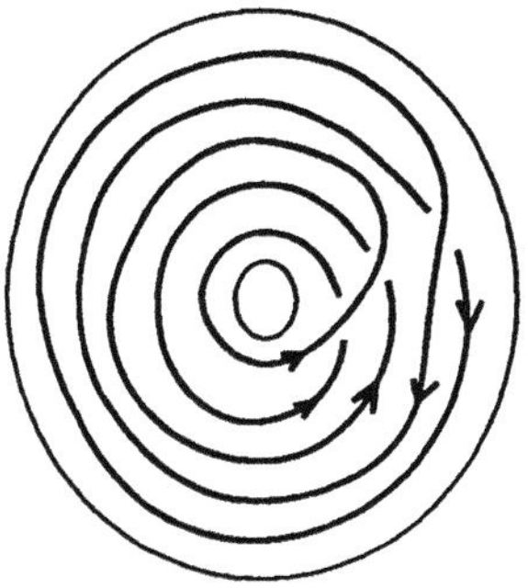

Oriented wiring diagrams can be used as before to induce linear maps between skeins. We may also decorate oriented link diagrams by elements of $\mathcal{C}$ and thus determine a multilinear map $P_D : \mathcal{C}^k \to \mathcal{S}_P(F)$ for any diagram D in F with k closed components. This map will be independent of D up to moves R_{II} and R_{III}, while changes of framing on a component of D, in other words alteration by moves R_I, can be accounted for by use of a framing change map $\mathcal{F} : \mathcal{C} \to \mathcal{C}$ defined as before by decorating the simple curl

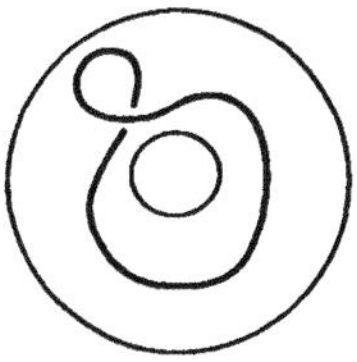

regarded as a diagram in the annulus.

In this way we can extend the Homfly polynomial to define invariants P_L of a framed oriented link L by decorating its components with elements of $\mathcal{C}$, so as to give a multilinear map $P_L : \mathcal{C} \times \ldots \mathcal{C} \to \mathcal{S}_P(\mathbf{R}^2)$, the 'total Homfly invariant' of L. The Homfly polynomial itself is recovered by evaluating the map P_L at $(\alpha_1, \ldots, \alpha_1)$, when $\mathcal{S}_P(\mathbf{R}^2)$ is identified with the ring of scalars. Other decorations give rise to further invariants of L, which I shall term 'satellite Homfly invariants' of L, as they are constructed from the Homfly polynomials of satellites of L.

6.4 Representing the Hecke algebra.

The closure wiring of a rectangle into the annulus induces a linear map $H_n \to \mathcal{C}$ for each n, with image $\mathcal{C}_n$ say. Every diagram in the annulus can be viewed as the closure of some tangle, but we cannot assume that the string orientations at the top of the tangle are all inputs, so the skein $\mathcal{C}$ is not necessarily the union of the subspaces $\mathcal{C}_n$. We can certainly recover the whole of $\mathcal{C}$ by considering tangles in which the boundary points at the bottom are divided into n inputs and p outputs, with the matching points at the top forming n outputs and p inputs, for varying n and p.

The algebra $\mathcal{C}$ is the product $\mathcal{C}_+ \times \mathcal{C}_-$ of the subalgebras generated respectively by $\{\alpha_i\}$ alone and by $\{\alpha_i^*\}$ alone. The image $\mathcal{C}_n$ of H_n lies in $\mathcal{C}_+$ for each n; it has a basis consisting of monomials in $\{\alpha_i\}$ of total weight n, where α_i has weight i. Its dimension is thus $\lambda(n)$, the number of partitions of n.

An alternative basis for $\mathcal{C}_n$ is suggested by the representation theory of H_n, which is a deformation of the group algebra $\mathbf{C}[S_n]$ of the symmetric group. For generic values of the parameter $z = s - s^{-1}$ (in fact for $s^{2r} \neq 1$, $r \leq n$) the algebra H_n is known to decompose as the direct sum of $\lambda(n)$ subalgebras, $\bigoplus M_\lambda$, each isomorphic to the algebra of $d_\lambda \times d_\lambda$ matrices for some d_λ. This decomposition is similar to the classical case of $\mathbf{C}[S_n]$; the subalgebras M_λ are traditionally indexed by the Young diagrams λ with n cells. Any such Young diagram is determined by a sequence of non-negative integers

$$\lambda_1 \geq \lambda_2 \geq \ldots \geq \lambda_n \geq 0$$

with $\lambda_1 + \ldots + \lambda_n = n$, and is commonly drawn diagramatically as an array of n cells with λ_i cells in row i. For example, the diagram

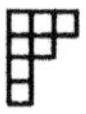

corresponds to the partition $3 \geq 2 \geq 1 \geq 1 \geq 0 \geq 0 \geq 0$ with $n = 7$.

Given the structure of H_n as a direct sum there will be a central idempotent $e_\lambda \in H_n$ for each λ, corresponding to the identity element of the subalgebra M_λ. These are orthogonal, in the sense that $e_\lambda e_\mu = 0$ if $\lambda \neq \mu$, while $e_\lambda^2 = e_\lambda$. The algebra H_n decomposes in this way, provided that the coefficient ring allows denominators $s^r - s^{-r}$ for $r \leq n$. The idempotents can be found explicitly, for example in [41]. The simplest of these are multiples of the elements a_n and b_n given above. They correspond to the two Young diagrams, each with $d_\lambda = 1$, which have n cells and just one row or just one column.

The closure $\widehat{e}_\lambda$ of the idempotents provide between them an alternative basis for C_n consisting of $\lambda(n)$ elements. They have the merit of all being eigenvectors of the framing change map $\mathcal{F} : C \to C$. This follows since any central element of H_n can be written as a linear combination of the idempotents $\sum c_\mu e_\mu$. The n-string curl Q_n, with appropriate orientation, which commutes up to R_{II} and R_{III} with all (n,n)-tangles, can then be written as $Q_n = \sum c_\mu e_\mu$. Orthogonality of the idempotents shows that $Q_n e_\lambda = c_\lambda e_\lambda$ and hence $\mathcal{F}(\widehat{e}_\lambda) = c_\lambda \widehat{e}_\lambda$. The elements $\widehat{e}_\lambda$ thus behave rather like the elements $w_i \in \mathcal{B}$.

Example. When $n = 2$ there are just two Young diagrams $\Box\Box$ and $\begin{smallmatrix}\Box\\\Box\end{smallmatrix}$, with corresponding idempotents

$$e_{\Box\Box} = (sv^{-1}\sigma_1 + 1)/(1+s^2), \quad e_{\begin{smallmatrix}\Box\\\Box\end{smallmatrix}} = (-s^{-1}v^{-1}\sigma_1 + 1)/(1+s^{-2}).$$

When $s = v = 1$ these are the symmetriser and skew-symmetriser respectively for the symmetric group S_2. It is easy to express each $\widehat{e}_\lambda$ in terms of the basis of monomials for C, noting that for $(2,2)$-tangles the closure of the identity braid 1 is α_1^2 and the closure of σ_1 is α_2.

Thus when K is the figure-eight knot with framing as shown

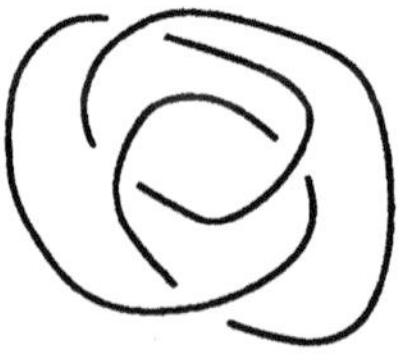

we have its satellite Homfly invariant

$$P_K(\widehat{e}_{\Box\Box}) = 1/(1+s^2)\left(v^{-1}s\, P_K(\alpha_2) + P_K(\alpha_1^2)\right)$$

given by calculating the Homfly polynomials of two 2-string satellites of K. The invariant can be written as

$$\frac{\delta}{s^2-s^{-2}}(v^{-5}s^5 - v^{-3}(s^7+s^5+s^{-1}) + v^{-1}(s^7+s^3+2s+s^{-5})$$
$$- v(s^{-7}+s^{-3}+2s^{-1}+s^5) + v^3(s^{-7}+s^{-5}+s) - v^5s^{-5}),$$

where $\delta = \dfrac{v^{-1}-v}{s-s^{-1}}$ is the Homfly invariant of the unknot. For compaiison the standard Homfly invariant of the figure-eight knot is $P_K(\alpha_1) = \delta(v^{-2} - s^{-2} + 1 - s^2 + v^2)$.

Similarly when $n = 3$ we can write down the two idempotents $e_{\square\square\square}$ and $e_{\begin{smallmatrix}\square\\ \square\\ \square\end{smallmatrix}}$ as above, using a_n and b_n. The remaining idempotent $e_{\begin{smallmatrix}\square\square\\ \square\end{smallmatrix}}$ can be found from the equation $1 = \sum e_\lambda$ in H_n. The closures of all three can be calculated in terms of the monomial basis, giving for instance

$$\widehat{e}_{\begin{smallmatrix}\square\square\\ \square\end{smallmatrix}} = \frac{2}{s^{-2}+1+s^2}\left(\alpha_1^3 + v^{-1}(s-s^{-1})\alpha_1\alpha_2 - v^{-2}\alpha_3\right).$$

This can also be written as $\dfrac{2}{s^{-2}+1+s^2}\widehat{d}$ where $d = 1 - \sigma_1^{-1}\sigma_2$.

Transition between the monomial basis and the basis $\{\widehat{e}_\lambda\}$ is not so convenient as in the case of $\mathcal{B}$, where the two bases of interest, $\{\alpha^j\}$ and $\{w_i\}$, are integrally related. In $\mathcal{C}_n$ we need a limited set of denominators of the form $s^r - s^{-r}$ and $vs^r - v^{-1}s^{-r}$ with $|r| \leq n$ to perform a complete transition. In principle, though, the information available from a link by taking its Homfly polynomial after decoration by elements of $\mathcal{C}_+$ is equivalent to knowing, on the one hand, its satellite Homfly polynomials when decorated by all possible monomials in the α_i and, on the other hand, the invariants when decorated by all possible $\widehat{e}_\lambda$ for Young diagrams λ. The connection with the quantum group $SU(k)_q$ invariants of the framed link L comes about through an identification of the quantum group invariants with the invariants above which use $\widehat{e}_\lambda$, as λ variės through Young diagrams restricted according to the value of k.

6.5 Unitary quantum groups.

The methods of Reshetikhin and Turaev [31] allow the quantum groups $\mathcal{G} = SU(k)_q$ to be used to represent oriented tangles whose components are coloured by $\mathcal{G}$-modules as $\mathcal{G}$-module homomorphisms. The scheme and necessary ingredients are similar to those outlined in section 4, with one additional feature, namely the use of the dual module V^* defined by means of the antipode in $\mathcal{G}$, (an antiautomorphism of $\mathcal{G}$ which is part of its structure as a Hopf algebra). When the components of the tangle are coloured by modules the tangle itself is represented by a homomorphism from the tensor product of the modules which colour the strings at the bottom to the tensor product of the modules which colour the strings at the top, provided that the string orientations are inwards at the bottom and outwards at the top. The dual module V^* comes into play in place of V when an arc of the tangle coloured by V has an output at the bottom or an input at the top.

For example, the $(4,2)$-tangle below, when coloured as shown, is represented by a homomorphism $U \otimes W^* \to U \otimes X^* \otimes X \otimes W^*$.

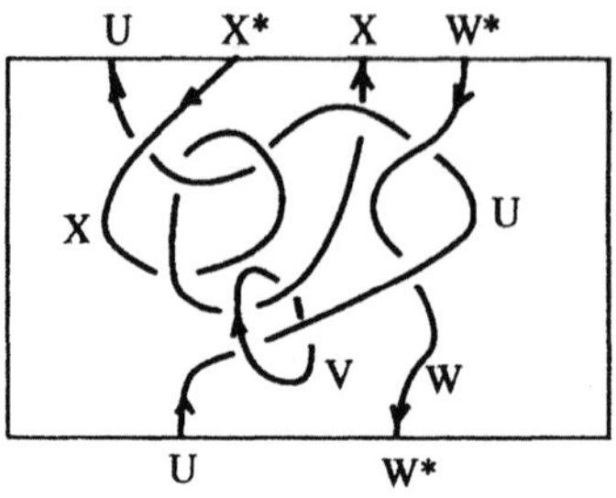

As in the earlier case it is possible [31] to build up the definition so that consistently coloured tangles are represented by the appropriate composite homomorphisms, starting from a definition of the homomorphisms for the elementary oriented tangles. Two cases, depending on the orientation, must be considered for both the local maximum and the local minimum, and a little care is needed here to ensure consistency. The final result is a definition of a homomorphism which is invariant when the coloured tangle is altered by R_{II} and R_{III}. When applied to an oriented k-component link diagram L regarded as an oriented $(0,0)$-tangle it gives an element $J(L; V_1, \ldots, V_k) \in \Lambda = \mathbf{Q}[[h]]$ for each colouring of the components of L by $\mathcal{G}$-modules, which is an invariant of the framed oriented link L. This element, apart from a simple factor, is an integer polynomial in $q = e^h$. A categorical account of the appropriate features needed to define an invariant in this way is given in [44].

As in section 4, this invariant $J(L)$ (for a fixed quantum group $\mathcal{G}$) is

(1) multilinear under direct sums of modules, and

(2) multiplicative on parallels.

We can use (1) to extend the definition of $J(L)$ to allow colouring by linear combinations of modules, and thus determine a multilinear map $J(L) : \mathcal{R} \times \ldots \times \mathcal{R} \to \Lambda$, where $\mathcal{R}$ is the representation ring of $\mathcal{G}$.

Definition. Refer to the map $J(L)$ as the *coloured invariants* of L, where the choice of quantum group $\mathcal{G}$ is clear. A *colouring* of L will mean a choice of an element of $\mathcal{R}$, (in other words, a linear combination of modules,) for each component of the link, and will determine an element of Λ by evaluation of $J(L)$.

Notation. Write $\mathcal{R}^{(k)}$ for the representation ring in the case when $\mathcal{G} = SU(k)_q$.

For generic q this ring is shown in [33] to be isomorphic to the classical representation ring of $SU(k)$. The irreducible modules of $SU(k)$ and hence of $SU(k)_q$ are also indexed by Young diagrams. There is an irreducible $SU(k)_q$-module V_λ for every Young diagram λ provided that λ is either the diagram with k rows and 1 column or otherwise has at most $k-1$ rows. Such Young diagrams are referred to later as 'admissible' for k. Among these modules there is a 'fundamental' irreducible module of dimension k, which is indexed by the Young diagram $\square$. Write $V_\square$ for this module. Each module V_λ whose Young diagram has n cells occurs as a summand of $V_\square^{\otimes n}$.

An early relation between the Homfly polynomial and the quantum invariants of a link was discovered by Jones and Turaev [34], when considering the invariant given by colouring all components with the fundamental module $V_\square$.

THEOREM 6.1 (Turaev, Jones). *For the quantum group $SU(k)_q$ the invariant $J(L; V_\Box, \ldots, V_\Box)$ of the framed oriented link L is, up to normalisation, the Homfly polynomial $P_L(v,z)$ with $z = s - s^{-1}$, $v = s^{-k}$ and $s = \sqrt{q} = e^{h/2}$. Assuming that $P_\phi = 1$ we have*

$$J(L; V_\Box, \ldots, V_\Box) = (xv^{-1})^{w(L)} P_L(v,z),$$

where $w(L)$ is the writhe of a correctly framed diagram of L, evaluated at $z = s - s^{-1}$, $v = s^{-k}$ and $x = s^{-1/k} = e^{-h/2k}$.

Proof: It is enough to show that $J(L; V_\Box, \ldots, V_\Box)$ satisfies a quadratic skein relation, and multiplies by a scalar under R_I, to identify it with some specialisation of P_L as at the beginning of this section. Turaev represents the $(2,2)$-tangle σ when coloured with the fundamental representation $V_\Box$ by a map $R : V_\Box \otimes V_\Box \to V_\Box \otimes V_\Box$ which satisfies the quadratic relation

$$R - R^{-1} = (s - s^{-1})\mathrm{Id} \ .$$

It is possible to deduce the existence of *some* quadratic relation for R from the fact that $V_\Box \otimes V_\Box$ is the sum of just two irreducible modules.

The $(1,1)$-tangle 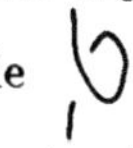 when coloured with any irreducible must be represented by a multiple of the identity, by Schur's lemma. Turaev shows that this multiple is s^{-k} when the k-dimensional fundamental module $V_\Box$ of $SU(k)_q$ is used. This would lead to the result of theorem 6.1, without the factor x. It appears, however, that a scalar multiple of Turaev's endomorphism is more appropriate, to permit a more consistent behaviour of the family of invariants $J(L)$ when evaluated on different modules. In the general construction of $J(L)$ this behaviour is ensured by the use of the universal R-matrix to determine the representation of the elementary tangle σ under each colouring. Since the universal R-matrix satisfies a non-homogeneous equation it is not possible to replace it by a scalar multiple of itself without losing the multiplicative behaviour of $J(L)$ on parallels. The endomorphism R used by Turaev is a non-trivial multiple of the one which arises from Drinfeld's universal R-matrix. The appropriate endomorphism R as given in [3] satisfies instead the relation

$$(*) \qquad x^{-1}R - xR^{-1} = (s - s^{-1})\mathrm{Id} \ ,$$

with $x = s^{-1/k}$.

Assuming that we use this endomorphism R to represent σ, equation $(*)$ enables us to define a function $\mathcal{S}'(\mathbf{R}^2) \to \Lambda$ from the variant skein $\mathcal{S}'(\mathbf{R}^2)$ with $z = s - s^{-1}$, $v = s^{-k}$ and $x = s^{-1/k}$ by taking the diagram L to $J(L; V_\Box, \ldots, V_\Box)$. Since $L = (xv^{-1})^{w(D)} P_L(v,z) \times \phi$ in $\mathcal{S}'(\mathbf{R}^2)$ and the value of J on the empty diagram ϕ is 1 we have the equation

$$J(L; V_\Box, \ldots, V_\Box) = s^{(k-1/k)w(L)} P_L(s^{-k}, s - s^{-1}) \ ,$$

where the Homfly polynomial P_L is normalised to have value 1 on the empty diagram, and $w(L)$ is the writhe of any diagram of L which realises the chosen framing. □

Given a Young diagram λ there is a corresponding $SU(k)_q$-module V_λ for each k, which should properly be distinguished from each other as k varies. It is, however, possible to organise things systematically so as to handle all the unitary quantum group invariants with colouring V_λ at once, by finding a 2-variable function of v and s depending on C and λ, from which the substitution $v = s^{-k}$ allows us to recover the invariant $J(C; V_\lambda)$ for the quantum group $SU(k)_q$, as shown in [41]. In the case when $\lambda = \square$ the theorem above shows that the Homfly polynomial of C itself provides a suitable function. For general λ we use a satellite Homfly polynomial of C; in fact we can use the closure $\hat{e}_\lambda$ derived from the Hecke algebra idempotent for the same Young diagram λ as the element of $\mathcal{C}$ to provide the satellite decoration.

We shall see that it is possible to realise all possible colourings of C as linear combinations of invariants which arise by varying the decoration P while restricting the colouring of P to the fundamental module $V_\square$. Thus all coloured invariants of C for the unitary quantum groups will arise, by the theorem of Jones and Turaev, as linear combinations of the Homfly polynomials of satellites of C, in which the variable v has been specialised to $v = s^{-k}$ for $SU(k)_q$.

There is a satellite theorem for the quantum invariants $J(L)$ of a satellite link L. This allows us to express the invariants of the link given when a companion knot C is decorated by some pattern P in the annulus in terms of the invariants of C and of the pattern P. From the point of view of constructing invariants of C we may choose the decorating pattern, and then choose a colouring of $P = \hat{T}$ to determine a coloured invariant of the satellite; this is an invariant of the original C, and the satellite theorem shows how to realise this as a coloured invariant of C itself, in other words as the value of $J(C)$ for some colouring of C.

Suppose that the pattern P, and hence the satellite, has r components, which we colour by modules $U_1, \ldots, U_r$. The tangle T, forming a subdiagram of P will then itself be coloured by these modules so that the top and bottom endpoints are represented by the same tensor product of modules, W say, drawn from $\{U_i, U_i^*\}$. The tangle T is represented by an endomorphism $T(\mathbf{U})$ of the module W. Write W as a direct sum $\bigoplus V_{\lambda_i}$ of irreducible modules, and choose $v_i \neq 0$ in V_{λ_i} for each i. The endomorphism $T(\mathbf{U})$ then determines $a_{ij} \in \Lambda$ with $T(\mathbf{U})(v_j) = \sum a_{ij} v_i$. Define a weighted trace $\mathrm{Tr}(T(\mathbf{U})) \in \mathcal{R}^{(k)}$ by setting

$$\mathrm{Tr}(T(\mathbf{U})) = \sum b_\lambda V_\lambda\ , \quad \text{where } b_\lambda = \sum_{V_{\lambda_i} \cong V_\lambda} a_{ii}.$$

SATELLITE THEOREM 6.2. *Let L be the framed oriented satellite of C with pattern $P = \hat{T}$ and let $\mathbf{U} = (U_1, \ldots, U_r)$ denote a colouring of its components. Then*

$$J(L;\mathbf{U}) = J(C; \mathrm{Tr}(T(\mathbf{U}))).$$

The proof can be constructed with care from [31]. Notice that $\mathrm{Tr}(T(\mathbf{U}))$ depends only on P and the colouring, and not on the companion C. It provides a multilinear map $J(P) : \mathcal{R}^{(k)} \times \ldots \times \mathcal{R}^{(k)} \to \mathcal{R}^{(k)}$ whose value on $(U_1, \ldots, U_r)$ is $\mathrm{Tr}(T(\mathbf{U}))$.

THEOREM 6.3. *Let C be an oriented framed knot, let λ be any Young diagram and let V_λ be the corresponding irreducible $SU(k)_q$-module. Then, with the convention that $J(C;V_\lambda) = 0$ if λ is not an admissible shape for $SU(k)_q$, we have*

$$d_\lambda J(C;V_\lambda) = (xv^{-1})^{|\lambda|^2 w(D)} P_C(\hat{e}_\lambda),$$

as functions of $s = \sqrt{q}$, when the variable v on the right-hand side is replaced by s^{-k} and x by $s^{-1/k}$. Here d_λ, independent of k, is the degree of the matrix algebra M_λ in the appropriate Hecke algebra, $|\lambda|$ is the number of cells in the Young diagram λ and $w(D)$ is the writhe of a diagram for C with the chosen framing.

A corresponding result holds for oriented framed links, dealing with each component independently.

Proof: An outline of the proof follows. Apart from the normalising factor this result is given in [41]; some further discussion will be found in [19]. Suppose that the given Young diagram λ has n cells, so that $|\lambda| = n$. We shall make use of a representation of H_n on $W = V_\square^{\otimes n}$ which carries the idempotent e_λ to the projection of W to the 'isotypic' submodule for V_λ, namely the submodule isomorphic to $\bigoplus V_{\lambda_i}$ for which $V_{\lambda_i} \cong V_\lambda$.

Any oriented (n,n)-tangle T determines an endomorphism $f(T)$ of W by colouring each of its components with the module $V_\square$. Because of the relation $(*)$ among the endomorphisms $f(T)$ as T varies, the map f induces a representation of the variant skein $\mathcal{S}'_P(R_n^n)$, with $v = s^{-k}$, $x = s^{-1/k}$, on W. Using the isomorphism of $H_n = \mathcal{S}_P(R_n^n)$ with this variant skein gives an explicit homomorphism

$$\varphi_k : H_n = \mathcal{S}_P(R_n^n) \to \mathrm{End}(W)$$

induced by $\varphi_k(T) = (x^{-1}v)^{w(T)} f(T)$, where again v and x are replaced appropriately when dealing with $SU(k)_q$.

Now decorate the diagram of C with the pattern $\widehat{T}$ to form a link diagram L, and colour all components of L with $V_\square$. By the satellite theorem we can calculate

$$J(L; V_\square, \ldots, V_\square) = J(C; \mathrm{Tr}\, f(T)) = (xv^{-1})^{w(T)} J(C; \mathrm{Tr}\, \varphi_k(T)) .$$

On the other hand, theorem 6.1 shows that

$$J(L; V_\square, \ldots, V_\square) = (xv^{-1})^{w(L)} P_L = (xv^{-1})^{w(L)} P_C(\widehat{T}) ,$$

where $v = s^{-k}$. Now the writhe of the decorated diagram L can readily be given as $w(L) = w(T) + n^2 w(C)$, since each crossing in C will give n^2 crossings of the same sign in L where the groups of n parallel strings cross. We can then write

$$J(C; \mathrm{Tr}\, \varphi_k(T)) = (xv^{-1})^{n^2 w(C)} P_C(\widehat{T}) , \text{ with } v = s^{-k}.$$

We may now replace T by any linear combination of (n,n)-tangles to get a similar result. In particular the idempotent e_λ in H_n can be written in this way, and then we have

$$(xv^{-1})^{|\lambda|^2 w(C)} P_C(\widehat{e}_\lambda) = J(C; \mathrm{Tr}\, \varphi_k(e_\lambda)) , \text{ with } v = s^{-k}, x = s^{-1/k}.$$

The proof of theorem 6.3 can then be completed by showing that $\varphi_k(e_\lambda)$ is the projection of W to the isotypic submodule for V_λ which is isomorphic to d_λ copies of V_λ. The trace of this projection is $d_\lambda V_\lambda$ so that the right-hand side in the equation above becomes $d_\lambda J(C; V_\lambda)$ as claimed. □

In the proof above the identification of $\varphi_k(e_\lambda)$ with the projection to one of the isotypic submodules of W remains to be established. A deeper understanding of the structure both of H_n and of the modules $W = V_\square$ for different k can be achieved by use of the representation φ_k. This representation gives a direct analogue of the setting for classical invariant theory of the symmetric group, where the Hecke algebra corresponds to the group algebra of the symmetric group S_n and the quantum groups to the special linear groups. By drawing on work of Wassermann [39] and Wenzl [41] it can be shown that the following generalisations of the classical results hold in this context.

THEOREM 6.4. *The homomorphism* $\varphi_k : H_n \to \mathrm{End}_{SU(k)_q} V_\square^{\otimes n}$ *is*

(1) *surjective for all* k,

(2) *injective when* $k \geq n$.

The first part shows that every module endomorphism of W can be represented as the linear combination of some tangles coloured with $V_{\square}$. In particular the projection to any submodule of W must be representable in this way; the choice of the element e_λ is then simply one explicit way to realise $J(C;V_\lambda)$ by means of a satellite Homfly polynomial. Indeed the element e_λ is generally rather complicated and it is usually possible to find a simpler combination with the same closure in $\mathcal{C}$.

The isomorphism of H_n with the endomorphism ring for large enough k permits us to extend the classical correspondence between the idempotent e_λ and the projection to the corresponding isotypic submodule in this case as well. It is also possible to describe readily the kernel of φ_k when $k < n$ as the ideal generated by those idempotents e_λ whose Young diagram has too many rows to be admissible for k, again exactly as in the classical case.

The most striking consequence of the approach using the skein of the annulus is the existence of the 2-variable invariant of C indexed by λ whose specialisations at $v = s^{-k}$ provide the quantum invariants $J(C;V_\lambda)$ for all $SU(k)_q$ at once. Links L can be treated in essentially the same way, taking the satellite Homfly polynomial when each component is decorated independently by some $\widehat{e}_\lambda$, multiplied by a suitable power of v, to specialise to the corresponding quantum invariant $J(L)$. It is interesting to note that when the orientation of one component is reversed the quantum invariant of the new link can be recovered from that of the old link by replacing the module on that component with its dual. The dual of the irreducible module V_λ is again irreducible, but its Young diagram λ^* depends on k as well as λ so it is not possible to give a similar universal treatment to handle string reversals for satellite Homfly invariants.

By way of example, the dual of the fundamental module $V_{\square}$ has Young diagram λ^* with a single column and $k-1$ cells. In the case of $SU(2)_q$ the fundamental module is then self-dual, as are all the other irreducibles, which accounts for the insensitivity of the bracket invariant to string orientation. For $SU(3)_q$ the calculation $J(C;V_{\begin{smallmatrix}\square\\\square\end{smallmatrix}})$ will then give $J(\overline{C};V_{\square}) = (xv^{-1})^{w(C)}P_{\overline{C}}$ with $v = s^{-3}$, where $\overline{C}$ is C with the opposite orientation. The Homfly polynomial of a knot is unchanged by string reversal, so we see that

$$(xv^{-1})^{4w(C)}P_C(\widehat{e}_{\begin{smallmatrix}\square\\\square\end{smallmatrix}}) = J(C;V_{\begin{smallmatrix}\square\\\square\end{smallmatrix}}) = J(\overline{C};V_{\square}) = (xv^{-1})^{w(C)}P_{\overline{C}},$$

and so $P_C(\widehat{e}_{\begin{smallmatrix}\square\\\square\end{smallmatrix}}) = (x^{-1}v)^{3w(C)}P_C$, the standard Homfly polynomial, with $v = s^{-3}$ and $x = s^{-1/3}$. This gives $P_C(\widehat{e}_{\begin{smallmatrix}\square\\\square\end{smallmatrix}}) = s^{-8w(C)}P_C$, when $v = s^{-3}$.

It is also possible to identify the module $V_{\begin{smallmatrix}\square\\\square\end{smallmatrix}}$ for $SU(4)_q$ with the fundamental module for $SO(6)_q$ and so relate $P_C(\widehat{e}_{\begin{smallmatrix}\square\\\square\end{smallmatrix}})$ with $v = s^{-4}$ to an evaluation of Kauffman's Dubrovnik polynomial, [20].

6.6 Remarks.

The satellite theorem provides a multiplicative homomorphism from $\mathcal{C}$ to $\mathcal{R}^{(k)}$ for each k, which is most readily defined on the variant skein of the annulus by taking each pattern $P = \widehat{T}$ to the trace of T when coloured entirely with $V_{\square}$. On $\mathcal{C}_+$ this description is independent of k and can be given on the basis $\{\widehat{e}_\lambda\}$ by $\widehat{e}_\lambda \mapsto d_\lambda V_\lambda$, so that after suitable writhe adjustment the functions J_L and P_L agree. The map carries the element α_1 to

$V_{\square}$ and α_2 to $vsV_{\square\square} - vs^{-1}V_{\begin{smallmatrix}\square\\\square\end{smallmatrix}}$, while on the other hand α_1^* is mapped to $V_{\square}^*$, which will depend on k as noted above.

The skein map P_L on the algebra $\mathcal{C}$, or even its restriction to the subalgebra $\mathcal{C}_+$, carries the information for all the total invariants J_L as k varies. Unlike the case earlier where we compared the algebra $\mathcal{B}$ for the bracket invariant and the representation ring of $SU(2)$, we have here a single algebra $\mathcal{C}$ arising from the Homfly polynomials and a whole series of non-trivial quotients $\mathcal{R}^{(k)}$ of $\mathcal{C}$ which organise the quantum invariants.

In fact the ring $\mathcal{R}^{(k)}$ is the quotient of $\mathcal{C}_+$ by the ideal generated by $X_n = \widehat{e}_\lambda$ for $n > k$, where λ is the Young diagram with one column and n cells. The corresponding module V_λ is the n-th exterior power of the fundamental module $V_{\square}$. It is possible to draw on classical knowledge of the representation rings $\mathcal{R}^{(k)}$ as polynomial rings in the exterior powers of the fundamental module to give alternative constructions for the general basis element $\widehat{e}_\lambda$ in $\mathcal{C}_+$ as a polynomial in the elements $\{X_n\}$. The element X_n is noted above to be $X_n = (\varphi_b(a_n))^{-1}\widehat{a}_n$. Equally the elements $Y_n = (\varphi_a(b_n))^{-1}\widehat{b}_n$, corresponding to the symmetric powers of $V_{\square}$, can be used to generate $\mathcal{C}_+$ as a polynomial ring.

An attempt to deal with 3-manifold invariants by means of $\mathcal{C}_+$, on the lines of the treatment in section 3, has the corresponding feature that when calculating with $v = s^{-k}$ and $s^{2(k+l)} = 1$, the invariant $P_{\mathcal{C}}(Y_n) = 0$ for $n = l, l+1, \ldots, l+k-1$. When the ideal generated by the k elements corresponding to Y_n, $n = l, \ldots, l+k-1$ is factored out from $\mathcal{R}^{(k)}$, the quotient is a finite-dimensional algebra (a Verlinde algebra), which gives an analogue to $\mathcal{R}_r$ in the case of the $SU(2)$ invariant, with $r = k + l$. It corresponds closely with the ingredients used by Turaev and Wenzl [37] in their construction of a 3-manifold invariant of level l based on $SU(k)_q$. It would be interesting to consider this approach via $\mathcal{C}$ in more detail, with enough care about the denominators in the ring of scalars to ensure that the substitutions of variables cause no problems.

Acknowledgements.

These notes have evolved from a series of seminars given in Liverpool during 1991, and represent one of many ways to choose a path through the wealth of material available. I am grateful to my colleagues and co-workers for providing a willing and critical audience, and particularly to Peter Cromwell and Ian Nutt for helpful criticism of the manuscript.

I must acknowledge the considerable influence of others, notably Raymond Lickorish and Lou Kauffman from conversations and lectures as well as from the articles cited here. I must also thank CIMAT in Guanajuato for the opportunity to lecture on this material in December 1991, and the organisers of the NATO ASI in Erzurum in September 1992 for the incentive to assemble it in its present form.

References.

1. Artin, E. 'Theorie der Zopfe', *Abh. Math. Sem. Univ. Hamburg*, 4 (1925), 47-72.
2. Blanchet, C., Habegger, N., Masbaum, G. and Vogel, P. 'Three-manifold invariants derived from the Kauffman bracket', to appear in *Topology*.
3. Drinfeld, V.G. 'Quantum groups', in Proceedings of the International Congress of Mathematicians, Berkeley 1986, (Amer. Math. Society 1987), 798-820.
4. Elrifai, E.A. and Morton, H.R. 'Algorithms for positive braids', preprint, Liverpool 1988, to appear in Oxford Quarterly Journal.

5. Fenn, R.A. and Rourke, C.P. 'On Kirby's calculus of links', *Topology,* 18 (1979), 1-15.
6. Freyd, P., Yetter, D., Hoste, J., Lickorish, W.B.R., Millett, K.C. and Ocneanu, A. 'A new polynomial invariant of knots and links', *Bull. Amer. Math. Soc.* 12 (1985), 239-246.
7. Jones, V.F.R. 'Hecke algebra representations of braid groups and link polynomials', *Annals of Math.* 126 (1987), 335-388.
8. Kauffman, L.H. 'State models for knot polynomials.' *Topology,* 26 (1987), 395-407.
9. Kauffman, L.H. 'An invariant of regular isotopy', *Trans. Amer. Math. Soc.* 318 (1990), 417-471.
10. Kauffman, L.H. 'Knots and Physics', World Scientific (1991).
11. Kirby, R. 'A calculus for framed links in S^3', *Invent. Math.* 45 (1978), 35-46.
12. Kirby, R. and Melvin, P. 'The 3-manifold invariants of Witten and Reshetikhin-Turaev for $Sl(2, \mathbf{C})$', *Invent. Math.* 105 (1991), 473-545.
13. Kirillov, A.N. and Reshetikhin, N.Y. 'Representations of the algebra $U_q(Sl(2))$, q-orthogonal polynomials and invariants of links,' in 'Infinite dimensional Lie algebras and groups', ed. V.G.Kac, World Scientific (1989), 285-342.
14. Kohno, T. 'Topological invariants for 3-manifolds using representations of the mapping class group I', *Topology,* 31 (1992), 203-230.
15. Lickorish, W.B.R. 'Linear skein theory and link polynomials', *Topology and its Applications,* 27 (1987), 265-274.
16. Lickorish, W.B.R. 'Three-manifold invariants from the combinatorics of the Jones polynomial', *Pacific J. Math.* 149 (1991), 337-347.
17. Lickorish, W.B.R. 'Three-manifolds and the Temperley-Lieb algebra', *Math. Ann.* 290 (1991), 657-670.
18. Lickorish, W.B.R. 'Skeins and handlebodies', preprint, Cambridge (1992).
19. Morton, H.R. 'Unitary knot invariants', preprint in preparation, Liverpool 1992.
20. Morton, H.R. 'Quantum invariants given by evaluation of knot polynomials', preprint, Liverpool 1992, to appear in Journal of Knot Theory and its Ramifications.
21. Morton, H.R. and Short, H.B. 'The 2-variable polynomial of cable knots.' *Math. Proc. Camb. Philos. Soc.* 101 (1987), 267-278.
22. Morton, H.R. and Short, H.B. 'Calculating the 2-variable polynomial for knots presented as closed braids', *J. Algorithms,* 11 (1990), 117-131.
23. Morton, H.R. and Strickland, P.M. 'Jones polynomial invariants for knots and satellites', *Math. Proc. Cambridge Philos. Soc.* 109 (1991), 83-103.
24. Morton, H.R. and Strickland, P.M. 'Satellites and surgery invariants', in 'Knots 90', Walter de Gruyter (1992), 798-820.
25. Morton, H.R. and Traczyk, P. 'The Jones polynomial of satellite links around mutants', in 'Braids', ed. Joan S. Birman and Anatoly Libgober, *Contemporary Mathematics* 78, Amer. Math. Soc. (1988), 587-592.
26. Morton, H.R. and Traczyk, P. 'Knots and algebras', in 'Contribuciones Matematicas en homenaje al profesor D. Antonio Plans Sanz de Bremond', ed. E. Martin-Peinador and A. Rodez Usan, University of Zaragoza, (1990), 201-220.
27. Neil, J. 'Combinatorial calculation of the various normalisations of the Witten invariants for 3-manifolds', preprint, Portland State, 1991.

28. Przytycki, J. and Traczyk, P. 'Invariants of links of Conway type', *Kobe J. Math.* 4 (1987), 115-139.
29. Reshetikhin, N. Y. 'Quantized universal enveloping algebras, the Yang-Baxter equation and invariants of links I and II ', preprint, LOMI E-4-87, 1987.
30. Reshetikhin, N. Y. 'Quasitriangular Hopf algebras and invariants of links', *Algebra i Analiz,* 1 (1989), 169-188.
31. Reshetikhin, N. Y. and Turaev, V. G. 'Ribbon graphs and their invariants derived from quantum groups', *Comm. Math. Phys.* 127 (1990), 1-26.
32. Reshetikhin, N. Y. and Turaev, V. G. 'Invariants of 3-manifolds via link polynomials and quantum groups', *Invent. Math.* 103 (1991), 547-597.
33. Rosso, M. 'Finite dimensional representations of the quantum analog of the enveloping algebra of a complex simple Lie algebra', *Comm. Math. Phys.* 117 (1988), 581-593.
34. Turaev, V.G. 'The Yang-Baxter equation and invariants of links', *Invent. Math.* 92 (1988), 527-553.
35. Turaev, V. G. 'The Conway and Kauffman modules of a solid torus', *Zap. Nauchn. Sem. Leningrad. Otdel. Mat. Inst. Steklov.* (LOMI), 167 (1988), Issled. Topol., 6, 79-89,
36. Turaev, V. G. 'The category of oriented tangles and its representations', *Funktsional. Anal. i Prilozhen,* 23 (1989), 93-94; translation in *Functional Anal. Appl.* 23 (1989), 254-255 (1990).
37. Turaev, V.G. and Wenzl, H. 'Quantum invariants of 3-manifolds associated with classical simple Lie algebras', preprint, San Diego (1991).
38. Walker, K. 'On Witten's 3-manifold invariants', preprint 1990.
39. Wassermann, A.J. 'Coactions and Yang-Baxter equations for ergodic actions and subfactors', *London Math. Soc. Lecture Notes* 136, ed. Evans and Takesaki, 203-236.
40. Wenzl, H. 'Hecke algebras of type A_n and subfactors', *Invent. Math.* 92 (1988), 349-383.
41. Wenzl, H. 'Representations of braid groups and the quantum Yang-Baxter equation', *Pacific J. Math.* 145 (1990), 153-180.
42. Wenzl, H. 'Quantum groups and subfactors of Lie type B , C and D ', *Comm. Math. Phys.* 133 (1990), 383-433.
43. Witten, E. 'Quantum field theory and the Jones polynomial', *Comm. Math. Phys.* 121 (1989), 351-399.
44. Yetter, D. 'Quantum groups and representations of monoidal categories', *Math. Proc. Camb. Philos. Soc.* 108 (1990), 261-290.

CLASSICAL NUMERICAL INVARIANTS IN KNOT THEORY

By Kunio Murasugi
Department of Mathematics
University of Toronto
Toronto, Canada M5S 1A1

Abstract

The recent progress made toward solving the determination problems of the minimal crossing number and braid index of a knot is discussed. Some relationships among these invariants and the bridge number are also discussed.

Key Words knot, link, alternating link, minimal crossing number, braid index, bridge number, Jones polynomial, skein polynomial, Seifert circle, graph.

1. Introduction

In knot theory there are a number of easily defined geometrical invariants that are extremely hard to compute. The minimal crossing number, braid index and bridge number are a few examples of this type of invariants. These numerical invariants have been estimated by using familiar algebraic invariants like the Alexander polynomials, signatures or the homology groups of branched (on unbranched) covering spaces of knots or links. However, in the past several years, the Jones polynomials, or more generally, the skein polynomials (or Homfly polynomials), have been used very successfully and effectively to compute these numerical invariants.

In this talk, we will outline some recent solutions to these problems which have been proposed.

2. The polynomial of a graph

In the early years of knot theory, the graph was one of the main tools used to study knots in 3-space $\mathbf{R}^3$ (or 3-sphere S^3). The progress of algebraic topology since the early 1920's, however, helped to establish knot theory as one of the major branches of low dimensional manifold topology. As a result, the topic of research changed from the knot K itself to the knot complement $S^3 - K$ (or knot manifold). The knot complement in fact determines the knot, as proven very recently [12].

In 1984, V.F.R. Jones defined a new polynomial invariant for knots or links. This discovery opened a new era in knot theory. The invariant was unexpectedly defined through operator algebras, but its combinatorial description indicated that through graph theory knot theory could most benefit from this new invariant.

M.E. Bozhüyük (ed.), Topics in Knot Theory, 157–194.

In this talk we will consider two types of graphs associated with each link diagram and define a few of their invariants.

2.1. Signed graphs

Let G be a graph. Let $V(G)$ and $E(G)$ be the sets of vertices and edges, respectively. We restrict ourselves to finite graphs, that is, graphs for which $V(G)$ and $E(G)$ are both finite. In this talk, however, slightly more general graphs shall be considered.

A graph G is said to be *signed* if either $+1$ or -1, called a *sign,* is assigned to each edge. More precisely, G (or (G, f_G)) is a *signed graph* if G is a graph equipped with a sign function $f_G : E(G) \to \{1, -1\}$. For convenience, we call an edge e *positive* if $f_G(e) = +1$ and *negative* otherwise. Since a positive graph may be considered as an unsigned (or an ordinary undirected) graph, our results can be applied to ordinary undirected graphs.

A subgraph H of G has induced sign function $f_H = f_G|E(H)$. A subgraph H is a *spanning* subgraph if $V(H) = V(G)$. In particular, a spanning subgraph consisting of only vertices is called the spanning vertex graph and is denoted by V_G.

Throughout this talk, what is meant by a *graph* is frequently the geometric realization of a graph as a finite 1-dimensional CW-complex in $S^3 = \mathbf{R}^3 \cup \infty$.

For a set $X, |X|$ denotes the cardinality of X. $\beta_i(G)$ denotes the ith Betti number of a graph G as a 1-complex.

In graph theory, $p_0(G)$ and $p_1(G)$ have been used instead of $\beta_0(G)$ and $\beta_1(G)$. $p_0(G)$ denotes the number of connected components of G, and $p_1(G)$ is called the cyclomatic number of G.

Let H and K be two graphs, both of which have at least one edge. Then the *one-point union* of H and K will be denoted by $H * K$.

For $V \subset V(G)$ and $E \subset E(G)$, $G - (V, E)$ denotes the maximal subgraph of G which does not contain vertices in V and edges in E. In particular, $G - e$ is the subgraph of G consisting of all vertices of G and all edges but e. Therefore $G - e$ is the subgraph obtained from G by deleting e. For a vertex v, $G - v$ is the subgraph consisting of all vertices but v and edges of G except those which are incident to v.

A graph G is said to be *separable* if there are two subgraphs H and K such that $G = H \cup K$ and $H \cap K = \{v_0\}$, where H and K both have at least one edge and v_0 is a vertex. Otherwise, G is non-separable. The vertex v_0 is called a *cut vertex.* If G has no loops, then G is separable when $\beta_0(G) < \beta_0(G - v)$ for some vertex v.

A *block* is a maximal non-separable connected subgraph of G. A connected graph is decomposed into finitely many blocks. Therefore, if $G_1, G_2, \ldots, G_k$ are blocks of G, we can write $G = G_1 * G_2 * \cdots * G_k$ and G is called the *block sum* of $G_1, G_2, \ldots, G_k$.

G is called *reduced* if G has neither loops nor isthmuses. An *isthmus* is an edge e such that $\beta_0(G) < \beta_0(G - e)$.

If two or more edges have the same ends, these edges are called *multiple-edges.* On the other hand, if two distinct vertices are joined by exactly one edge e, then e is called a *singular* edge of G. A loop is not a singular edge.

A two-vertex graph G is called a *multiple-edge graph* (or a *single-edge graph*) if all edges have the common (distinct) ends and $|E(G)| \geq 2$ (or $|E(G)| = 1$).

Let G be a graph and v a vertex of G. *star* v is the smallest subgraph containing v and all edges of G which are incident to v. If X is a connected subset of G, then G/X is defined as the subgraph obtained from G by identifying all points in X to one point.

For convenience, for subgraphs H and K of a graph G, we define H/K as $H/(H \cap K)$. Therefore, if $H \cap K = \phi$, then H/K is H itself. For an edge e, $G/(e)$ constructed from $G - e$ by identifying the ends of e is said to be obtained by ***contracting*** e. If e is a loop, then $G - e = G/(e)$.

An alternate sequence of vertices v_i and edges e_i: $v_0, e_1, v_1, \ldots, u_{n-1}, e_n, v_n$ is called a ***chain*** (connecting v_0 and v_n) of G if v_i and v_{i+1} are ends of the edge e_{i+1}, for $i = 0, 1, \ldots, n-1$. The *length* of the chain is n.

A chain C is called a *cycle* if $v_n = v_0$. The length of C, denoted by $|C|$, is n. A chain or a cycle is called *simple* if $e_i \neq e_j$ and $v_i \neq v_j$ for any i and j, $i \neq j$, except possibly $v_n = v_0$. For simplicity, a cycle of length n will be called an n-cycle. A chain (or a cycle) in which all the edges are distinct is called a ***trail*** (or a ***closed trail***).

A graph G is said to be ***bipartite*** if any cycle of G has an even length. A bipartite graph cannot have a loop. A graph is called an ***even*** graph if every vertex has an even valency. A vertex of valency 1 is called a ***stump***. A ***twig*** is a vertex of valency 2.

A graph G is called ***planar*** if G can be embedded into $S^2 = \mathbf{R}^2 \cup \infty$ as a 1-complex. G is called a ***plane*** graph if G is a graph embedded in S^2.

If G is a connected plane graph, we can define the dual graph G^*. $V(G^*)$ and the set $F(G)$ of domains in $S^2 - G$ are in one-to-one correspondence, and, $E(G^*)$ and $E(G)$ are in one-to-one correspondence in such a way that $e^* \in E(G^*)$ and its partner have exactly one point, not a vertex, in common. We define the sign of e^* as the opposite of its partner. If G is a plane disconnected graph, then G^* is a disjoint union of graphs dual to connected components of G.

Example 2.1

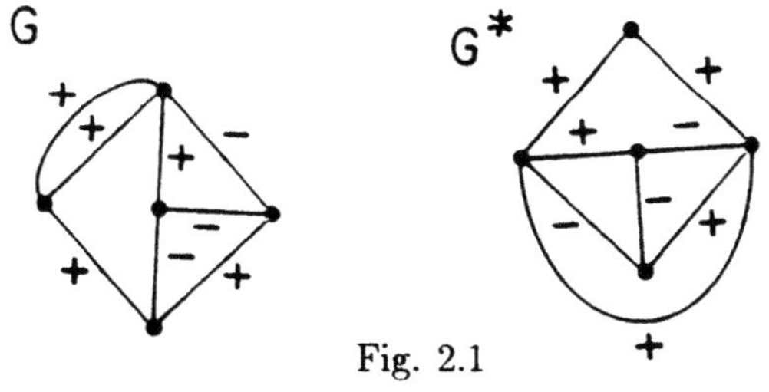

Fig. 2.1

2.2. Polynomial of a graph

Now, we fix some notations, before we define an integer polynomial $F_G(x, y, z)$ for any finite signed graph G.

Let H be a subgraph of G. Denote $p(H)$ and $n(H)$, respectively, the number of positive and negative edges in H. The ***maximal positive spanning subgraph*** P is the spanning subgraph that contains all positive edges but no negative edges. Analogously, the *maximal* negative spanning subgraph N is defined. Therefore, $P \cup N = G$ and $P \cap N = V_G$. We reserve P and N for these subgraphs.

Example 2.2

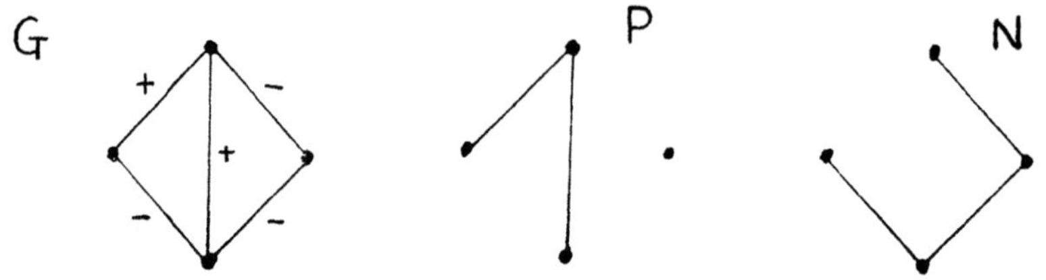

Fig. 2.2

Let $\mathcal{S}_G(r,s)$ be the set of all spanning subgraphs H of G such that $\beta_0(H) = r+1$ and $\beta_1(H) = s$. Therefore $\mathcal{S}_G(0,0)$ is, in particular, the set of all spanning trees in G.

Definition 2.3. We define

$$F_G(x,y,z) = \sum_{r,s} \Big\{ \sum_{H \in \mathcal{S}_G(r,s)} x^{p(H)-n(H)} \Big\} y^r z^s \tag{2.1}$$

where the second summation runs over all spanning subgraphs H in $\mathcal{S}_G(r,s)$. $F_G(x,y,z)$ will be called the *polynomial* of a graph G.

From the definition, we have immediately

Proposition 2.4.
(1) If G has n connected components $G_1, \cdots, G_n$, then

$$F_G(x,y,z) = y^{n-1} \prod_{i=1}^{n} F_{G_i}(x,y,z)$$

(2) If G is connected and is the block sum of m blocks $G_1, \cdots, G_m$, then

$$F_G(x,y,z) = \prod_{i=1}^{m} F_{G_i}(x,y,z) .$$

Since positive graphs are important in both graph theory and its application to knot theory, it is worth writing $F_G(x,y,z)$ more precisely for a positive graph.

Proposition 2.5. *If G is a positive (connected) graph, i.e. $n(G) = 0$, then*

$$F_G(x,y,z) = x^{v-1} \sum_{r,s} |\mathcal{S}_G(r,s)| \Big(\frac{y}{x}\Big)^r (xz)^s. \tag{2.2}$$

where $v = |V(G)|$.

Proof. For $H \in \mathcal{S}_G(r,s)$, $p(H) = v - 1 - r + s$ and $n(H) = 0$. Therefore, (2.1) is reduced to

$$F_G(x,y,z) = \sum_{r,s}\{\sum_{H \in \mathcal{S}_G(r,s)} (\frac{y}{x})^r (xz)^s\} x^{v-1} = x^{v-1}\sum_{r,s}|\mathcal{S}_G(r,s)|(\frac{y}{x})^r (xz)^s . \qquad \square$$

Remark 2.6. It is possible to define a similar polynomial $F_M(x,y,z)$ for a (circuit) matroid M.

$F_G(x,y,z)$ is invariant under 2-isomorphism. Two graphs G_1 and G_2 are said to be *2-isomorphic* if one is obtained from the other by applying the following two operations Ω_1 and Ω_2 finitely many times. Let G be the one-point union of two subgraphs H and K which meet at a vertex v. Then $\Omega_1(G)$ is another one-point union of H and K which meet at a different vertex v'. To define $\Omega_2(G)$, suppose that G is obtained from two disjoint graphs H and K by identifying vertices u_1 and u_2 of H with v_1 and v_2 of K, respectively. $\Omega_2(G)$ is a new graph obtained from H and K by modifying the identification so that $u_1 = v_2$ and $u_2 = v_1$. (Cf. [28].) See Fig. 2.3.

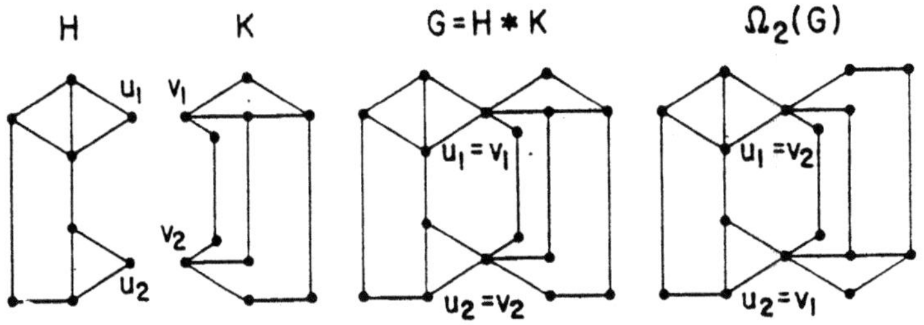

Fig. 2.3

Example 2.7. (1) If G consists of only one vertex, then $F_G(x,y,z) = 1$.
(2) If $G =$  , then $F_G(x,y,z) = x^3z + 3x^2 + 3xy + y^2$.

The dual graph depends on an imbedding of a planar graph G in S^2, but the polynomial of the dual graph is uniquely determined no matter how G imbeds in S^2 and we have

Proposition 2.8. *If G^* is a dual graph of a plane graph G, then*

$$x^{p(G)-n(G)} F_{G^*}(x,z,y) = F_G(x,y,z) \tag{2.3}$$

Warning: Variables y and z are interchanged in the left-hand side.
Proof. We need the following lemma.

Lemma 2.9. *Let G be a connected plane graph. Then for any r and s, $\mathcal{S}_G(r,s)$ and $\mathcal{S}_{G^*}(s,r)$ are in one-to-one correspondence.*

Proof. Take $H \in \mathcal{S}_G(r,s)$ and consider the spanning complement $\hat{H}$ of H in G, i.e. $\hat{H}$ is a spanning subgraph of G such that $E(\hat{H}) = E(G) - E(H)$. Let $\hat{H}^*$ be the dual of $\hat{H}$. $\hat{H}^*$ is a spanning subgraph of G^*. Now by the duality theorem, $\hat{H}^* \in \mathcal{S}_{G^*}(s,r)$, i.e. $\beta_0(\hat{H}^*) = s + 1$ and $\beta_1(\hat{H}^*) = r$. Therefore $H \leftrightarrow \hat{H}^*$ is what we want. $\square$

Now we return to a proof of (2.3). By Lemma 2.9, $S_G(r,s)$ and $S_{G^*}(s,r)$ are in one-to-one correspondence by $H \leftrightarrow \hat{H}^*$. The term associated with H in $F_G(x,y,z)$ is $x^{p(H)-n(H)}y^r z^s$, while the term associated with $\hat{H}^*$ in $F_{G^*}(x,y,z)$ is $x^{-(p(\hat{H}^*)-n(\hat{H}^*))}y^s z^r$. Since $n(\hat{H}^*)+p(H)=p(G)$ and $p(\hat{H}^*)+n(H)=n(G)$, we have $x^{p(G)-n(G)}x^{p(\hat{H}^*)-n(\hat{H}^*)} = x^{p(H)-n(H)}$. This proves (2.3). □

2.3. Recursion formulas

In practice, $F_G(x,y,z)$ can be calculated recursively using the following formulas.

(2.4) (I) If $E(G)=\phi$ and $|V(G)|=v$, then $F_G(x,y,z)=y^{v-1}$.

(II) Let $e \in E(G)$ and $f_G(e)=\varepsilon$.

(i) If e is not a loop, then $F_G(x,y,z)=F_{G-e}(x,y,z)+x^{\varepsilon}F_{G/(e)}(x,y,z)$.

(ii) If e is a loop, then $F_G(x,y,z)=(1+x^{\varepsilon}z)F_{G/(e)}$.

Proof. (I) is obvious, since $S_G(r,s)=\phi$ except for $S_G(v-1,0)=V_G$.

(II) Let $S_1(r,s)=\{H \in S_G(r,s) \mid H \not\ni e\}$ and $S_2(r,s)=\{H \in S_G(r,s) \mid H \ni e\}$. $S_G(r,s)$ is the disjoint union of $S_1(r,s)$ and $S_2(r,s)$.

(i) Suppose that e is not a loop. Then $S_1(r,s)$ and $S_2(r,s)$, respectively, correspond to $S_{G-e}(r,s)$ and $S_{G/(e)}(r,s)$, in one-to-one fashion. Therefore, we have

$$\begin{aligned}
&F_{G-e}(x,y,z)+x^{\varepsilon}F_{G/(e)}(x,y,z)\\
&=\sum_{r,s}\sum_{S_{G-e}(r,s)\ni H} x^{p(H)-n(H)}y^r z^s \sum_{r,s}\sum_{S_{G-e}(r,s)\ni H} x^{p(H)-n(H)+\varepsilon}y^r z^s\\
&=\sum_{r,s}\sum_{S_1(r,s)\ni H_1} x^{p(H_1)-n(H_1)}y^r z^s \sum_{r,s}\sum_{S_2(r,s)\ni H_2} x^{p(H_2)-n(H_2)+\varepsilon}y^r z^s\\
&=\sum_{r,s}\sum_{S_G(r,s)\ni H} x^{p(H)-n(H)}y^r z^s\\
&=F_G(x,y,z).
\end{aligned}$$

(ii) A proof is completely analogous. □

Example 2.10. Let $\varepsilon_i = f_G(e_i)$.

(1) If G consists of k loops $e_1,\ldots,e_k$ then $F_G(x,y,z)=\prod_{i=1}^{k}(1+x^{\varepsilon_i}z)$.

(2) If G is a tree with ℓ edges $e_1,\ldots,e_\ell$, then $F_G(x,y,z)=\prod_{i=1}^{\ell}(y+x^{\varepsilon_i})$.

$F_G(x,y,z)$ is closely related to other polynomials which appear in graph theory. In fact, we have

Proposition 2.11. *Let G be a connected positive graph without multiple edges. Let $C_G(y)$ denote the chromatic polynomial of G. Then*

$$F_G(-1,y,1) = C_G(y)y^{-1}.$$

Proposition 2.12. *Let G be a connected positive graph. Let $\chi(G;y,z)$ denote Tutte's dichromate of G. Then*

$$F_G(1, y-1, z-1) = \chi(G; y, z) .$$

Proofs of these propositions follow immediately if we compare the recursion formulas needed to evaluate $C_G(y)$ or $\chi(G;y,z)$ and (2.4) (I)-(II).

2.4. Degree of $F_G(x,y,z)$

For a Laurent polynomial $f = \sum_{i_1\cdots i_n} a_{i_1\cdots i_n} x_1^{i_1}\cdots x_n^{i_n} \in \mathbb{Z}[x_1^{\pm 1},\cdots,x_n^{\pm 1}]$, we define $\max\deg_{x_j} f = max\{i_j | a_{i_1\cdots i_n} \neq 0\}$ and $\max\deg f = max\{i_1+\cdots+i_n | a_{i_1\cdots i_n} \neq 0\}$. $\min\deg_{x_j} f$ and $\min\deg f$ are defined analogously. Furthermore, we denote $x_j - span$ $f = \max\deg_{x_j} f - \min\deg_{x_j} f$ and *span* $f = \max\deg f - \min\deg f$.

Now we will evaluate $\max\deg F_G(x,y,z)$ and $\min\deg F_G(x,y^{-1},z^{-1})$, using a simple combinatorial argument on spanning subgraphs.

Theorem 2.13. *For any signed graph G,*

$$\begin{aligned} \max\deg F_G(x,y,z) &= p(G) + \beta_0(P) + \beta_1(P) - 1 , \qquad \textit{and} \\ \min\deg F_G(x,y^{-1},z^{-1}) &= -\{n(G) + \beta_0(N) + \beta_1(N) - 1\}, \end{aligned}$$

where P and N denote, respectively, the maximal positive and negative spanning subgraphs.

Proof. A spanning subgraph H belongs to some $S_G(r,s)$. For convenience, we say $H \in S_G(r_H, s_H)$, i.e. $r_H = \beta_0(H) - 1$ and $s_H = \beta_1(H)$.

First we note that it suffices to prove the theorem for a connected graph.

Now, to each spanning subgraph H of G, there is associated a term $x^{p(H)-n(H)} y^{r_H} z^{s_H}$ in $F_G(x,y,z)$ and $x^{p(H)-n(H)} y^{-r_H} z^{-s_H}$ in $F_G(x,y^{-1},z^{-1})$. For convenience, the degrees of these terms are called the *degrees* of H, and are denoted by $\deg H = p(H) - n(H) + r_H + s_H$ and $\deg^* H = p(H) - n(H) - r_H - s_H$. Since $\deg P = p(G) + r_P + s_P$ and $\deg^* N = -n(G) - r_N - s_N$ and since $F_G(x,y,z)$ and $F_G(x,y^{-1},z^{-1})$ are positive polynomials, i.e. all non-zero coefficients are positive, we only need to show that for any spanning subgraph H,

$$\text{(2.5)} \qquad \deg H \leq \deg P \quad \text{and} \quad \deg^* H \leq \deg^* N .$$

We use the following lemma.

Lemma 2.14. *For any spanning subgraph H of G,*

$$\begin{aligned} \text{(2.6)} \quad &(1) \quad r_H - \{p(G) - p(H)\} \leq r_P \quad \textit{and} \quad s_H - n(H) \leq s_P . \\ &(2) \quad r_H - \{n(G) - n(H)\} \leq r_N \quad \textit{and} \quad s_H - p(H) \leq s_N . \end{aligned}$$

Proof. Since P contains all positive edges in G, any positive edge in H is contained in P. Therefore, H is obtained from P by removing $p(G)-p(H)$ positive edges and then by adding $n(H)$ negative edges. Let B be a spanning subgraph of G and B' be the subgraph obtained from B either by removing one edge or by adding one edge. Then we see easily that

$$|r_B - r_{B'}| \le 1 \qquad \text{and} \qquad |s_B - s_{B'}| \le 1\,,$$

and hence

$$r_H \le r_P + p(G) - p(H) \qquad \text{and} \qquad s_H \le s_P + n(H)\,.$$

This proves (2.6) (1). A proof of (2) is analogous by taking P instead of N, and hence, is omitted. □

Now, by using Lemma 2.14, we have

$$\begin{aligned} \deg P &= p(G) + r_P + s_P \ge p(H) + r_H + s_H - n(H) = \deg H\,, \qquad \text{and} \\ \deg^* N &= -n(G) - r_N - s_N \le -n(H) - r_H + p(H) - s_H = \deg^* H\,. \end{aligned}$$

This proves (2.5) and hence Theorem 2.13. □

When G is reduced, we can slightly strengthen Theorem 2.13 to Theorem 2.15 below. These theorems are of a fundamental importance for applications to link theory.

Theorem 2.15. *If G is a reduced connected positive graph, then for any spanning subgraph H, not P or N.*

(2.7) $$\deg H \lneqq \deg P \qquad \textit{and} \qquad \deg^* H \gneqq \deg^* N\,.$$

Proof. Since G is positive, $P = G$ and $N = V_G$, and hence $n(G) = 0$, $r_P = 0$, $r_N = |V(G)|-1$ and $s_N = 0$. Therefore, we have from (2.6) $s_H \le s_P$ and $s_H - p(H) \le 0$. Using these inequalities, we have $\deg H = p(H) + r_H + s_H \le p(H) + r_H + s_P$, and $\deg P = p(G) + s_P$, and moreover, $\deg^* H = p(H) - r_H - s_H \ge -r_H$ and $\deg^* N = p(H) - r_N - s_N = -\{|V(G)|-1\}$. Therefore, it suffices to show that

(2.8) (1) $p(H) + r_H \lneqq p(G)$, and

(2) $r_H \lneqq |V(G)| - 1$.

Let $p(G) - p(H) = q \ge 1$. Then H is obtained from G by removing q edges, $e_1, e_2, \ldots, e_q$ say. Let $G_j = G - \{e_1, e_2, \ldots, e_j\}$, $0 \le j \le q$. Note that $G_0 = P$ and $G_q = H$. Since G has no isthmuses, G_1 is still connected, i.e. $r_{G_1} = 0$. Inductively, we can prove that $r_G \le j-1$ for $1 \le j \le q$, and hence $r_H = r_{G_q} \le q-1 = p(G) - p(H) - 1$. This implies that $p(H) + r_H < p(G)$, which proves (2.8) (1).

To prove (2.8) (2), we note that N is the spanning vertex graph V_G and H is obtained from N by adding $p(H)$ positive edges $e'_1, \ldots, e'_{p(H)}$, say. Let $N_j = N \cup \{e'_1, \ldots, e'_j\}$, $0 \le j \le p(H)$. Note that $N_0 = N$ and $N_{p(H)} = H$, and $p(H) \ge 1$. Since N does not have a loop, e'_1 is not a loop and hence, $r_{N_1} = |V(G)| - 2$ and inductively, we can prove that $r_{N_j} \le |V(G)| - j - 1$ for $1 \le j \le p(H)$. Therefore, $r_H = r_{N_{p(H)}} \le |V(G)| - p(H) - 1 \lneqq |V(G)| - 1$. □

Theorem 2.16. *Let G be a connected graph. Let H be a spanning subgraph of G and H' be the spanning complement of H. Then*

$$(2.9) \qquad r_H + s_H + r_{H'} + s_{H'} \leq |E(G)|\,.$$

Furthermore, equality holds in (2.9) iff H is a union of some (not necessarily all) blocks of G (and hence H' is a union of the remaining blocks).

Proof. Use Mayer-Vietoris homology sequence:

$$\begin{array}{cccccc} \to & H_1(H \cap H') & \to & H_1(H) \oplus H_1(H') & \xrightarrow{i_*} & H_1(H \cup H') \\ \to & H_0(H \cap H') & \to & H_0(H) \oplus H_1(H') & \to & \cdots \end{array}$$

Since $H \cap H' = V_G$, it follows that $H_1(G \cap H') = 0$ and i_* is a monomorphism. Since very homology group is free abelian, we see that $\beta_1(H) + \beta_1(H') = s_H + s_{H'} \leq \beta_1(H \cup H') = |E(G)| - \{|V(G)| - 1\}$. Since $r_H + 1 - s_H = |V(H)| - |E(H)|$ for any spanning subgraph H, it follows that $r_H + s_H + r_{H'} + s_{H'} = s_H - |E(H)| + |V(G)| - 1 + s_H + s_{H'} - |E(H')| + |V(G)| - 1 + s_{H'} = 2(s_H + s_{H'}) - |E(G)| + 2|V(G)| - 2 \leq 2(|E(G)| - |V(G)| + 1) - |E(G)| + 2|V(G)| - 2 = |E(G)|$.

Furthermore, equality holds in (2.9) iff i_* is an isomorphism, and hence, H must be a union of blocks of G. □

2.5. Signature of a graph

In this section, we introduce another invariant of a signed graph, and estimate it in terms of P and N.

Definition 2.17. Let G be a (signed) graph. The *adjacency* matrix of G is a $|V(G)| \times |V(G)|$ integer matrix $A_G = \|a_{ij}\|$ defined as follows.

(2.10) (1) For any $i \neq j$, a_{ij} is the number of those positive edges minus negative edges which have two ends v_i and v_j.

(2) $a_{ii} = 0$ for $i = 1, 2, \ldots, |V(G)|$.

Let $Q_G = \|q_{ij}\|$ be a diagonal matrix of order $|V(G)|$, where $q_{ii} = \sum_{j=1}^{|V(G)|} a_{ij}$, $i = 1, 2, \ldots, |V(G)|$. Then the matrix $B_G = Q_G - A_G$ is called the *matrix* of G. We should note that if G is positive then B_G is the known matrix of an (unsigned) graph. B_G is not affected by the presence of loops in G.

Now since B_G is symmetric, we can define the signature of B_G which is called the *signature* of G and is denoted by $\sigma(G)$. From the definition we have

Proposition 2.18.

(1) $\sigma(G)$ is an invariant of a graph.

(2) If G is a connected positive graph, then $\sigma(G) = |V(G)| - 1$.

Other elementary properties of the signature are:

(2.11) (1) Let $G_1, \ldots, G_m$ be connected components of G. Then

(i) $\sigma(G) = \sum_{i=1}^{m} \sigma(G_i)$, and

(ii) $|\sigma(G)| \leq |V(G)| - m$.

Equality holds in (ii) if all G_i are positive or all negative.

(2) Let G be a connected graph and $G_1, \ldots, G_k$ be the blocks of G. Then $\sigma(G) = \sum_{i=1}^{k} \sigma(G_i)$.

(3) If G is a tree with $p(G)$ positive and $n(G)$ negative edges, then $\sigma(G) = p(G) - n(G)$.

We note that the signature is an invariant for 2-isomorphic graphs. Now we prove the following

Theorem 2.19. *For a connected graph G,*

$$r_N + s_N - n(G) \leq \sigma(G) \leq p(G) - (r_P + s_P)\,. \tag{2.12}$$

Equalities hold in (2.12) simultaneously iff each block of G is either a positive or negative graph.

Proof. The second statement follows from 2.16.

We will prove (2.12). Since $r_N - s_N = \beta_0(N) - 1 - \beta_1(N) = |V(N)| - |E(N)| - 1 = |V(G)| - n(G) - 1$, we have $s_N = r_N - |V(G)| + n(G) + 1$, and hence, $r_N + s_N - n(G) = 2r_N - \big(|V(G)| - 1\big)$. Similarly, since $s_P = r_P - |V(G)| + p(G) + 1$, we have $p(G) - (r_P + s_P) = \big(|V(G)| - 1\big) - 2r_P$. Therefore, (2,12) is equivalent to

$$2r_N - |V(G)| + 1 \leq \sigma(G) \leq |V(G)| - 1 - 2r_P\,. \tag{2.13}$$

To prove (2.13), we use the following lemma

Lemma 2.20. *Let G be a connected graph. Let G_+ (or G_-) be the graph obtained from G by adding finitely many, h say, positive (or negative) edges, each of which has the same ends, a and b, say. Then*

$$\sigma(G) \leq \sigma(G_+) \leq \sigma(G) + 2\,, \qquad \text{and} \qquad \sigma(G) - 2 \leq \sigma(G_-) \leq \sigma(G)\,.$$

This lemma follows easily by comparing three symmetric matrices B_G, B_{G_+} and B_{G_-} needed to evaluate the signatures.

Now we return to a proof of Theorem 2.19. Suppose that P has k connected components $P_1, P_2, \ldots, P_k$. Since G is connected, there are $(k-1)$ negative isthmuses $e_1, \ldots, e_{k-1}$, each of which connects two components of P and $G_0 = P \cup \{e_1, \ldots, e_{k-1}\}$ is connected. Let N_i be the graph consisting of e_i and its ends. Then (2.11) (2) shows that

$$\sigma(G_0) = \sum_{i=1}^{k} \sigma(P_i) + \sum_{i=1}^{k-1} \sigma(N_i)\,.$$

Since $\sigma(P_i) = |V(P_i)| - 1$ and $\sigma(N_i) = -1$, it follows that

$$\sigma(G_0) = \sum_{i=1}^{k}\{|V(P_i)| - 1\} - (k-1) = \sum_{i=1}^{k}|V(P_i)| - 2k + 1 = |V(G)| - 2k + 1 .$$

Now in order to obtain G from G_0, we have to add the remaining negative edges to G_0. However, by Lemma 2.20, $\sigma(G) \leq \sigma(G_0) = |V(G)| - 2k + 1$. Since $r_P = k - 1$, we have $\sigma(G) \leq \sigma(G_0) = |V(G)| - 1 - 2r_P$. This proves the right-side inequality in (2.12).

A proof of the other inequality is completely analogous. □

There is a "mixed" graph G for which one of the equalities holds but not both. See Fig. 2.4.

Example 2.21.

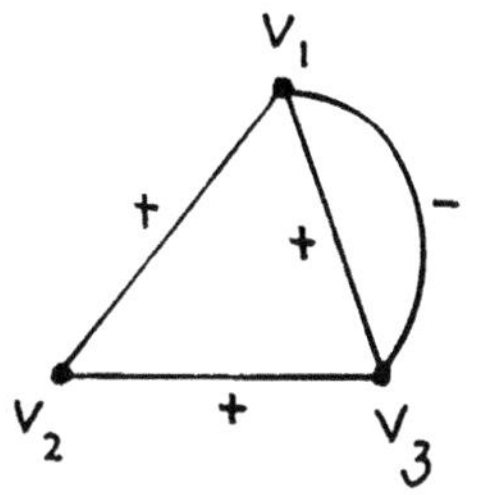

Fig. 2.4

The matrix of G is $B_G = \begin{bmatrix} 1 & -1 & 0 \\ -1 & 2 & -1 \\ 0 & -1 & 1 \end{bmatrix}$. Note that $\sigma(G) = 2$, $v = 3$, $r_P = s_N = 0$, $s_P = r_N = 1$, $p(G) = 3$ and $n(G) = 1$. Therefore, $r_N + s_N - n(G) = 0 \neq \sigma(G) = 2 = p(G) - r_P - s_P$.

3. Jones polynomials of links

3.1. Graph of link

To utilize the theorems proved in the previous Chapter, it is necessary to construct an appropriate graph from a link. In fact, there are several ways to associate a graph with a link. In this talk, we will use two different types of graph. The first graph, defined below, was used by C. Bankwitz [2] in 1930 to study alternating knots. We call it the *graph* of a link L (or more precisely the *graph of a diagram* D of L).

Let D be a link diagram of a link L. D divides S^2 into a finite number of domains $R_1, R_2, \cdots, R_n$ which will be classified as shaded or unshaded. There are no common edges between two shaded or unshaded domains. There are exactly two such classifications (or shadings) of $S^2 - D$. For convenience, one is called the *opposite* of the other. Now we construct a signed plane graph G from D and shading ρ. Take a point v_i from each unshaded domain R_i. These points form a set of vertices of G. Suppose that two unshaded domains R_i and R_j meet at a crossing c_k. If R_i is different from R_j, then we join v_i and v_j by an edge e_k passing through c_k. If R_i and R_j are the same domain, then we form a loop e_k passing through c_k. See Fig. 3.1.

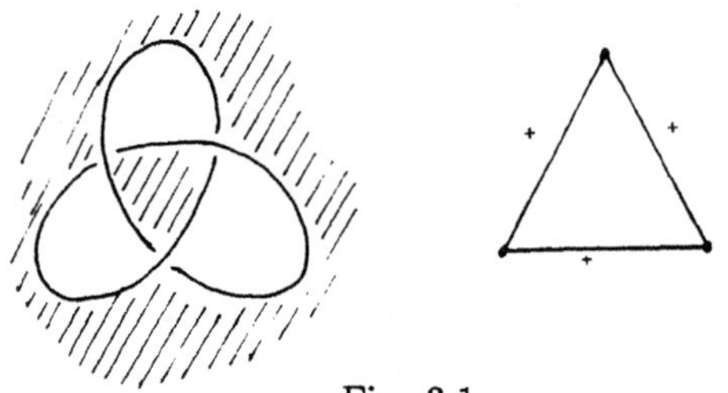

Fig. 3.1

Each edge of G, therefore, corresponds to a crossing of D. Furthermore, each edge e_i is signed with $+1$ or -1 according to whether the twist at the crossing is positive or negative. (See Fig. 3.2)

$sign(e) = +1$ $\qquad\qquad$ $sign(e) = -1$

Fig 3.2

The resulting signed plane graph is called the ***graph*** of a link w.r.t. D and shading ρ, and is denoted by $G(D,\rho)$ (or $G(D)$, G_D or simply G). $G(D,\rho)$ depends not only on D, but also on shading. If we use the opposite shading ρ', the resulting graph $G(D,\rho')$ is the dual of $G(D,\rho')$.

Conversely, given a signed plane graph G, one can construct uniquely the link diagram D of a link and can choose shading ρ so that $G(D,\rho)$ is exactly G.

Now each vertex and each edge of $G(D,\rho)$ correspond, respectively, to an unshaded domain and a crossing of D. If $G(D,\rho)$ is a reduced graph, a diagram D is said to be ***reduced.*** The diagram in Fig. 3.1 is not reduced. For an alternating diagram D, $G(D,\rho)$ is either a positive or negative graph.

3.2. Jones polynomial

Now, given a link L, we can define the polynomial $F_{G(D,\rho)}$ (x,y,z), or simply denoted by $F_D(x,y,z)$, of the graph $G(D,\rho)$. Since a link L can have many different link diagrams, we may have many different polynomials. In other words, the polynomial obtained from $G(D,\rho)$ is not necessarily unique to each L. However, we can define a ***unique*** polynomial to each L no matter which diagram we use. This polynomial depends on the orientation of a link L.

Suppose that we are given an oriented link L. The orientation of L induces the orientation of a diagram D. We then defined the second index $w(c)$, called the ***twist*** or *writhe*, at each crossing c as follows

Fig. 3.3

A crossing c with $w(c) = 1$ (or $w(c) = 1$) is called a positive (or negative) crossing. The sum of all writhes on D, $w(c) = \sum_{c\in D} w(c)$ is called the *writhe* or the *Tait number* of D. We should note that the writhe is independent of shading.

Let $F_{G(D,\rho)}(x,y,z)$ be the polynomial of the graph $G(D,\rho)$ associated with an (oriented) diagram D.

Theorem 3.1. *Let $p = p(G(D,\rho))$ and $n = n(G(D,\rho))$. Then*

$$\mathbf{V}_L(x) = (-1)^{p-n} x^{-\frac{p-n-3w(D)}{2}} F_G(x, -(x+x^{-1}), -(x+x^{-1}))$$

is independent of a diagram D and ρ, and depends only on the ambient isotopy type of L. In other words, $\mathbf{V}_L(x)$ is an invariant of L.

For a proof, we refer to [14]. We should note that $\mathbf{V}_L(x)$ does note depend on shading.

Now since $p - n - 3w(D) \equiv p + n - w(D) \equiv 0 \pmod 2$, $\mathbf{V}_L(x)$ is a polynomial on $x^{\pm 1}$, and the polynomial $V_L(x)$:

$$(3.1) \qquad V_L(x) = \mathbf{V}_L(x^{\frac{1}{2}}) = (-1)^{p-n} x^{-\frac{p-n-3w(D)}{4}} F_D(x^{\frac{1}{2}}, -(x^{\frac{1}{2}} + x^{-\frac{1}{2}}), -(x^{\frac{1}{2}} + x^{-\frac{1}{2}}))$$

is in fact the *Jones polynomial* of a link L. This (due to Kauffman) is an important second definition of the Jones polynomial [13].

For example, the following theorem due to M.B. Thistlethwaite [26] is an immediate consequence of this definition of $V_L(x)$.

Theorem 3.2. *The Jones polynomial of a non-split alternating link L is alternating. In other words, if $V_L(x) = \sum_{-\infty<k<\infty} c_k x^k$, then $c_k c_{k+1} \leq 0$ for any k, $-\infty < k < \infty$.*

3.3. Estimate of the degree of $V_L(X)$

We have already had two indices $\operatorname{sign}(c)$ and $w(c)$ at each crossing c. We now need the third index $\eta_\rho(c)$ at c.

Let D be an oriented diagram of a link L and ρ a shading of $S^3 - D$. Let $\eta_\rho(c) = w(c)\delta_{\mathrm{sign}(c),w(c)}$, where δ denotes Kronecker's data. We define $\eta_\rho(D) = \Sigma\eta_\rho(c)$, where the summation runs over all crossings in D. $\eta_\rho(D)$ depends not only on shading ρ, but also on orientation of D. Therefore, $\eta_\rho(D)$ is not determined by the graph $G(D,\rho)$ alone. To be more precise, let $c_+(D)$ and $c_-(D)$ be, respectively, the number of positive and negative crossings in D. Note that $w(D) = c_+(D) - c_-(D)$ and $c_+(D) + c_-(D)$ is the total number of crossings in D which will be denoted by $c(D)$.

Lemma 3.3. $p\big(G(D,\rho)\big) = c_-(D) + \eta_\rho(D)$ and $n\big(G(D,\rho)\big) = c_+(D) - \eta_\rho(D)$.

Proof. There are four cases to be considered. A broken line in Fig. 3.4 indicates an edge in the graph of D.

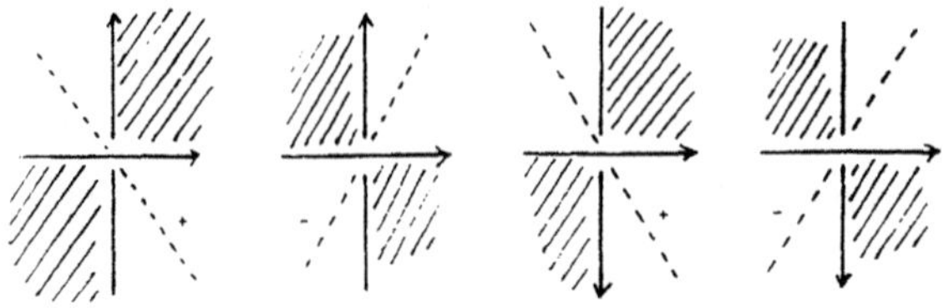

$w = 1$	$w = 1$	$w = -1$	$w = -1$
sign $= 1$	sign $= -1$	sign $= 1$	sign $= -1$
$\eta = 1$	$\eta = 0$	$\eta = 0$	$\eta = -1$

Fig. 3.4

From Fig. 3.4, we see easily that $p(G) = \big|\{c \in D \mid w(c) = -1 \text{ and } \eta_\rho(c) = 0\}\big| + \big|\{c \in D \mid w(c) = 1 \text{ and } \eta_\rho(c) = 1\}\big|$ and hence $p(G) = c_-(D) + \eta_\rho(D)$. Similarly, $n(G) = \big|\{c \in D | w(c) = 1 \text{ and } \eta_\rho(c) = 0\}\big| + \big|\{c \in D \mid w(c) = -1 \text{ and } \eta_\rho(c) = -1\}\big|$ and hence $n(G) = c_+(D) - \eta_\rho(D)$. □

The following lemma is proved in [11]. See also [4].

Lemma 3.4. $\sigma(L) = \sigma\big(G(D,\rho)\big) - \eta_\rho(D)$.

Using these lemmas and Theorem 2.9, we can now prove the following

Theorem 3.5. *For any link diagram D of a link L,*

$$\text{(3.2)} \qquad \begin{cases} \max\deg V_L(x) & \le \ \ c_+(D) - \frac{1}{2}\sigma(L) \\ \min\deg V_L(x) & \ge -c_-(D) - \frac{1}{2}\sigma(L). \end{cases}$$

Both inequalities hold in (3.2) simultaneously iff D is a reduced alternating diagram or the connected sum of these.

Proof. Let P and N be, as usual, the maximal positive and negative subgraphs of G. Then by Theorems 2.13 and 3.1, we have

$$\text{(3.3)} \qquad \begin{cases} \max\deg \mathbf{V}_L(x) & \le p(G) + r_P + s_P - \frac{p(G)-n(G)-3w(D)}{2} \\ \min\deg \mathbf{V}_L(x) & \ge -(n(G) + r_N + s_N) - \frac{p(G)-n(G)-3w(D)}{2} \end{cases}$$

Therefore, to prove (3.3), it suffices to show

$$\text{(3.4)} \quad (1) \quad p(G) + r_P + s_P - \frac{p(G) - n(G) - 3w(D)}{2} \le 2c_+(D) - \sigma(L)$$

$$(2) \quad -(n(G) + r_N + s_N) - \frac{p(G) - n(G) - 3w(D)}{2} \ge -2c_-(D) - \sigma(L).$$

Since $w(D) = c_+(D) - c_-(D)$ and $p(G) + n(G) = c_+(D) + c_-(D)$, a straightforward computation shows that (3.4) is equivalent to (3.5):

$$\text{(3.5)} \quad (1) \quad \sigma(L) \le c_-(D) - (r_P + s_P)$$

$$(2) \quad r_N + s_N - c_+(D) \le \sigma(L).$$

However, since $\sigma(L) = \sigma(G) - \eta_\rho(D)$ by Lemma 3.4, (3.5) becomes

$$(3.6) \qquad r_N + s_N - c_+(D) + \eta_\rho(D) \le \sigma(G) \le c_-(D) + \eta_\rho(D) - (r_P + s_P)$$

or, by Lemma 3.3,

$$(3.7) \qquad r_N + s_N - n(G) \le \sigma(G) \le p(G) - (r_P + s_P)\ .$$

(3.7) is exactly what we have proved in Theorem 2.19, and equalities hold in (3.2) simultaneously iff D is a reduced alternating diagram or the connected sum of these. □

3.4. Other estimates of the degree of $V_L(x)$

The formula (3.2) is a good estimate for $\max\deg V_L(x)$ and $\min\deg V_L(x)$ for an oriented link L, but V. Jones also gives another estimate for these degrees. In fact, he proves the following proposition [13].

Proposition 3.6. *Let β be an n-braid and $\hat{\beta}$ the closure of β. Let e_+ and e_- denote, respectively, the sum of positive and negative exponents in β. Write $e = e_+ - e_-$. Then*

$$(3.8) \quad (a) \quad \max\deg V_{\hat{\beta}}(x) \le e_+ + \frac{e+n-1}{2}$$

$$(b) \quad \min\deg V_{\hat{\beta}}(x) \ge -e_- + \frac{e-n+1}{2}$$

T. Fiedler [7] recently obtained a much better estimate for a closed braid $\hat{\beta}$ as follows

Proposition 3.7. *Let $\delta^+(\beta)$ (and $\delta^+(\beta)$) denote the number of distinct (Artin) braid generators appearing in β with positive (and negative) exponent. Then*

$$(3.9) \quad (1) \quad \max\deg V_{\hat{\beta}}(x) \le e_+ + \frac{e+n-1}{2} - \delta^+(\beta)$$

$$(2) \quad \min\deg V_{\hat{\beta}}(x) \ge -e_- + \frac{e-n+1}{2} + \delta^-(\beta)$$

Since at least one of $\delta^+(\beta)$ or $\delta^-(\beta)$ is positive, (3.9) is a considerable improvement of (3.8). In this section, we will show that (3.9) is a consequence of (3.3).

Let L be the closure of an n-braid β and D the link diagram obtained naturally from a braid representation of L. Let G be a plane graph in S^2 associated with D. We classify the domains of $S^2 - G$ into shaded or unshaded domains in such a way that the unbounded domain i.e. the domain containing ∞, is shaded. We only prove (3.9)(1), since a proof of (3.9)(2) is analogous.

Now in order to prove (3.9)(1), it suffices to show the following inequality

$$(3.10) \quad \frac{1}{2}\left\{p(G) + r_P + s_P - \frac{p(G) - n(G) - 3w(D)}{2}\right\} \le e_+ + \frac{1}{2}(e+n-1) - \delta^+(\beta)\ .$$

First we express $p(G)$, $n(G)$ and $w(D)$ in terms of e_+, e_- and e. Let $\sigma_1, \sigma_2, \ldots, \sigma_{n-1}$ be Artin's generators of the n-braid group B_n. Define $p(\sigma_i)$ and $n(\sigma_i)$, respectively, as the sum of positive and negative exponents of σ_i appearing in β. Then $e_+ = \sum_{i=1}^{n-1} p(\sigma_i)$ and $e_- = \sum_{i=1}^{n-1} p(\sigma_i)$. It is then easy to see

$$(3.11)\ p(G) = \sum_{i=\text{odd}} n(\sigma_i) + \sum_{j=\text{even}} p(\sigma_j) \quad \text{and} \quad n(G) = \sum_{i=\text{odd}} p(\sigma_i) + \sum_{j=\text{even}} n(\sigma_j).$$

Furthermore, noting that $w(D) = e = e_+ - e_-$ and $p(G) + n(G) = e_+ + e_-$, (3.10) is now equivalent to

$$(3.12) \qquad r_P + s_P \leq e_+ + n - 1 - 2\delta^+(\beta).$$

Note that $r_P + 1 - s_P = |V(G)| - p(G)$ and $|V(G)| = \sum_{i=\text{odd}} \{p(\sigma_i) + n(\sigma_i)\} + \varepsilon$, where $\varepsilon = 0$ or 1 depending on whether n is even or odd. See Figure 3.5. Therefore, (3.12) is reduced to

$$(3.13) \qquad s_P + \delta^+(\beta) \leq \sum_{j=\text{even}} p(\sigma_j) + \left[\frac{n}{2}\right].$$

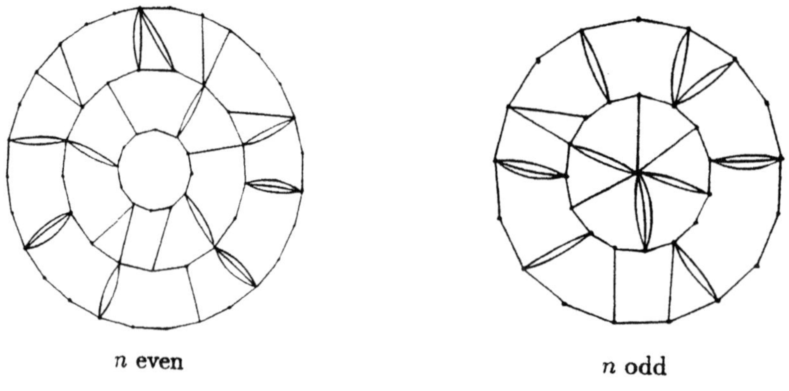

Fig. 3.5

Now, by definition, $s_P + 1$ is the number of domains in which S^2 is divided by the maximal positive subgraph P of G. Note that G consists of $\left[\frac{n+1}{2}\right]$ concentric circles C_i, $i = 1, 2, \ldots, \left[\frac{n+1}{2}\right]$, and edges joining vertices on adjacent two circles C_i and C_{i+1}, $i = 1, 2, \ldots, \left[\frac{n+1}{2}\right] - 1$. Obviously, $\sum_{j=\text{even}} p(\sigma_j)$ gives the number of those positive edges which do not belong to any circle C_i.

Now we evaluate $s_P + 1$. First consider the special case where all the circles C_i consist of positive edges alone. In this case, no generator σ_{2i+1} appears in β with a positive exponent. Since $\delta^+(\beta)$ is, in our case, the number of $p(\sigma_{2_j})$ such that $p(\sigma_{2_j}) \geq 1$ and since each positive edge corresponding to σ_{2_i} in β divides the annulus bounded by

C_i and C_{i+1}, we see easily that $s_P + \delta^+(\beta) = \sum_{j=\mathbf{even}} p(\sigma_j) + [\frac{n}{2}]$. This proves (3.13) for the special case.

Next, we consider the general case. A graph G is obtained from the special graph by replacing some positive edges in C_i by negative edges. If we replace one positive edge of C_i by a negative edge, then δ^+ increases by at most one, but s_P decreases by at least one, and therefore

$$s_P + \delta^+(\beta) \leq \sum_{j=\mathbf{even}} p(\sigma_j) + [\frac{n}{2}] .$$

This completes our proof. □

Remark 3.8. We compare two estimates (3.2) and (3.9) for a non-alternating torus knot of type (r, q), where $3 \leq q < r$ and g.c.d $(r, q) = 1$. If we represent L as the closure of a q-braid $\beta = (\sigma_1 \sigma_2 \dots \sigma_{q-1})^r$, then $c_+(D) = r(q-1)$ and $c_-(D) = 0$. Furthermore, $e_+ = r(q-1)$, $e_- = 0$, $e = e_+$, $\delta^+(\beta) = q - 1$ and $\delta^-(\beta) = 0$. Since $-\sigma(L) \leq (r-1)(q-1)$, a simple computation shows that

(3.14) (1) $$c_+(D) - \frac{\sigma(L)}{2} \leq e_+ + \frac{e+n-1}{2} - \delta^+(\beta)$$

(2) $$-c_-(D) - \frac{\sigma(L)}{2} \leq -e_- + \frac{e-n+1}{2} + \delta^-(\beta)$$

Therefore, (3.2) is a better estimate than (3.9)(1) for $\max \deg V_L(x)$, while (3.9)(2) is better than (3.2) for $\min \deg V_L(x)$.

4. Minimal crossing number

4.1. History

A link diagram D has a finite number of double points which is denoted by $c(D)$. The minimal crossing number of a link L is defined as $\min c(D)$, where the minimum is taken over all link diagrams of L. $c(L)$ is obviously a link invariant. For example, $c(L) = 0$ if L is trivial, and conversely. There are no *knots* L with $c(L) = 1$ or 2, but there is one knot (up to mirror image) with $c(L) = 3$ that is called a trefoil knot and one *link* with $c(L) = 2$ called the Hopf link. (See Fig. 4.1.)

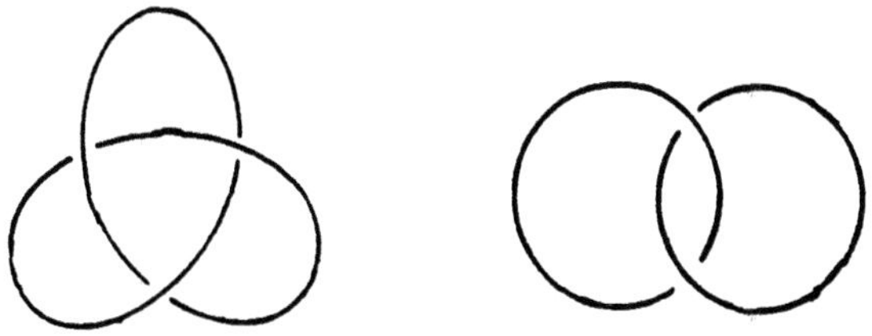

Fig. 4.1

One of the classical problems is:

Problem 4.1. Given a link L, determine $c(L)$.

This problem is extremely hard. Before 1984, there are only a few results involving $c(K)$ in the literature. One of the earliest results is the following theorem due to Bankwitz [2].

Theorem 4.2. *Let $\Delta_K(t)$ be the Alexander polynomial of a knot K. If K is an alternating knot, then $c(K) \leq |\Delta_K(-1)|$.*

Later, this theorem was improved considerably by R.H. Crowell [5].

Theorem 4.3. *If K is an alternating knot that is not a torus knot of type $(n,2)$, then*

$$c(K) \leq \frac{|\Delta_K(-1)| + 3}{2}.$$

If K is a torus knot of type $(n,2)$, then $c(K) = |\Delta_K(-1)|$. In general these theorems cannot be used to determine $c(K)$.

Now the Jones polynomial or, more generally, the skein polynomial has played a crucial role to solve Problem 4.1 for many links including alternating links.

4.2. Minimal crossing number for alternating links

The first general solution to Problem 4.1 is obtained as an immediate consequence of Theorem 3.5.

A link diagram that has exactly $c(L)$ crossings is called a *minimal diagram* of L.

For any diagram D of L, we have from (3.2) that

$$\max\deg V_L(x) - \min\deg V_L(x) \leq c_+(D) + c_-(D) = c(D)\,. \tag{4.1}$$

Therefore, $\operatorname{span} V_L(x)$ gives us a lower bound for the number of crossings any diagram of L can have. This proves the following

Theorem 4.4. *If L is not a split link, then $\operatorname{span} V_L(x) \leq c(L)$.*

In particular, if L is alternating and D is a reduced alternating diagram, then equalities in (3.2) give us the following theorem that solves Problem 4.1 for alternating link.

Theorem 4.5. *If L if a (non-split) alternating link, then*

$$\operatorname{span} V_L(x) = c(L)\,.$$

Furthermore, for any reduced alternating diagram D, $\operatorname{span} V_L(x) = c(D)$. Therefore, any reduced alternating diagram is a minimal diagram.

The second statement of Theorem 3.5 implies

Theorem 4.6. span $V_L(x) = c(L)$ *iff* L *is a connected sum of alternating links.*

Moreover, if L is alternating and D is a reducted alternating diagram, then (3.2) yields

$$\text{(4.2)} \qquad \max\deg V_L(x) + \min\deg V_L(x) \;=\; c_+(D) - c_-(D) - \sigma(L) \;=\; w(D) - \sigma(L)\,.$$

Since $\max\deg V_L(x)$, $\min\deg V_L(x)$ and $\sigma(L)$ are all link invariants, so is $w(D)$. It verifies a conjecture by P.G. Tait. Namely, we have

Theorem 4.7. *If* D_1 *and* D_2 *are reduced alternating diagrams of the same alternating link, then* $w(D_1) = w(D_2)$. *In other words, the writhe is independent of the diagram insofar as we consider reduced alternating diagrams.*

Remark 4.8. Theorems 4.4–4.7 are also proved by M.B. Thistlethwaite with a slightly different method. L. Kauffman also proved Theorems 4.4–4.6.

From Theorem 4.5, we can show that for large n, the number $\lambda(n)$ of prime knots K (up to mirror image) with $c(K) = n$ is at least n^2 [6].

4.3. Adequate links

Although every reduced alternating diagram is a minimal diagram, there are many other non-alternating reduced diagrams which are also minimal. One of the notable examples has been introduced in [15].

Let D be an unoriented diagram of a link L. If we split D at each crossing c in either way as is shown in Fig. 4.2 (a) or (b)

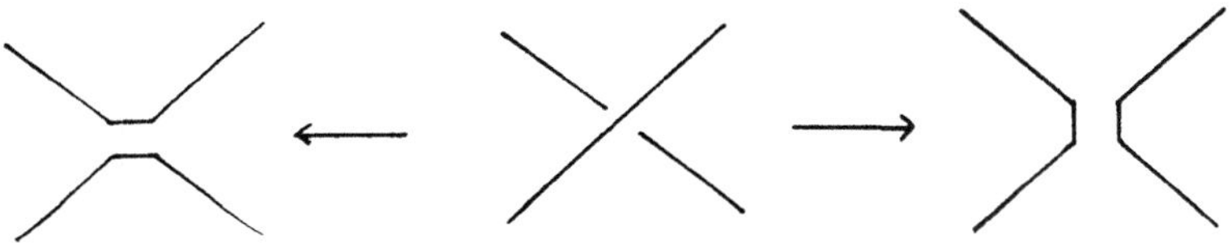

(a) Fig. 4.2 (b)

D is decomposed into finitely many circles on S^2. For convenience, we call the first split a *positive split* and the other *negative split*. By applying these splits, the original crossing c_i "splits" into two points c_i', c_i'' lying on each branch (cf. Fig. 4.3)

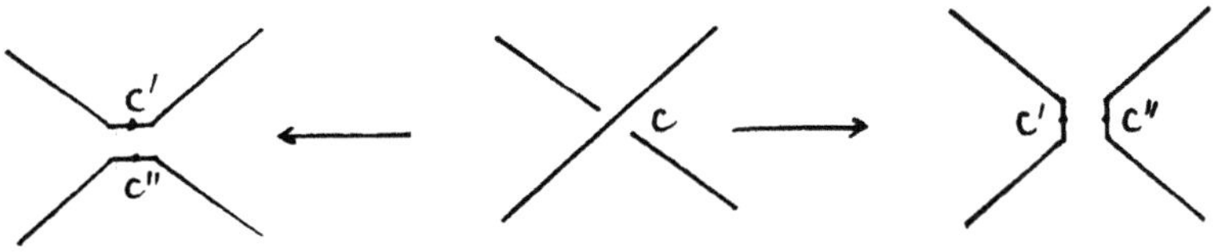

Fig. 4.3

A link diagram D is called a $+$ adequate (or $-$ adequate) if in the diagram D_+ (or D_-) obtained from D by applying positive splits (or negative splits) on all crossings, new points created in D_+ (or D_-) always occur in the different components. D is called *adequate* if D is both $+$ adequate and $-$ adequate. A link is adequate if it admits an adequate diagram. Non-reduced diagrams are never adequate.

Example 4.9.

(1) A reduced alternating diagram is adequate, and hence an alternating link is adequate.

(2) A pretzel link of type $(n_1, n_2, \ldots, n_k)$ is adequate if $k \geq 2$ and $|n_i| \geq 2$ for $i = 1, 2, \ldots, k$. Fig. 4.4 depicts an unoriented pretzel link of type (2,3, – 2,4).

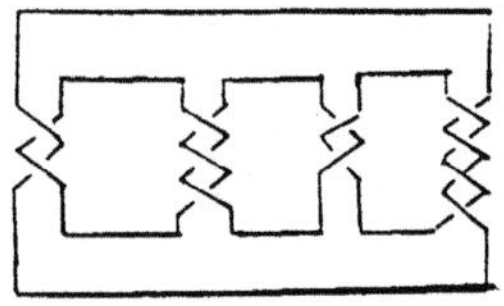

Fig. 4.4

(3) The standard torus knot diagram is not adequate.

Now Problem 4.1 is completely solved for adequate links. In fact, the following theorem is proven in [15].

Theorem 4.10. *An adequate diagram is a minimal diagram.*

Although the standard diagram of a torus knot K of type (p, q) is not adequate if $|p|, |q| > 2$, its minimal crossing number has been determined. See 5.7.

4.4. Semi-alternating links

Another type of links for which Problem 4.1 may be solvable is a link type introduced by H. Terasaka in terms of the graph of a link diagram.

Let G be a signed plane graph. For a vertex v of G, let $e_1, e_2, \ldots, e_n$ be a sequence of all edges emerging from v in the counter-clockwise order. Set $\zeta(v) = \sum_{i=1}^{n} \text{sign}(e_i)\ \text{sign}(e_{i+1})$, $e_{n+1} = e_1$. Then $\gamma(v) = \frac{1}{2}\{n + \zeta(v)\} \geq 0$ is called the *alternation-index* at v. The alternation-index of G is defined as $\gamma(G) = \min_v \gamma(v)$, where the minimum is taken over all vertices of G.

Example 4.11.

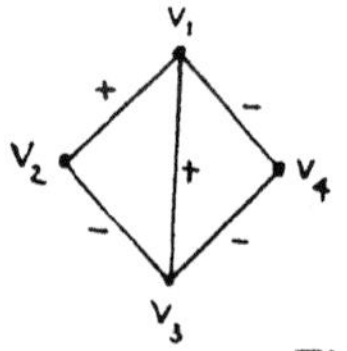

$\gamma(v_1) = 1$

$\gamma(v_2) = 0$

$\gamma(v_3) = 1$

$\gamma(v_4) = 2$

Fig. 4.5

$\gamma(v)$ is the number of pairs of consecutive edges $\{e_i, e_{i+1}\}$ such that $\text{sign}(e_i)\,\text{sign}(e_{i+1}) = 1$. Next, we will construct two graphs $\Gamma_0(G)$ and $\Gamma_1(G)$ from G.

First we subdivide G by adding one vertex w_j to each edge e_j of G so that e_j is divided into two edges e'_j, e''_j. The resulting graph is denoted by G'. By defining $\text{sign}(e'_j) = \text{sign}(e''_j) = \text{sign}(e_j)$, G' becomes a signed plane graph. Therefore, $V(G') = V(G) \cup \{w_1, \ldots, w_m\}$ and $D(G') = \{e'_1, e''_1, \ldots, e'_m, e''_m\}$, where $m = |E(G)|$.

Secondly, two subgraphs $\Gamma_0(v)$ and $\Gamma_u(v)$ are associated with a vertex v of G as follows. If all the edges of G' emerging from v have the same sign, $\Gamma_0(v) = \Gamma_u(v) = \mathit{star}\, v$. Consider the general case. Suppose $e_{11}, \ldots, e_{1n_1}, e_{21}, \ldots, e_{2n_2}, \ldots, e_{2k1}, \ldots, e_{2kn_{2k}}$ be the edges of G' emerging from v, where $\text{sign}\, e_{ij} = (-1)^{i+1}$, $i = 1, 2, \ldots, 2k$. Let w_{ij} be one end ($\neq v$) of e_{ij}. Then $\Gamma_0(v)$ is the plane (unsigned) graph consisting of $\sum_{i=1}^{2k} n_i - k$ edges, $e_{11}, \ldots, e_{1,n_1-1}, \bar{e}_1, e_{22}, \ldots, e_{2n_2}$, $e_{31}, \ldots, e_{3n_3-1}, \bar{e}_3, e_{42}, \ldots, e_{4,n_4}, \ldots, \bar{e}_{2k-1}, e_{2k,2}, \ldots, e_{2k,n_{2k}}$ and their ends. On the other hand, $\Gamma_u(v)$ consists of edges of $e_{12}, \ldots, e_{1,n_1}, e_{21}, \ldots, e_{2n_2-1}, \bar{e}_2$, $e_{32}, \ldots, e_{3,n_3}, e_{41}, \ldots, e_{4n_4-1}\bar{e}_4, \ldots, e_{2k,1}, \ldots, e_{2k,n_{2k}-1}, \bar{e}_{2k}$, and their ends. Here, $\bar{e}_\ell (\ell = 1, 2, \ldots, 2k)$ is the edge joining two vertices w_{ℓ,n_ℓ} and $w_{\ell+1,1}$, where $w_{2k+1,1} = w_{1,1}$. Finally, $\Gamma_0(G)$ and $\Gamma_u(G)$ are obtained from G by replacing every *star* v, $v \in V(G)$, by $\Gamma_0(v)$ and $\Gamma_u(v)$, respectively. $\Gamma_0(G)$ and $\Gamma_u(G)$ are called the ***over-graph*** and ***under-graph*** of G, respectively.

Example 4.12

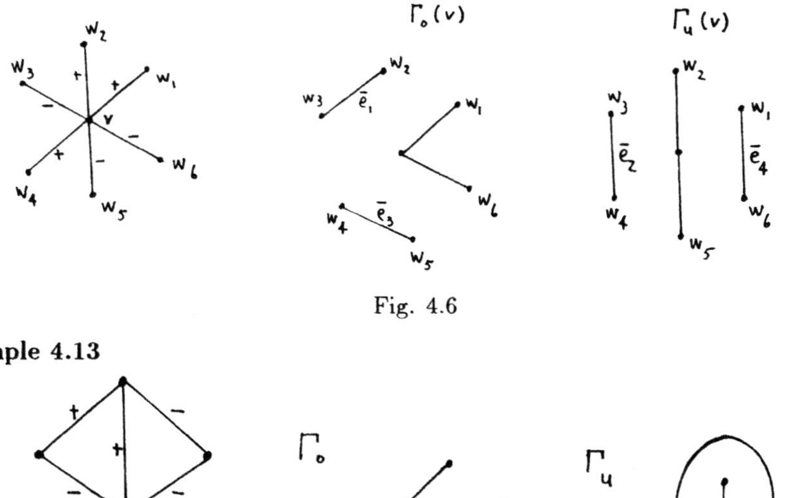

Fig. 4.6

Example 4.13

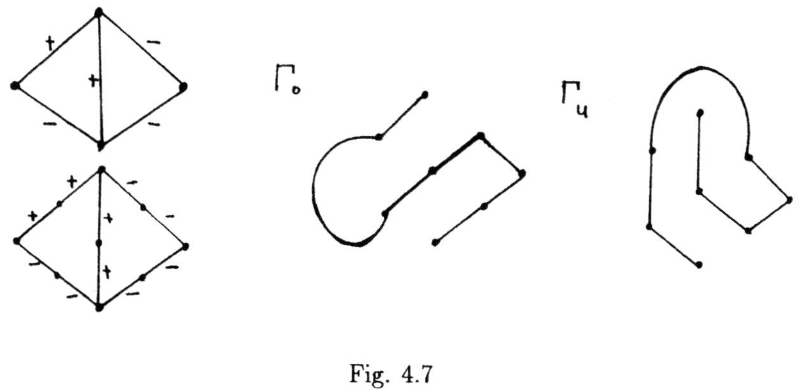

Fig. 4.7

Although this conjecture was recently disproved in [18], we will show that equality in (5.3) holds for many links, including 2-bridge links, alternating fibred links and alternating pretzel links [22]. However, there is an alternating link for which the equality is false [22]. The simplest example is the alternating link depicted in Fig. 5.2.

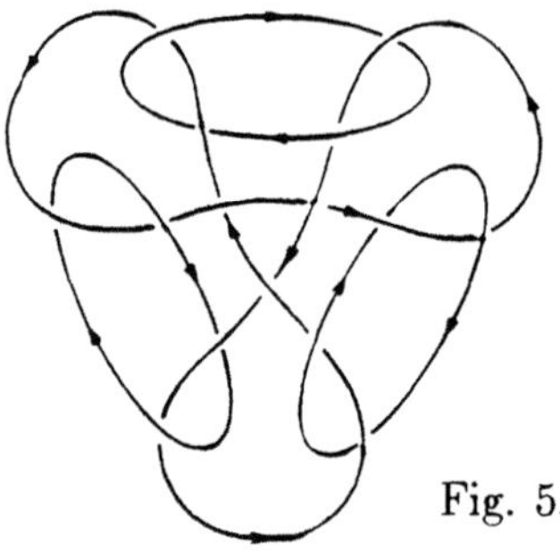

Fig. 5.2

A calculation by a computer shows that $v-\text{span}\, P_L(v,z) = 8$, but $\mathbf{b}(L)$ is indeed 6.

5.2. Seifert graphs

Now to prove equality in (5.3) or to establish some relationship between v-span $P_L(v,z)$ and $\mathbf{b}(L)$, we use the second type of graph called the Seifert graph of a link diagram.

Let L be a link and D a diagram of L. If we split D at each crossing along the orientation of D (Fig. 5.3), D is decomposed into finitely many circles on S^2, called *Seifert circles.*

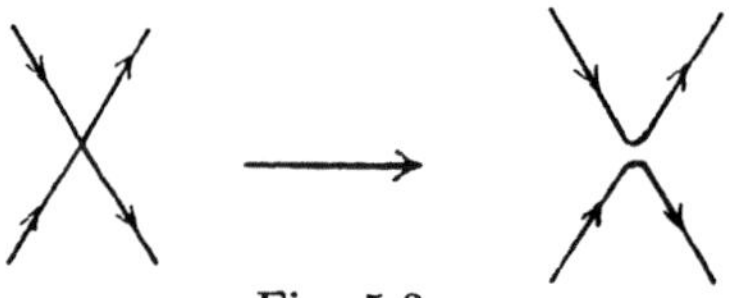

Fig. 5.3

Let $s(D)$ denote the number of Seifert circles in D and $c(D)$ the number of crossings in D. The *Seifert graph* Γ_D (associated with D) is a graph with $s(D)$ vertices $v_1, \cdots, v_{s(D)}$ and $c(D)$ edges $e_1, \cdots, e_{c(D)}$. Each vertex corresponds to a Seifert circle and each edge corresponds to a crossing. Two distinct vertices v_i and v_j are connected by e_k iff two Seifert circles S_i and S_j (corresponding to v_i and v_j) are joined by a crossing c_k (corresponding to e_k). Furthermore, each edge is given the same sign as that of the corresponding crossing in D. (See Fig. 3.3) Therefore, the Seifert graph is a signed (plane) graph. A Seifert graph has no loops. In contrast to the graphs discussed in §2.1, a Seifert graph does not represent a link type. In other words, the exact same graph may represent different links. In fact, the figure eight knot (Fig. 5.4 (a)) and a 3-component

Let L_+, L_- and L_0 be links which are identical except in the neighborhood of a crossing, where they look like

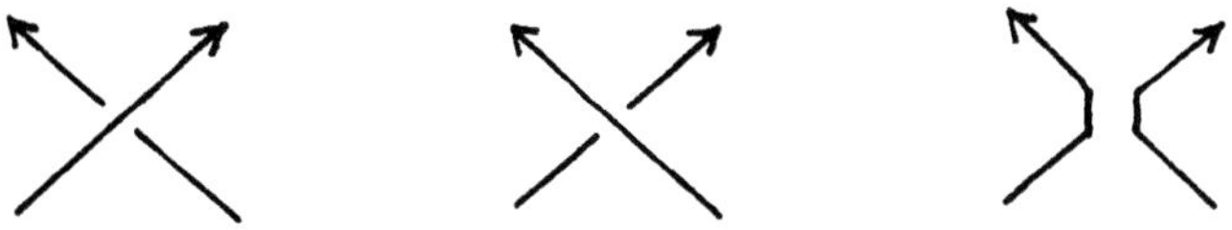

L_+ $\qquad$ L_- $\qquad$ L_0

Fig. 5.1

Then $P_L(v,z)$ satisfies the following formula

$$\frac{1}{v}P_{L_+}(v,z) - vP_{L_-}(v,z) = zP_{L_0}(v,z)\ . \tag{5.1}$$

If L is a trivial knot, then

$$P_L(v,z) = 1\ . \tag{5.2}$$

The integer polynomial uniquely defined by (5.1) and (5.2) will be called the *skein polynomial* of a link L. The skein polynomial is a generalization of the Jones polynomial, since $V_L(x) = P_L(x, x^{\frac{1}{2}} - x^{-\frac{1}{2}})$.

Example 5.2.

P  $(v,z) = (v^{-1} - v)z^{-1}$

Since $P_L(v,z)$ involves two variables v and z, we can define the v-span $P_L(v,z)$ and the z-span $P_L(v,z)$. As is suggested by Theorem 5.3 below, however, v-span $P_L(v,z)$ is more interesting and important.

Theorem 5.3. *[9, 17] For any link L,*

$$v\text{-span } P_L(v,z) \leq 2(\mathbf{b}(L) - 1). \tag{5.3}$$

Surprisingly, equality holds in (5.3) for many links, although inequality is sharp for some knots. One of the earliest conjecture on $\mathbf{b}(L)$ was the following

Conjecture 5.4. [9] If L is the closure of a positive braid, then

$$v\text{-span } P_L(v,z) = 2(\mathbf{b}(L) - 1), \tag{5.4}$$

where a positive braid is a braid in which every Artin's generator σ_i appears with a non-negative exponent.

Obviously, if G is a positive or negative graph, then $\Gamma_0(G) = \Gamma_u(G)$ and moreover, $\Gamma_0(G)$ is homeomorphic to G as topological spaces, and conversely.

Definition 4.14. A signed graph G is said to be *semi-alternating* if (1) $\gamma(G) \geq 2$ and (2) both $\Gamma_0(G)$ and $\Gamma_u(G)$ are connected and non-separable. A link diagram D is said to be semi-alternating if the signed graph G_D associated with D or its dual G_D^* is semi-alternating. A link L is semi-alternating if L admits a semi-alternating diagram.

Example 4.15.
(1) A reduced alternating diagram is semi-alternating.
(2) Non reduced diagrams are never semi-alternating.
(3) A standard torus knot diagram need not be semi-alternating.
(4) A standard pretzel link diagram need not be semi-alternating.

Terasaka proved that a semi-alternating link is not trivial. It is not known whether a semi-alternating diagram is adequate, but the following conjecture seems plausible.

Conjecture 4.15. A semi-alternating diagram D is a minimal diagram of the link represented by D.

Finally we mention one interesting question that remains unsolved

Conjecture 4.16. $c(K_1 \# K_2) = c(K_1) + c(K_2)$.

5. The Braid index

The theory of braids has played a fundamental role in the discovery of the new polynomials [13]. The original Jones' theory depends on the fact that every oriented link is represented as a closure of an n-braid for some n [1]. The minimum number of strings n needed to represent L as a closed n-braid, is called the *braid index* of L, denoted by $\mathbf{b}(L)$. Obviously, the braid index is a knot invariant.

Problem 5.1. Given a link, determine $\mathbf{b}(L)$.

This problem is also very difficult, and before the new polynomial was discovered, it was almost impossible to evaluate $\mathbf{b}(L)$ for L, except for a few limited types of links. For example, it is shown in [24] that if $K(p,q)$ is a torus link of type (p,q), $p,q \geq 1$, then $\mathbf{b}(K(p,q)) = \min\{p,q\}$.

However the recent development of the new polynomials revealed a strong connection between these polynomials and the braid index of a link. And now the evaluation of the braid index is possible at least for certain type of links. In this chapter, we will discuss the recent progress on this problem.

5.1 Skein polynomials

We begin with a definition of a new integer polynomial $P_L(v,z)$ in variables v and z [10].

link (Fig. 5.4 (b)) have the same Seifert graph Γ (Fig. 5.4 (c)).

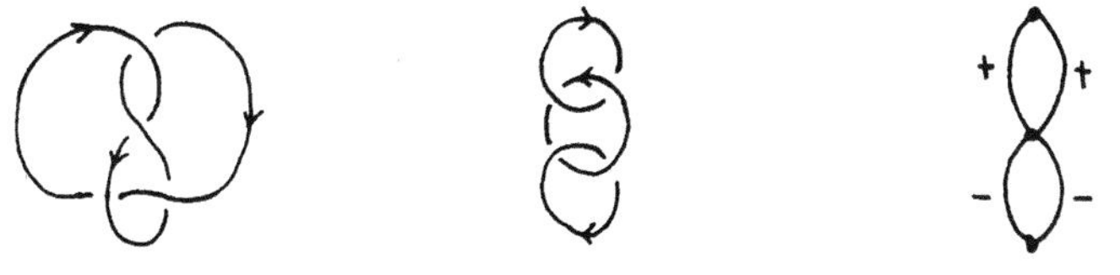

(a) (b) (c)

Fig. 5.4

If Γ_D is not separable, D is called a ***special*** diagram, and Γ_D uniquely represents a link. In this case, Γ_D is in fact the graph $G(D,\rho)$ associated with D defined in §2.1 for some shading. Therefore there is no ambiguity in recovering the original link diagram from a non-spearable Seifert graph. A Seifert graph is a spine of some orientable spanning surface of L, and hence it is ***bipartite***. Particularly interesting examples of Seifert graphs arise from ***braids***.

A Seifert graph of a closed braid is the block sum of multiple-edge graphs (see Fig. 5.5).

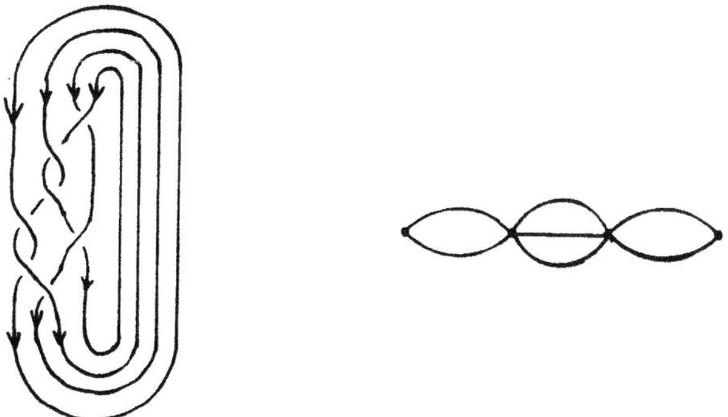

Fig. 5.5

Also, each block of the Seifert graph of an alternating link diagram is either a positive or negative graph.

As we have seen above, a Seifert graph of a closed n-braid is the block sum of $n-1$ multiple edge graphs and hence the natural diagram of a closed n-braid has exactly n Seifert circles. (See Fig. 5.5.) Therefore, any link has at least one diagram D_0 for which $s(D_0) = \mathbf{b}(L)$, and we have $min\ s(D) \leq \mathbf{b}(L)$, where the minimum is taken over all diagrams of L. In 1987, Yamada proved the reverse inequality. In fact, he proved

Theorem 5.5. *[29] For any diagram D of L, $s(D) \geq \mathbf{b}(L)$.*

This theorem suggests that for any link diagram D of L, the study of the ***surplus*** $s(D) - \mathbf{b}(L)$ would eventually lead to the determination of $\mathbf{b}(L)$. We may ask, for example, for what diagram D, is $s(D) - \mathbf{b}(L)$ equal to 0? If D is a reduced alternating diagram, $s(D) - 1$ equals the degree of the reduced Alexander polynomial of L and hence $s(D) - \mathbf{b}(L)$ is a link type invariant. Our study of the surplus $s(D) - \mathbf{b}(L)$ leads to a new invariant of a graph G, called the *index* of G, which is a topic of the next section.

5.3. Index of a graph

Let G be a graph. A family $\mathcal{F} = \{e_1, \ldots, e_k\}$ of edges of G is said to be ***independent*** if (i) all e_j $(j = 1, 2, \ldots, k)$ are singular and (ii) there is an edge e_i in $\mathcal{F}$ and a vertex v, one of the ends of e_i, such that $\{\phi(e_1), \ldots, \phi(e_{i-1}), \phi(e_{i+1}), \ldots, \phi(e_k)\}$ is an independent set of $k-1$ edges in the graph $G/star\ v$, where $\phi : G \to G/star\ v$ is the collapsing map. (In the rest of this talk, we do not distinguish between e_i and $\phi(e_i)$ unless confusion arises.) We define that the empty set of edges is independent. $ind\,(G)$ is defined to be the maximal number of independent edges in G. If G is a signed graph, then $ind_+(G)$ (respectively $ind_-(G)$) is defined to be the maximal number of independent edges $\{e_1, \ldots, e_k\}$ in G, where all $e_j (j = 1, \ldots, k)$ are positive (respectively negative) and singular in G.

It is obvious that $ind\,(G) \leq ind_+(G) + ind_-(G)$.

Example 5.6. For the graph G depicted in Fig. 5.6, $ind\,(G) = 1$, $ind_+(G) = 1$ and $ind_-(G) = 1$.

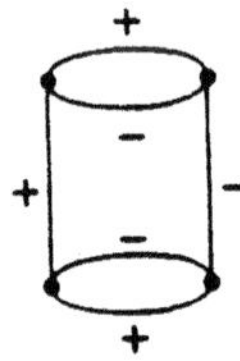

Fig. 5.6

From the definition, we have immediately the following

Proposition 5.7. If two graphs G_1 and G_2 are disjoint, then

$$ind\,(G_1 \cup G_2) = ind\ G_1 + ind\ G_2,$$

$$ind_+(G_1 \cup G_2) = ind_+\ (G_1) + ind_+\ (G_2), \text{ and}$$
$$ind_-(G_1 \cup G_2) = ind_-\ (G_1) + ind_-\ (G_2).$$

One of the important properties of the index is the following theorem.

Theorem 5.8. *Let G be a connected bipartite graph. If G consists of blocks $G_1, G_2, \ldots, G_k$ then*
(1) $ind\ G = ind\,(G_1) + \cdots + ind\,(G_k)$.
Furthermore, if G is a signed graph, then
(2) $ind_+(G) = \sum_{i=1}^{k} ind_+(G_i)$ and $ind_-(G) = \sum_{i=1}^{k} ind_-(G_i)$.

The original proof of this theorem has been simplified considerably after P. Traczyk proved our conjecture on bipartite graphs. Therefore, we will postpone the proof to the next section.

5.4. Cycle index

Since the determination of the index of a graph is by no means straightforward, we define a *cycle index* of a graph G, as the first approximation of $ind\ G$. Usually, the determination of the cycle index is much easier than that of the index and therefore, it provides a quite effective method to determine the index of a graph.

Definition 5.9. Let $S = \{e_1, \ldots, e_n\}$ be a set of n distinct edges in a graph G.

(1) S is said to be *cyclically independent* if no k edges in $S (1 \leq k \leq n)$ occur on a simple cycle of length at most $2k$. Otherwise S is called *cyclically dependent.*

(2) The cycle index of G, denoted by $\alpha(G)$, is defined as the maximal number of cyclically independent edges of G.

In the Definition 5.11 (1), a simple cycle can be replaced by a closed trail.

Example 5.10. For a graph G depicted in Fig. 5.7, $\alpha(G) = 3$, but $indG = 2$.

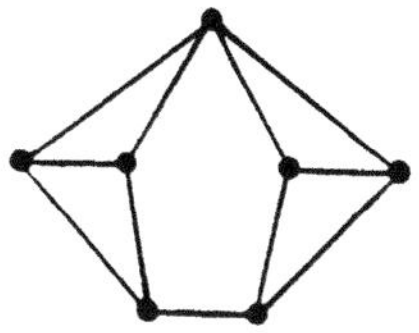

Fig. 5.7

Theorem 5.11. *For a graph G, $ind(G) \leq \alpha(G)$.*

Proof. We proceed by induction on $\alpha(G)$. If $\alpha(G) = 0$, then G has no singular edges and hence, $ind\ G = 0$.

Let $\alpha(G) = n \geq 1$ and assume that Theorem 5.11 holds for a graph H with $\alpha(H) < n$. Let $v(e)$ be an end of a singular edge e in G. First we will show that $\alpha(G/star\, v(e)) \leq n - 1$. Take a family $S = \{e_1, \ldots, e_n\}$ of n distinct edges in $G/star\, v(e)$. S gives rise to a family S' of $n+1$ edges in G by adding e to S. Since $\alpha(G) = n$, S' is cyclically dependent in G. Therefore, there are, say k, edges $e_{i_1}, \ldots, e_{i_k}$ in S' such that these edges occur on a simple cycle C in G of length at most $2k$. Let $U = \{e_{i_1}, \ldots, e_{i_k}\}$. We consider two cases.

Case (1) $e \notin U$. Then U is also a family of k edges, all of which occur on the closed trail $C/star\, v(e)$ in $G/star\, v(e)$, where $|C/star\, v(e)| \leq |C| \leq 2k$.

Case (2) $e \in U$. Then $U - \{e\}$ is a family of $k-1$ edges, all of which occur on the closed trail $C/star\, v(e)$ in $G/star\, v(e)$, where $|C/star\, v(e)| \leq |C| - 2 \leq 2k - 2$.

In either case S is cyclically dependent in $G/star\, v(e)$, and therefore, $\alpha(G/star\, v(e)) \leq n - 1$. But the inductive assumption yields $ind\ (G/star\, v(e)) \leq \alpha(G/star\, v(e)) \leq n - 1$ and hence, $ind\ (G) \leq n$. □

The proof of the following corollary is elementary.

Corollary 5.12.

(1) If $ind\ (G) \leq 1$, then $ind\ (G) = \alpha(G)$. In particular, $ind\ G = 1$ iff G has singular edges and each pair of singular edges in G occurs on a simple 3 - or 4 - cycle in G.

(2) Suppose that there are no simple 3-cycles in G. Then $ind\ G = 2$ iff $\alpha(G) = 2$.

Corollary 5.12 (2) is false if G has a 3-cycle. See the graph G in Fig. 5.7.

Very recently Traczyk proved the reverse inequality in Theorem 5.11 for bipartite graphs. He proves [27].

Theorem 5.13. *If G is a bipartite graph, then $ind(G) \geq \alpha(G)$, and hence $indG = \alpha(G)$.*

This theorem is false if G is not bipartite. See Example 5.10.

Since $\alpha(G)$ is additive w.r.t the block sum, Theorem 5.8 is an immediate consequence of Theorem 5.13. Furthermore, the index of a bipartite graph is a 2-isomorphic invariant, but it is not for non-bipartite graphs.

Example 5.14. G is 2-isomorphic to G', but $ind\, G = 4$ and $ind\, G' = 3$.

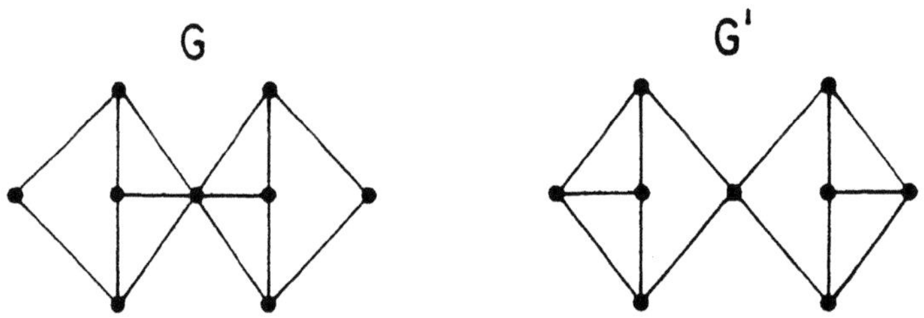

Fig. 5.8

5.5 Improvement of Morton-Frank-Williams inequalities

Now we return to link theory. Let D be a link diagram of L and Γ_D (or simply Γ) the Seifert graph associated with D. Then the index of D, $ind\, D$, is defined as the index of Γ_D. Recall that Γ_D is a block sum of bipartite graphs and hence $ind\, D$ is the sum of indices of these blocks.

As was mentioned in 5.1 the maximal and minimal v-degree of $P_L(v,z)$ have been studied in [9, 17] and the following theorem was proven which estimated their upper and lower bounds in terms of the numbers the following theorem was proven which implies (5.3).

Theorem 5.15. *For any link diagram D of a link L,*

$$(5.5)\quad w(D) - s(D) + 1 \leq min\, deg_v\, P_L(v,z) \leq max\, deg_v\, P_L(v,z) \leq w(D) + s(D) - 1.$$

However, these inequalities have been improved considerably by the following theorem.

Theorem 5.16. *For any link diagram D and the associated Seifert graph $\Gamma(D)$, we have*

$$\text{(5.6)} \qquad \begin{aligned} &max\ deg_v\ P_L(v,z) \leq w(D) + s(D) - 1 - 2\ ind_+\Gamma(D), \qquad \textit{and} \\ &min\ deg_v\ P_L(v,z) \geq w(D) - s(D) + 1 + 2\ ind_-\Gamma(D) \end{aligned}$$

and hence

$$\text{(5.7)} \qquad v\text{-}span\ P_L(v,z) \leq 2\{s(D) - 1 - ind_+\Gamma(D) - ind_-\Gamma(D)\}.$$

Now, to prove Theorem 5.16, the following lemma is crucial.

Lemma 5.17. *Given an oriented link diagram D of a link L, there are new link diagrams D', D'' and D''' of L such that*

$$\text{(5.8)} \qquad \begin{aligned} &(1) \quad w(D') = w(D) - ind_+(D) \quad \textit{and} \quad s(D') = s(D) - ind_+(D), \\ &(2) \quad w(D'') = w(D) + ind_-(D) \quad \textit{and} \quad s(D'') = s(D) - ind_-(D) \\ &(3) \quad s(D''') = s(D) - ind(D) \end{aligned}$$

and hence

$$\text{(5.9)} \qquad \mathbf{b}(L) \leq s(D) - ind(D)\ .$$

Remark 5.18. (1) It may not exist a diagram D' such that $s(D') = s(D) - (ind_+(D) + ind_-(D))$. (2) If D is an alternating diagram, then $ind\ D = ind_+D + ind_-D$ and we can choose D''' so that $w(D''') = w(D) - ind_+\Gamma(D) + ind_-\Gamma(D)$.

Proof of Lemma 5.17. A proof is to find a new diagram D' associated with the graph $\Gamma(G)/star\ v$. Suppose that there exists a singular positive edge e and a vertex v, one of two ends of e, such that $ind_+\big(\Gamma(D)/star\ v\big) = k - 1$. e corresponds to a crossing c of D.

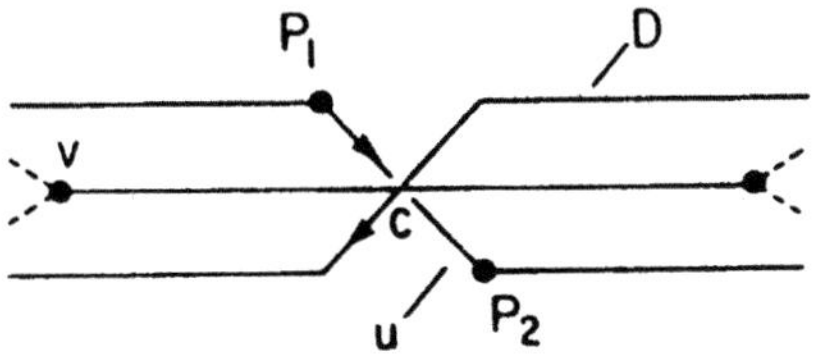

Fig. 5.9

Let u be a small part of D that crosses under the other part of D at c. Let P_1 and P_2 be the end points of u. See Fig. 5.9. We will deform isotopically the short path u to a long under-crossing path ℓ.

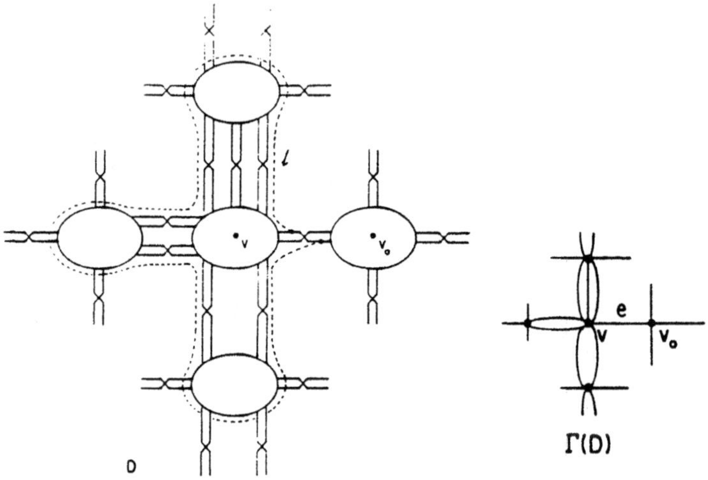

Fig. 5.10

ℓ is depicted by a dotted line in Fig. 5.10. ℓ crosses under those "bands" which are *not* connected to v. In this new diagram D', two Seifert circles represented by v and v_0 are amalgamated to one circle and hence $s(D') = s(D) - 1$.

Now we see that $\Gamma(D')$ is the one-point union of $\Gamma(D)/star\ v$ and some multiple-edge graph K, and $ind_+\Gamma(D') = ind_+\Gamma(D) - 1$. We repeat the same argument so that finally $\Gamma(D)$ is reduced to the block sum of $\Gamma(D^{(k)})$ and multiple-edge graphs where $ind_+\Gamma(D^{(k)}) = 0$.

The final link diagram $\hat{D}$ corresponding to this graph has $s(\hat{D}) = s(D) - ind_+(D)$ and $w(\hat{D}) = \tilde{w}(D) - ind_+(D)$. $\hat{D}$ is what we sought. □

We are now in position to prove Theorem 5.16. Using the diagrams D' and D'' in Lemma 5.17, we have from Theorem 5.15

$$\max \deg_v\ P_L(v,z) \leq w(D') + s(D') - 1 = w(D) + s(D) - 1 - 2\ ind_+\Gamma(D), \quad \text{and}$$
$$\min \deg_v\ P_L(v,z) \geq w(D'') - s(D'') + 1 = w(D) - s(D) + 1 + 2\ ind_-\Gamma(D)\,.$$

This proves Theorem 5.16. □

As a consequence of Theorem 5.16, we have

Corollary 5.19. *Suppose that the equalities hold in (5.6). Then if* $ind\ D = ind_+D + ind_-D$, *we have*

$$\mathbf{b}(L) = s(D) - ind\ D.$$

Proof. It follows from Theorem 5.13 that v-*span* $P_L(v,z) \leq 2\{\mathbf{b}(L)-1\}$, and hence $s(D)-1-ind\ D \leq \mathbf{b}(L)-1$, i.e. $s(D)-ind\ D \leq \mathbf{b}(L)$. However, Lemma 5.17 shows that there is a diagram D''' of L such that $s(D''')=s(D)-ind\ D$, and hence, $s(D''') \leq \mathbf{b}(L)$. Since $\mathbf{b}(L) \leq s(\tilde{D})$ for any diagram $\tilde{D}$ of L, it follows that $\mathbf{b}(L) = s(D''') = s(D)-ind\ D$. □

Theorem 5.20. *Let L be an alternating link. If the equalities hold in (5.6), then we have*

$$\mathbf{b}(L) = s(D) - ind\ D.$$

Proof. Because $ind\,D = ind_+ D + ind_- D$.

5.6. Determination of the braid index

Using Corollary 5.19, we can determine the braid index of many links, particularly, alternating links.

Theorem 5.21. *For the links listed in (5.10), the braid index is determined by the formula:* $\mathbf{b}(L) = s(D) - indD$,

(5.10)(1) *L is an alternating link that has a reduced alternating diagram D of index 0. In particular, L is an alternating fibred link.*

(2) *L is a (not necessarily alternating) link that has a special diagram D with at most one singular edge.*

(3) *L is a 2-bridge link. (For the precise definition, see the next section.)*

(4) *L is an alternating pretzel links.*

(5) *L is an alternating link such that the leading coefficient of the Alexander polynomial is $\pm 1, \pm 2$, or ± 3.*

We should note that the assumption L being alternating in (1), (3)-(5) cannot be dropped. (A part of Theorem 5.23 is also proved by Y. Yokota [30] using *one variable* skein polynomial $P_L(q^m, q-q^{-1})$ obtained by substitutions $v=q^m$ and $z=q-q^{-1}$.)

For a 2-bridge link, it is possible to give a precise formula that evaluates its braid index.

Theorem 5.22. *Let L be a 2-bridge link of type (α,β), where $0<\beta<\alpha$ and β is odd. Let $[2n_{1,1}, 2n_{1,2}, \ldots, 2n_{1,k_1} - 2n_{2,1}, \ldots, -2n_{2,k_2}, \ldots, (-1)^{t-1}2n_{t,1}, \ldots, (-1)^{t-1}2n_{t,k_t}]$ be a continued fraction form of $\frac{\alpha}{\alpha-\beta}$ where $n_{i,j}>0$ for all i,j. Then*

$$\mathbf{b}(L) \ = \ \sum_{i=1}^{t}\sum_{j=1}^{k_i}(n_{ij}-1)+t+1\ .$$

Example 5.23. K is a 2-bridge knot of type (241, 183). (See Fig. 5.11.) The continued fraction form of $\frac{241}{58}$ is $[4,-6,2,-4]$ and hence $\mathbf{b}(K) = 1+2+1+5 = 9$. On the other hand, $s(D)=13$ and $ind\,D = 4$ and $s(D)-ind\,D = 9$.

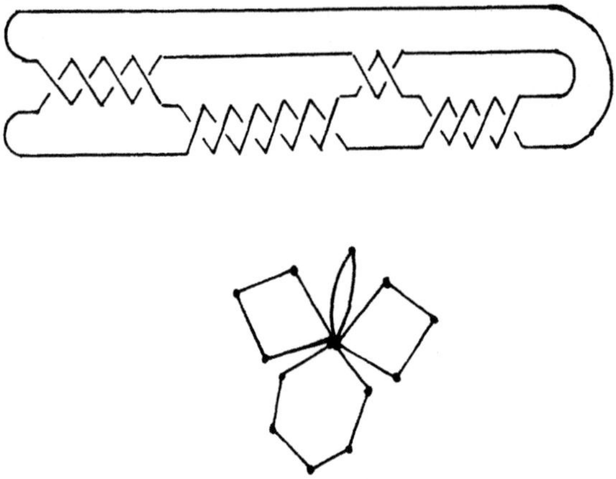

Fig. 5.11

At present, if equality in (5.6) does not hold for a link L, the determination of $\mathbf{b}(L)$ is still a hard problem. However, a recent result announced in [16] shows that both $s(D)$ and $ind\,D$ are invariants for a reduced alternating diagram D, and so is $s(D) - ind\,D$. We conjecture that this number gives the braid index for an alternating link.

Conjecture 5.24. If L is an alternating link and D is a reduced alternating diagram of L, then $\mathbf{b}(L) = s(D) - ind\,D$.

This conjecture is true for for the link depicted in Fig. 5.2. Finally, we note that the additivity of $\mathbf{b}(L)$ w.r.t # is proved in [3].

5.7. Minimal crossing number (II)

Rather surprisingly, some results obtained in the previous sections will be used to determine the minimal crossing number $c(L)$ of some non-alternating links L.

Proposition 5.25. *Let L be an oriented link. Suppose that L has a diagram D such that*

$$(5.11) \qquad \begin{aligned} &(1) \quad s(D) = \mathbf{b}(L) \\ &(2) \quad \max \deg_z P(v,z) = c(D) - s(D) + 1\,. \end{aligned}$$

Then $c(L) - c(D)$.

Proof. Let D' be a link diagram of L. Since $\max\deg_z P(v,z)$ is bounded above by $n(D') - s(D') + 1$, we see from (5.11) (2) that

$$c(D') - s(D') + 1 \geq c(D) - s(D) + 1\,.$$

Since $\mathbf{b}(L) \leq s(D')$, it follows from (5.11) (1) that $s(D) \leq s(D')$ and hence $c(D') - c(D) \geq s(D') - s(D) \geq 0$, i.e. $c(D') \geq c(D)$. Therefore $c(L) = c(D)$. □

Let L be a torus link of type (p,q), $p > q > 0$. Represent L as a closed q-braid $\beta = (\sigma_1\sigma_2 \dots \sigma_{q-1})^p$ and let D be the standard diagram of L. Then D satisfies (5.11) and hence we have

Theorem 5.26. *Let L be a torus link of type (p,q). Then*

$$c(L) = \min \{|p|(|q|-1), |q|(|p|-1)\} .$$

6. Bridge number

6.1. Problem

The concept of bridges of a link L was introduced by H. Schubert in 1954. An (unoriented) link L is said to have an n-bridge representation if $L \cap S^2$ consists of exactly $2n$ simple points $P_1, P_2, \dots, P_{2n}$ that divide L into $2n$ simple arcs $A_1, A_2, \dots, A_{2n}$ in such a way that (i) $A_i \cap A_{i+1} = P_i$, $i = 1, 2, \dots, 2n$, where $A_{2n+1} = A_1$, (ii) n arcs $A_1, A_3, \dots, A_{2n-1}$ are in the upper half space $S^3_+ = \{(x,y,z) \mid x \geq 0\} \cup \{\infty\}$ while $A_2, A_4, \dots, A_{2n}$ are in the lower half space $S^3_- = \{(x,y,z) \mid x \leq 0\} \cup \{\infty\}$, and (iii) each arc A_i is unknotted, i.e. A_i and an arc B_i in S^2 span a disk D_i in S^3_+ or S^3_-, and (iv) $D_i \cap D_j = \emptyset$ whenever $i \equiv j$ (mod 2). (See Fig. 6.1.)

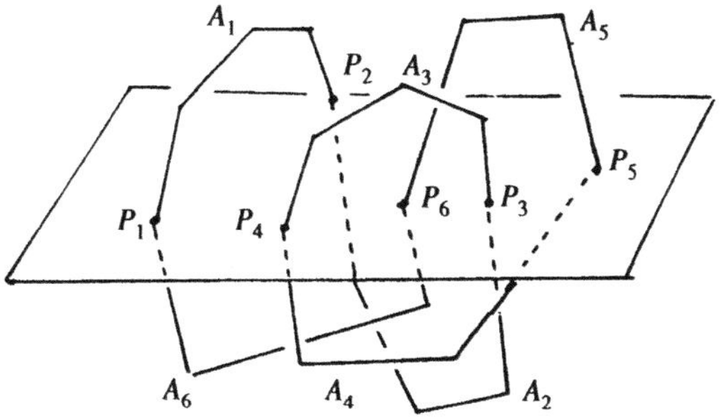

Fig. 6.1

Since each arc A_{2k-1}, $k = 1, 2, \dots, n$, has at least one local maximal point, an n-bridge representation of L has at least n local maxima. A link L has the *bridge number* (or *index*) n, $bg(L) = n$, if L has an n-bridge representation, but cannot have a fewer bridge representation. An unknotted knot has the bridge index 1 and conversely. The bridge index of a μ-component link is at least μ. For $\mu \geq 2$, there are many non-trivial μ-component links L with $bg(L) = \mu$.

The bridge number, $bg(D)$, of a link diagram D is the number of local maximal points on D. From the definition we have

Proposition 6.1. $bg(K) = \min_D bg(D)$, *where the minimum is taken over all diagrams.*

It is well-known that 2-bridge links are alternating.

Problem 6.2. Given a knot L, determine $bg(L)$.

There are no algorithms with which one can compute $bg(L)$, but the following proposition is one of the earliest results due to Schubert [24].

Proposition 6.3. *The bridge number of a torus link of type* (p,q) *is* $\min\{|p|, |q|\}$.

It is easily seen that $bg(L)$ has some connections with the number of Wirtinger generators of the link group of L.

Let $G(L) = \pi_1(S^3 - L)$ be the group of a link L, and let $\langle x_1, x_2, \ldots, x_m : r_1, r_2, \ldots, r_m\rangle$ be a Wirtinger presentation of $G(L)$ associated with a link diagram D. Each generator x_i, called a *Wirtinger generator*, represented by a meridian of L. For example, since a 2-bridge link L has a diagram with two local maxima, two Wirtinger generators x_1 and x_2 suffice to generate $G(L)$. In general, if L has an n-bridge representation, the group $G(L)$ is generated by n Wirtinger generators and therefore, we have

Proposition 6.4. *Let* $G(L)$ *be the group of a link* L. *Let* m *be the minimum number of Wirtinger generators that generates* $G(L)$. *Then we have*

$$m \leq bg(K). \tag{6.1}$$

It is not known whether equality holds in (6.1) for any link L. Obviously $m = bg(L)$ if $bg(L) = 1$ or 2. If a link L is represented as a closed n-braid, then its diagram has n bridges and hence we have immediately

Proposition 6.5. $bg(K) \leq \mathrm{b}(K)$.

The following additivity of $bg(K)$ is proved in [24].

Theorem 6.6.

$$bg(K_1 \# K_2) - 1 = \sum_{i=1}^{2} \{bg(K_i) - 1\}.$$

6.2. The rank of a graph

In this section, we will introduce a new concept called the *rank* of a graph which leads another definition of $bg(L)$.

Let G be a (finite) unsigned graph. Let W_0 be a collection of edges of G, i.e. $W_0 \subseteq E(G)$. An edge e is called a *consequence* of W_0 if (1) there is an end of e, say v_e, such that all edges but e in *star* v_e belong to W_0, or (2) there is a simple cycle C including e such that all edges but e in C belong to W_0.

Let $\tau(W_0)$ be the set of all consequences of W_0. $\tau(W_0)$ is a subset of $E(G)$. Define $W_1 = W_0 \cup \tau(W_0) \subset E(G)$, and then inductively, define $W_{k+1} = W_k \cup \tau(W_k)$, $k = 0, 1, 2, \ldots$. Thus we obtain an ascending sequence:

$$W_0 \subset W_1 \subset W_2 \subset \cdots \subset W_k \subset \ldots \subset E(G).$$

If, for some ℓ, $W_\ell = E(G)$, then W_0 is said to generate G and W_0 is called a set of generators of G. Since $E(G)$ itself generates G, every (finite) graph has at least one set of generators.

Definition 6.7. The rank of G, $r(G)$, is defined as $\min_W \{|W|\}$, where the minimum is taken over all sets of generators W of G.

Example 6.8.
(1) If $E(G) = \phi$ then $r(G) = 0$.
(2) If G is a tree, then $r(G)$ is at most the number of stumps minus one.
(3) For the graph G in Fig. 6.2, $r(G) = 3$, since $\{e_1, e_2, e_3\}$ generates G, but no two-edge set generates G.

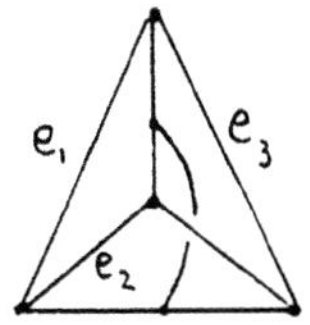

Fig. 6.2

From the definition we have immediately

Proposition 6.9.
(1) If $G_1 \cap G_2 = \phi$, then $r(G_1 \cap G_2) = r(G_1) + r(G_2)$
(2) $r(G_1 * G_2) \leq r(G_1) + r(G_2) + 1$
(3) If G^* is the dual of a plane graph G, then $r(G^*) = r(G)$.

If the graph G is associated with an n-bridge representation of a connected L, then $r(G_D) + 1 \leq bg(L)$ and we propose the following

Conjecture 6.10.

(6.2) $$bg(L) - 1 = \min_{G_D} r(G_D),$$

where the minimum is taken over all diagrams.

6.3. The Fox Conjecture

As is seen from many examples, two numerical invariants $bg(L)$ and $\mathbf{b}(L)$ are much smaller than $c(L)$. How small are they? In 1950, R.H. Fox suggested the following inequality [8]. For a knot K,

(6.3) $$3(bg(K) - 1) \leq c(K).$$

Difficulties of proving the statement like (6.3) are an involvement of the minimal crossing number $c(K)$. However, these difficulties are now eliminated at least for alternating knots or links. Since (6.3) is false for a link, we modify (6.3) as follows.

Conjecture 6.11. If L is a μ-component link with λ split components, then

(6.4) $$3(bg(L) - \lambda) \leq c(L) + \mu - \lambda.$$

Equality holds in (6.4) if L is a connected sum of trefoil knots and Hopf links.

This conjecture still remains unsolved, but we can prove (6.4) for the following type of links.

A link L is called *algebraic* if L has a minimal diagram D such that the graph associated with D or its dual is *reducible*. (A graph G is said to be *reducible* if G has a vertex v_0 such that $G - v_0$ is a tree.) Note that there is a plane graph that is reducible, but its dual is not. A 2-bridge link is algebraic.

Theorem 6.12. *If L is algebraic, then conjecture 6.11 holds.*

Proof. It suffices to prove (6.4) for $\lambda = 1$. Let G be a reducible (unsigned) graph of a minimal diagram D. For a reducible graph G, it is shown [19] that $bg(L) - 1 \leq r(G)$, and hence it is enough to prove

$$3r(G) \leq c(L) + \mu - 1 , \tag{6.5}$$

where μ is the number of components of L. Since $c(L) = c(D) = |E(G)|$ and $\mu =$ 2-null (G), the mod 2 nullity of the matrix of G, it follows that (6.5) is equivalent to

$$3r(G) \leq |E(G)| + \text{2-null}\,(G) - 1 . \tag{6.6}$$

To prove (6.6), first consider the simplest case G_0 where T and G_0 are of the form depicted in Fig. 6.3, i.e. $T = star\ v_{k+1}$.

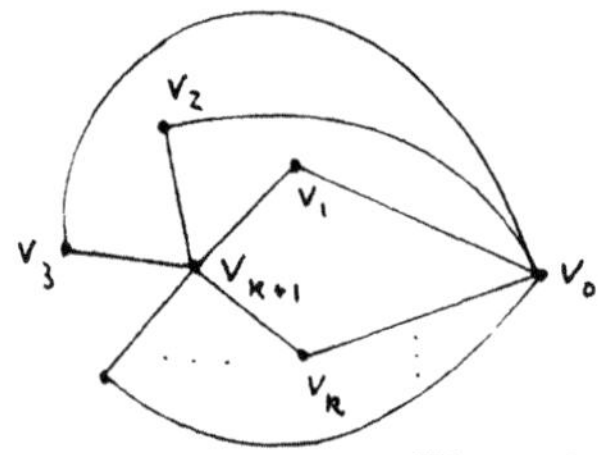

Fig. 6.3

Then $r(G_0) \leq k - 1$ and $|E(G)| = 2k$, while 2-null $(G_0) = k$, and hence (6.6) holds. Now G is obtained from G_0 first by expanding T by inserting edges at (non-stump) vertices, (see Fig. 6.4)

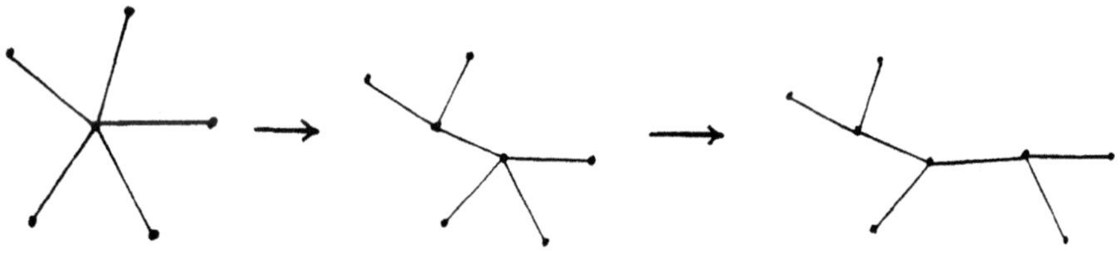

Fig. 6.4

and then by adding edges connecting v_0 and vertices on T. During these extensions, $r(G)$ remains unchanged. However, if we add the edge to a graph, 2-null (G) may increase or decrease by at most one, and hence, $|2\text{-null}(G) - 2\text{-null}(G_0)| \leq |E(G)| - |E(G_0)|$. Now $3r(G) = 3r(G_0) \leq |E(G_0)| + 2\text{-null}(G_0) - 1 \leq |E(G)| + 2\text{-null}(G) - 1$. □

In conjecture 6.11, $br(K)$ cannot be replaced by $\mathbf{b}(K)$, because the figure eight knot K has the braid index 3 but $c(K) = 4$. Recently, however, the following theorem is proven.

Theorem 6.13. *[23] For any knot K,*

$$2(\mathbf{b}(K) - 1) \leq c(K).$$

References

[1] J.W. Alexander, A lemma on systems of knotted curves, Proc. Nat. Acad. Sci., U.S.A., 9 (1923) 93–95.

[2] C. Bankwitz, Über die Torsionzahlen der altermierenden Knoten, *Math. Ann.* **103** (1930) 145–161.

[3] J. Birman-W. Menasco, Studying links via closed braids IV: composite and split links, *Invent. Math.* **102** (1990 115–139.

[4] G. Burde-H. Zieschang, *Knots*, de Gruyter (1985).

[5] R.H. Crowell, Non-alternating links, Ill. *J. Math.* **3** (1959) 101–120.

[6] C. Ernst - D.W. Sumners, The growth of the number of prime knots, *Math. Proc. Cambridge Phil. Soc.* **102** (1987) 303–315.

[7] T. Fiedler, On the degree of the Jones polynomial, *Topology* **30** (1991), 1-8.

[8] R.H. Fox, On the total curvature of some tame knots, *Ann. of Math.* **52** (1950) 258–260.

[9] J. Frank-R.F. Williams, Braids and the Jones polynomial, *Trans. Amer. Math. Soc.* **303** (1987) 97–108.

[10] P. Freyd, et al., A new polynomial invariant of knots and links, *Bull. Amer. Math. Soc.* **12** (1985) 103–111.

[11] C. McA Gordon - R.A. Litherland, On the signature of a link, *Invent. Math.* 47 (1978) 53–69

[12] C. McA. Gordon-J. Luecke, Knots are determined by their complements, *J. Amer. Math. Soc.* **3** (1989) 371–415.

[13] V. Jones, Hecke algebra representations of braid groups and link polynomials, *Ann. of Math.* **126** (1987) 335–388.

[14] L.H. Kauffman, State models and the Jones polynomial, *Topology* **26** (1987) 395–407.

[15] W.B.R. Lickorish-M. Thistlethwaite, Some links with non-trivial polynomials and their crossing-numbers, *Comment. Math. Helv.* **63** (1988) 527–539.

[16] W. Menasco-M.B. Thistlethwaite, The Tait flyping conjecture, *Bull. Amer. Math. Soc.* **25** (1991) 403–412.

[17] H.R. Morton, Seifert circles and knot polynomials, *Math. Proc. Cambridge Phil. Soc.* **99** (1986) 107–109.

[18] H.R. Morton-H.B. Short, The 2-variable polynomial of cable knots, *Math. Proc. Cambridge Phil. Soc.* **101** (1987), 267–278.

[19] K. Murasugi, On estimate of the bridge index of links, *Kobe J. Math.* **5** (1988) 75–86.

[20] —, On invariants of graphs with applications to knot theory, *Trans. Amer. Math. Soc.* **314** (1989) 1–49.

[21] —, On the braid index of alternating links, *Trans. Amer. Math. Soc.* **326** (1991) 237–260.

[22] K. Murasugi-J.H.Przytycki, The index of a graph with applications to knot theory. To appear in Memoirs, Amer. Math. Soc.

[23] Y. Ohyama, On the minimal crossing number and the braid index of links.

[24] H. Schubert, Über eine numerische Knoten invariante, *Math. Z.* **61** (1954) 245–288.

[25] H. Terasaka, On the non-triviality of some kinds of knots, *Osaka Math. J.* **12** (1960) 113–144.

[26] M. Thistlethwaite, A spanning tree expansion of the Jones polynomial, *Topology* **26** (1987) 287–309.

[27] P. Traczyk, On the index of graphs: Index versus cycle index, (preprint).

[28] H. Whitney, 2-isomorphic graphs, *Amer. J. Math.* **55** (1933) 236–244.

[29] S. Yamada, The minimal number of Seifert circles equals the braid index of a link, *Inv. math.* **89** (1987) 347–356.

[30] Y. Yokota, On the degree of the two-variable Jones polynomial of certain links.

THE QUEST FOR A KNOT WITH TRIVIAL JONES POLYNOMIAL: DIAGRAM SURGERY AND THE TEMPERLEY-LIEB ALGEBRA

DALE ROLFSEN
Mathematics Department,
The University of British Columbia,
Vancouver, B.C., V6T 1Z2
Canada

ABSTRACT. This article reviews several methods of altering knot and link diagrams without changing the Jones polynomial of the underlying link. The technique, which may be called diagram surgery or generalized mutation, involves removing a part of the diagram and replacing it in an altered form. In general, the resulting knot or link is different from the original. An important possible application of this technique would be to find a nontrivial knot with trivial Jones polynomial. Our point of view involves skein theory and the Temperley-Lieb algebra, and underlines the utility of these ideas.

1 Introduction

Since V. Jones introduced his new polynomial invariant of knots, about eight years ago, nobody has answered the following very basic question: Is there a nontrivial knot with trivial Jones polynomial?

I'll take this as the main motivation for discussing various methods of producing pairs of knots (or links) which have the same Jones polynomial. They are all generalizations of Conway's concept of "mutation." By analogy with gene splicing, you take part of a picture (= diagram) of a link, remove it and replace that part in a different way. Under certain circumstances, which we'll study, the new diagram represents a link which has the same Jones polynomial.

My hope is that, with patience or cleverness, one of you will find an example which gives an answer of "yes" to that question, by finding a knot whose appropriate generalized mutant is unknotted.

Some of the ideas I'll discuss also apply to the Alexander polynomial, the so-called HOMFLY polynomial, the Kauffman polynomial, etc. For simplicity, I'll concentrate on the Jones polynomial. You can check the literature, or figure out for yourself, which methods also apply to the other polynomials. Of course, it has been known for a long time that there are plenty of knots with Alexander polynomial equal to 1; J. H. C. Whitehead gave a general method (doubling) to produce such examples. Another pair of examples will be mentioned shortly.

M.E. Bozhüyük (ed.), Topics in Knot Theory, 195–210.

Another thing I'd like to emphasize is the beauty of the ideas of skein theory. When I first learned this idea, also due to Conway, of turning the set of tangles (parts of knot diagrams) into a vector space, I thought it was a pretty wierd bit of abstract nonsense. However, the idea has proven to be extremely useful. The very well-developed ideas of linear algebra can be used to great advantage, and this algebraization of the geometry actually enables us to make explicit calculations, which in turn have concrete geometric consequences. Since vector spaces have bases, and linear transformations and pairings correspond to matrices, one can use skein theory to reduce certain questions to a simple check of a finite number of things. This is the heart of several of our arguments. The 3-manifold invariants discussed by H. Morton in this meeting give an even more convincing illustration of the power of skein theory.

2 The Jones polynomial and the Kauffmann bracket

By now, you must be familiar with this, so a quick review should suffice. Kauffman's beautiful and elementary construction of the Jones polynomial is through the bracket $\langle D \rangle$ of a planar diagram of a link. I refer you to Kauffman's book [K] for further details. Consider the ring $Z[A^{\pm 1}]$ of (Laurent) polynomials in a variable A. Let D be a diagram of a link, which as usual is a curve or set of curves in the plane with only double-point self-crossings, and an indication at each such crossing of which strand is to be the underpass in the third dimension. As usual, we indicate this by a little gap. The bracket $\langle D \rangle$ is defined by the axioms:

Axiom 1: 

Axiom 2: $\langle D \amalg O \rangle = \delta \langle D \rangle$ where $\delta = -A^2 - A^{-2}$

together with the stipulation that the bracket of the empty diagram equals 1. These axioms insure that $\langle D \rangle$ is invariant, not only under ambient isotopy of the plane, but also the Reidemeister moves of type II and III, which (by definition) generate the relation of regular isotopy of diagrams.

The Reidemeister moves:

I 

II 

III 

An easy calculation shows that 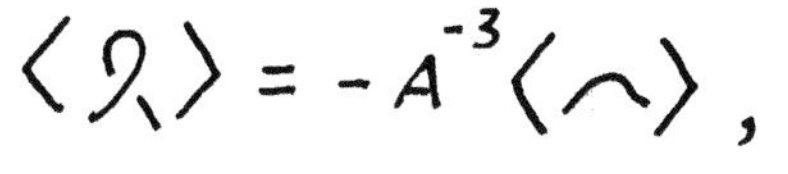

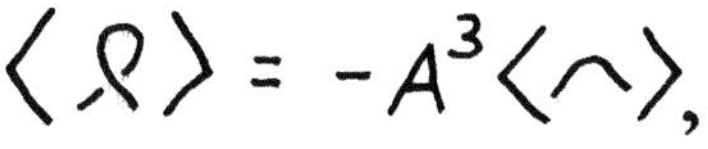

so we get an invariant of all three Reidemeister moves (and hence an ambient isotopy invariant of the link L that D represents) by *orienting* the curves and calculating the writhe $w(D)$ as the algebraic sum of the crossings, counting signs ϵ according to the convention

$\epsilon = +1$ $\qquad\qquad$ $\epsilon = -1$

and then defining $f_L(A) = (-A^{-3})^{-w(D)}\langle D\rangle$, the normalized bracket invariant. (From now on we will drop the distinction between a link L and a diagram D representing it.) Kauffmann showed that this is the same as the Jones polynomial $V_L(t)$, up to a change of variable $t = A^{-4}$ and a factor of δ. That is,

$$(-t^{1/2} - t^{-1/2})V_L(t) = f_L(t^{-1/4})$$

A key observation is that if a knot or (oriented) link diagram is modified in such a way that its bracket *and* its writhe are unchanged, then its Jones polynomial is also unchanged.

3 Skein theory, mutants and the Temperley-Lieb algebra

There are various skein theories for classical knots and links, corresponding to the Conway polynomial (the original version), the HOMFLY plynomial and other invariants which can be defined by "skein relations" such as the axioms defining the bracket. Moreover, the idea can be generalized to skein theory of 3-manifolds (maybe also higher dimensions) and bears a strong similarity with the topological quantum field theories which are currently being developed. Again for simplicity, I'll only discuss the skein theory corresponding to the Kauffman bracket, and planar diagrams of classical links.

A *room* R is a region of the plane (the boundary, assumed polygonal, may be empty or disconnected), together with an even number of marked points on the boundary. An *inhabitant* is a diagram in R (part of a link diagram) whose boundary is precisely the set of marked points. Two inhabitants of the room R are called *equivalent* if they are related by an isotopy of R fixed on the boundary and a finite number of Reidemeister moves of type II or III, i. e., regular isotopy within R. Let $\mathcal{M}(R)$ denote the free $Z[A^{\pm 1}]$-module generated by all equivalence classes of inhabitants of R. Define the skein module $\mathcal{S}(R)$ to be the quotient of $\mathcal{M}(R)$ modulo all equations among inhabitants, of the type stated in Axioms 1 and 2 of the previous section. Only now we imagine the brackets to be erased, and the equations asserting a relation among the inhabitants themselves, as "vectors" in $\mathcal{M}(R)$. That is, $\mathcal{S}(R) = \mathcal{M}(R)/\mathcal{I}(R)$, where $\mathcal{I}(R)$ is the 2-sided ideal generated by all elements which are differences between the left-hand and right-hand side of an equation given in Axiom 1 or 2. Axiom 1, as usual, involves inhabitants which are identical except in a neighbourhood of the crossing depicted. We interpret Axiom 2 as applying only if the unknotted curve O bounds a disk in R.

Examples: 1. If R is the entire plane, then $\mathcal{S}(R)$ is one-dimensional, with basis the empty link $\emptyset$. Any link L can be expressed $L = \langle L \rangle \emptyset$.

2. Similarly, if R_2 is a disk with 2 marked points, $\mathcal{S}(R_2)$ has a basis consisting of an arc in R connecting the two points.

3. If R_4 = the disk, with 4 marked points, then $\mathcal{S}(R_4)$ is the free module with basis consisting of the inhabitants:

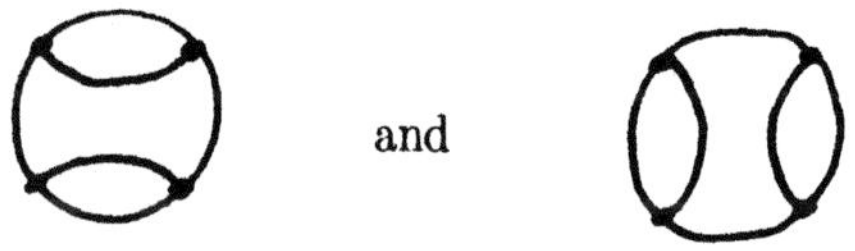

4. For R_6 = disk with 6 marked points, $\mathcal{S}(R_6)$ is 5-dimensional with basis

1 $\quad$ e_1 $\quad$ e_2 $\quad$ e_1e_2 $\quad$ e_2e_1

5. The skein theory of the disk R_{2n} with $2n$ marked points has basis consisting of all (equivalence classes of) inhabitants with no crossings, and its dimension is the Catalan number $C_n = 2n(2n-1)\cdots(n+2)/n!$

6. If R is an annulus with no marked points, then $\mathcal{S}(R)$ has a countably infinite basis, the k-th basis element consisting of k parallel copies of disjoint curves which go around the hole, $k = 0, 1, 2, \ldots$

Problem: Prove the formula for C_n = dimension of $\mathcal{S}(R_{2n})$ as follows. Consider $C_0 = 1$ and observe that $C_n = C_0C_{n-1} + C_1C_{n-2} + \cdots + C_{n-1}C_0$. Define a (formal) power series $f(x) = \sum_{i=0}^{\infty} C_i x^i$, and argue that $f(x) - 1 = xf(x)^2$, solve for f and deduce the form of its coefficients.

Sometimes it will be convenient to consider A to be a fixed nonzero complex number. If we do that, then $\mathcal{S}(R)$ may be defined as the *complex vector space* formed by taking all formal complex linear combinations of equivalence classes of inhabitants, modulo the relations given in Axioms 1 and 2. For "generic" A the dimensions of the above examples, as complex vector spaces, are as stated. However, there are exceptions: for example if A

is a fourth root of -1, then $\delta = 0$ and so the skein vector space of the plane becomes zero dimensional.

Consider a room R and its complementary room R', so that $R \cup R'$ is the whole plane and $R \cap R' = \partial R = \partial R'$ and the marked points on the boundary of the rooms are the same. Then if D and D' are inhabitants of R and R', respectively, $D \cup D'$ is a link diagram. By extending linearly to the skein modules (or vector spaces), this construction defines a bilinear pairing:

$$\mathcal{S}(R) \times \mathcal{S}(R') \to \mathcal{S}(\text{plane}) = \text{scalars}$$

Again consider a room R and a function F taking inhabitants of R to inhabitants of R. Suppose the function has the property that whenever inhabitants of R satisfy a skein relation, then also their images under F satisfy the same relation. Then this induces a linear transformation $F : \mathcal{S}(R) \to \mathcal{S}(R)$.

An example of the above is Conway's *mutation*. Let R_4 be a disk symmetric under 180 degree rotations in either the x-axis, y-axis, or z-axis (which we visualize in the usual way as pointing respectively to the right, upwards, or pointing out of the page towards us); moreover, suppose the four marked points are also chosen to be setwise invariant under these rotations. If F denotes the operation of rotating an inhabitant in any one of these three senses (with crossings changed under the x- or y-rotation, as if the diagram were three-dimensional) we get a linear transformation $F : \mathcal{S}(R) \to \mathcal{S}(R)$.

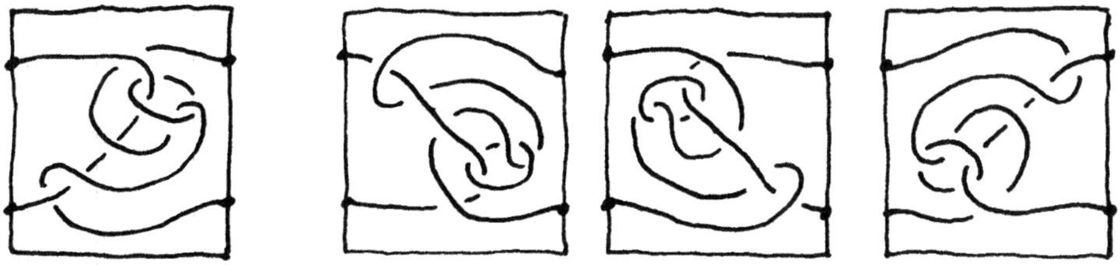

Rotating a tangle: about x-axis y-axis z-axis.

Since the two basis elements are invariant under each of these three rotations, we conclude that:

Proposition 1 *If $F : \mathcal{S}(R_4) \to \mathcal{S}(R_4)$ is induced by one of the three rotations described above, then F is the identity transformation.*

Mutation of a knot or link L consists of locating a room R_4 in a diagram for L, so that $T = L \cap R_4$ is an inhabitant of R_4 (also called a *tangle*), and replacing T in the diagram by the tangle $F(T)$, to form L'. L and L' are called *mutants*.

Example: A well-known pair of mutants are the Kinoshita-Terasaka knot and Conway's

11 crossing knot, both of which have Alexander polynomial equal to one:

The Conway knot

The Kinoshita-Terasaka knot

Their common Jones polynomial is $V(t) = t^{-6} - 2t^{-5} + 2t^{-4} - 2t^{-3} + t^{-2} + 2t - 2t^2 + 2t^3 - t^4$.

Because of proposition 1, and the fact that (when orientations are unchanged outside the room) the writhe does not change under mutation, we conclude the following.

Proposition 2 *Mutant links have the same Jones polynomial.*

As Conway observed, mutants also have the same Conway, or Alexander, polynomial as well. They also have equal HOMFLY and Kauffman polynomials, and the proof is essentially the same.

Alas, this result is useless for our strategy of altering an unknot to get a knot with trivial Jones polynomial, because of the following "folk" theorem.

Proposition 3 *Any mutant of an unknot is itself unknotted.*

One way to see this involves considering the knot K to be in S^3, and letting M_K^3 be the two-fold branched covering of S^3 branched over K. The preimage of a thickened room R_4 enclosing a tangle of K is a solid torus upstairs in M_K. Mutation from K to K' can be lifted to a surgery on M_K, but one can check that the surgery is really a trivial surgery. Therefore, M_K and M'_K are homeomorphic if K and K' are mutants. Proposition 3 follows from this observation, together with the fact (the $Z/2$ Smith conjecture proved by Waldhausen) that K is unknotted if and only if M_K is S^3.

Questions: What happens to $\mathcal{S}(R_4)$ under a *reflection* in an appropriate plane, instead of a rotation? If we performed such an operation — a "reflective" mutation — would the Jones polynomial be invariant?

One can define a multiplicative structure on $\mathcal{S}(R_{2n})$ by the following operation:

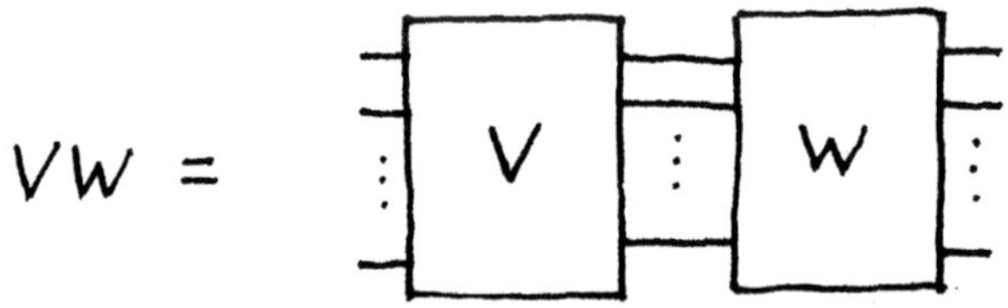

Clearly the element $\mathbf{1} =$ 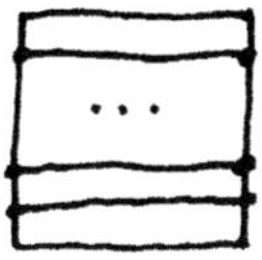 is an identity for this product.

Then, for any fixed nonzero complex value of A, we have that the complex vector space $\mathcal{S}(R_{2n})$ is an ***algebra***, which is called the nth Temperley-Lieb algebra, $\mathcal{TL}_n$. It's actually a family of algebras depending on the parameter A as well as the positive integer n. This version of the Temperley-Lieb algebras is due to Kauffmann (see [K]) and many of their fascinating properties are being discussed by H. Morton at this workshop. The multiplication in $\mathcal{TL}_1$ corresponds to connected sum of knots (and multiplication of their bracket invariants) and is commutative. You can also easily verify the following.

Proposition 4 *Multiplication in $\mathcal{TL}_2$ is also commutative.*

Problem: Show that $\mathcal{TL}_n, n \geq 3$ is a non-commutative algebra.

There are inclusions $\mathcal{TL}_n \subset \mathcal{TL}_{n+1}$, by adjoining an extra horizontal strand at (say) the top.

As an algebra, $\mathcal{TL}_n$ is generated by $\mathbf{1}$ and the $n-1$ elements:

$\mathbf{e}_1 =$  $\mathbf{e}_2 =$  $\cdots$ $\mathbf{e}_{n-1} =$ 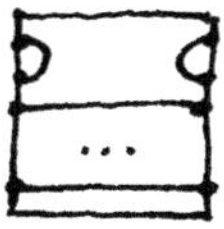

They satisfy the relations:

$$\begin{aligned} \mathbf{e}_i^2 &= \delta \mathbf{e}_i \\ \mathbf{e}_i \mathbf{e}_{i\pm1} \mathbf{e}_i &= \mathbf{e}_i \\ \mathbf{e}_i \mathbf{e}_j &= \mathbf{e}_j \mathbf{e}_i, \qquad |i-j| > 1 \end{aligned}$$

It can be shown that these relations in fact define $\mathcal{TL}_n$ abstractly as an algebra (with parameter $\delta = -A^2 - A^{-2}$).

Problem: Verify the following equations in $\mathcal{TL}_2$:

 $= A^2\mathbf{1} + (1 - A^{-4})\mathbf{e}_1$

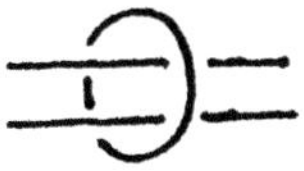 $= (-A^6 - A^{-6})\mathbf{1} + (1 - A^{-4})(1 - A^4)\mathbf{e}_1$

Calculate:

 and 

as sums of products of the generators in $\mathcal{TL}_2$ and $\mathcal{TL}_3$.

Proposition 5 *A dense set of elements of $\mathcal{TL}_2$ have multiplicative inverses.*

Here, dense refers to the topology as a complex vector space. In fact, a typical element W of $\mathcal{TL}_2$ can be expressed $W = w_0\mathbf{1} + w_1\mathbf{e}_1$. If $X = x_0\mathbf{1} + x_1\mathbf{e}_1$, then we calculate

$$WX = w_0x_0\mathbf{1} + (w_0x_1 + w_1x_0 + \delta w_1x_1)\mathbf{e}_1.$$

We can solve $WX = 1$ by setting $x_0 = 1/w_0$ and $x_1 = -w_1/w_0(w_0 + \delta w_1)$. To do this, of course, one must avoid having $w_0 = 0$ or $w_0 + \delta w_1 = 0$, but those are just lines in complex 2-space.

This proposition really needs the use of coefficients in a field, rather than the ring $Z[A^{\pm 1}]$, which has very few invertible elements. We can also see that, assuming $w_0 \neq 0$, the invertibility of W is assured for all but a finite number of values of the parameter A. Note also that if W is represented by an actual tangle, its inverse will probably not be a tangle, but rather will be a formal linear combination of tangles. Nevertheless, this algebraic device will have strictly geometric consequences, as we shall see later.

Question: Is Proposition 5 true for $\mathcal{TL}_n, n > 2$?

4 Rotants

I will now describe another method, a generalized mutation, discovered by R. Anstee, J. Przytycki and myself [APR], for constructing pairs of knots with the same Jones polynomial. Anstee is a combinatorist who showed us a trick of W. T. Tutte to produce different graphs with equal polynomial invariants — the chromatic and Tutte polynomials. We adapted Tutte's method as follows. Forgetting the Temperley-Lieb algebra for the moment, picture the room $R_{2n}, n \geq 3$ as a regular n-gon with a pair of marked points on each edge, so that the figure is symmetric under rotation by $2\pi/n$, as well as the dihedral flip. Consider an inhabitant D of this room which is also symmetric under the rotation, but *not* necessarily under the flip. Let E be *any* inhabitant of the room R'_{2n} which is complementary to R_{2n} so that $L = D \cup E$ is (a diagram of) a link in the plane. Borrowing Tutte's terminology, we call D a *rotor* and E a *stator*. Let E' denote the result of flipping over the stator, by turning it over 180 degrees (in 3-space) about a line of symmetry of R_{2n}. Let $L' = (L - E) \cup E'$ denote the resulting link; we call L' a *rotant* of L of order n. (Maybe "flippant" would be a better term, but that sounds too... well, flippant.) It is crucial to note that the choice of axis for flipping is immaterial, up to ambient isotopy, due to the symmetry of the rotor. Also note that we could flip over the rotor instead and the result would be ambient isotopic to L'.

Example 4.1:

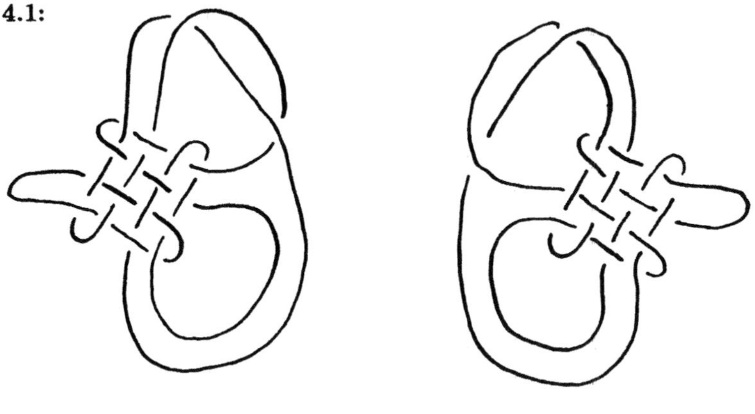

Rotants of order four

Proposition 6 *If the links L and L' are rotants of order n, $3 \leq n \leq 5$, then they have the same bracket invariant. If L and L' are oriented and have the same writhe, then their Jones polynomials agree.*

If an orientation of L orients the rotor in a rotationally invariant manner, then writhe is preserved automatically under the flip. The proof of the proposition, using skein theory, goes as follows. Fix a rotor D, and let $G : \mathcal{S}(R'_{2n}) \to \mathcal{S}(\text{plane})$ denote the linear map defined by taking an inhabitant E of R' and forming the inhabitant $G(E) = D \cup E$. Similarly define $H : \mathcal{S}(R'_{2n}) \to \mathcal{S}(\text{plane})$ by defining $H(E) = D \cup E'$, where E' is E after a dihedral flip as above. We argue that G and H are equal as linear maps, by checking on a basis of $\mathcal{S}(R'_{2n})$. Indeed, if you examine a basis for $\mathcal{S}(R'_{2n})$, you will see that every basis element has a dihedral symmetry, provided $n \leq 5$. For example, the fourteen basis elements of $\mathcal{S}(R'_8)$ are of the following six types, together with their rotated versions (the stator R'_8 is turned inside out for easier visualization):

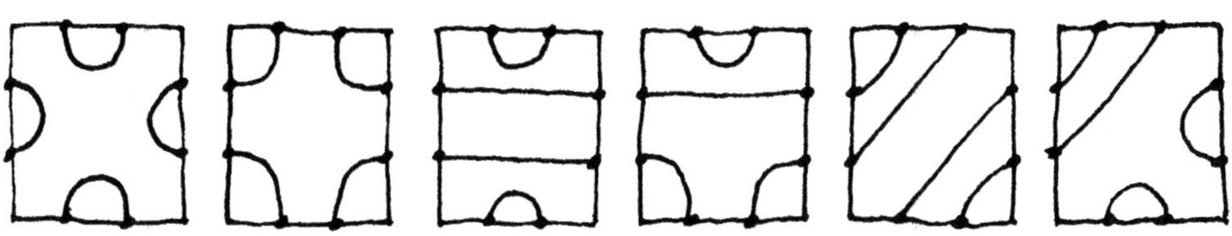

Example 4.1, continued The rotants L and L' pictured above, which are each 2-component links, have Jones polynomial $V_L(t) = V_{L'}(t) =$

$$-t^{-41/2} + 7t^{-39/2} - 26t^{-37/2} + 68t^{-35/2} - 139t^{-33/2}$$
$$+ 237t^{-31/2} - 348t^{-29/2} + 450t^{-27/2} - 518t^{-25/2}$$
$$+ 533t^{-23/2} - 494t^{-21/2} + 410t^{-19/2} - 302t^{-17/2}$$
$$+ 195t^{-15/2} - 109t^{-13/2} + 50t^{-11/2} - 19t^{-9/2} + 5t^{-7/2} - t^{-5/2}.$$

However, their Kauffmann polynomials are different, showing that one can not go from L to L' by a sequence of mutations in the sense of Conway (section 2).

Problem: Verify that all 42 basis elements of $\mathcal{S}(R'_{10})$ have an axis of symmetry. Find, on the other hand, a generator of $\mathcal{S}(R'_{12})$ which cannot be represented by a tangle with a symmetry axis, and so the above argument breaks down.

It is shown in [APR] that Proposition 6 holds for the HOMFLY polynomial if $n \leq 4$ and the Kauffmann polynomial for $n = 3$. The links of Example 4.1, and other examples presented in a paper [JR] by G. T. Jin and myself show these bounds on n are the best possible.

Problem: I don't know the answer to this one or the next — they're good questions for you graduate students to have a crack at. If L and L' are rotants of *arbitrary* order n, then do their Alexander polynomials agree?

Problem: Is a rotant of an unknot necessarily unknotted (at least for $n \leq 5$)? If so, using rotants as a strategy for producing a knot with $V(t) = 1$ is thwarted.

Here are examples of 6-rotants, from [JR], which have different Jones polynomials, although they are tantalizingly close, being:

$$t^{-1} - 8 + 42t - 168t^2 + 552t^3 - 1555t^4 + 3846t^5 - 8481t^6$$
$$+ 16863t^7 - 30459t^8 + 50275t^9 - 76164t^{10} + 106279t^{11}$$
$$- 136966t^{12} + 163352t^{13} - 180517t^{14} + 184917t^{15}$$
$$- 175495t^{16} + 154062t^{17} - 124748t^{18} + 92778t^{19}$$
$$- 63004t^{20} + 38756t^{21} - 21367t^{22} + 10408t^{23} - 4392t^{24}$$
$$+ 1561t^{25} - 448t^{26} + 97t^{27} - 14t^{28} + t^{29}$$

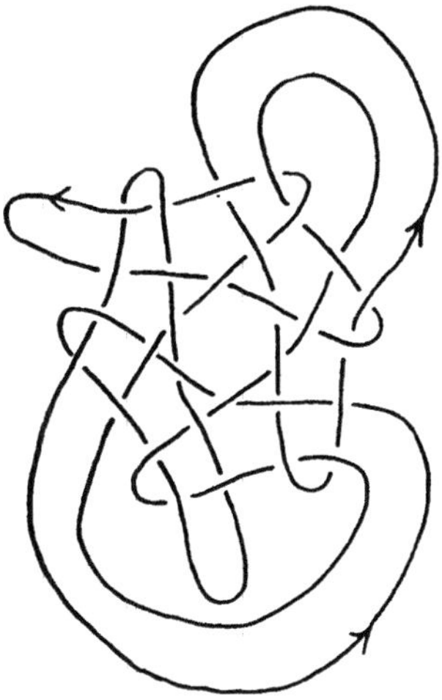

$$
\begin{aligned}
&t^{-1} - 8 + 42t - 168t^2 + 552t^3 - 1555t^4 + 3845t^5 - 8478t^6 \\
&+ 16856t^7 - 30445t^8 + 50253t^9 - 76134t^{10} + 106247t^{11} \\
&- 136939t^{12} + 163337t^{13} - 180519t^{14} + 184934t^{15} \\
&- 175524t^{16} + 154095t^{17} - 124778t^{18} + 92800t^{19} \\
&- 63016t^{20} + 38761t^{21} - 21368t^{22} + 10408t^{23} - 4392t^{24} \\
&+ 1561t^{25} - 448t^{26} + 97t^{27} - 14t^{28} + t^{29}
\end{aligned}
$$

5 Jones' trick

In a very recent paper [J], Jones described yet another generalization of mutation, and showed that this can be used to explain duplications in the Jones polynomial of several examples with quite low crossing number.

Proposition 7 *Let V and W be tangles, i.e., inhabitants of the room R_4. Then there exists an element X of $\mathcal{S}(R_4)$, such that the following diagrams are equal, as elements of $\mathcal{S}(R_6)$:*

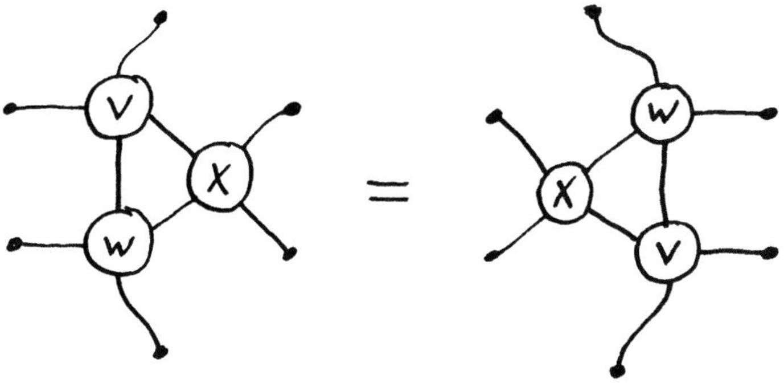

You might want to prove it yourself, at this point, although I'll show shortly that it is a special case of a more general result. Now here's the great idea. If we have tangles V and W stacked as shown below, and we want to interchange them, we *introduce* the solution X

to the above, *together with its inverse* $\overline{X}$ in $\mathcal{TL}_2$, and notice the equalities in $\mathcal{S}(R_6)$:

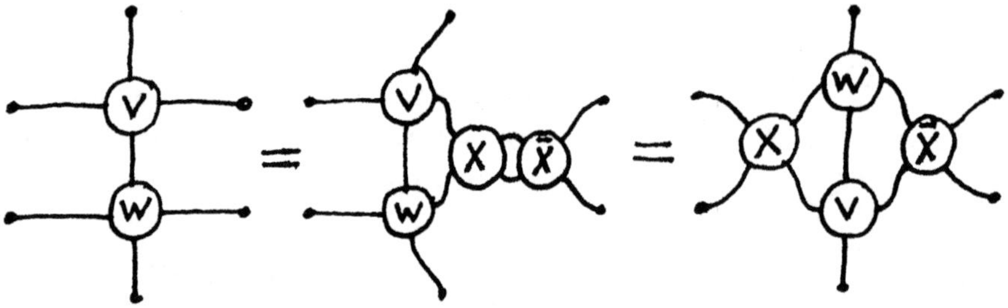

Now suppose the rest of the diagram is arranged so that X can slide around by an ambient isotopy around to the other side. Then it ends up next to $\overline{X}$ again and they annihilate, again not affecting the skein class. Also note that in its journey, X can pass through other V over W configurations, interchanging them as well, and that X can also pass through any tangle in its path, as multiplication in $\mathcal{TL}_2$ is commutative. This argument establishes the following.

Proposition 8 (Jones) *Consider a link of the form:*

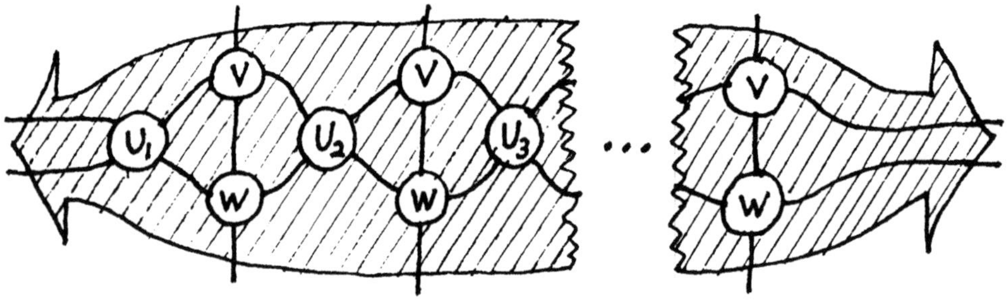

where the V, W and the U_i are arbitrary tangles. Suppose the shaded region is connected to itself, as indicate by large arrows, to form a band (which may itself be knotted) and that the other loose ends of the picture are connected in 3-space in any manner, provided that they do not pass through the band. Then one can interchange all the V tangles with all the W tangles, and the resulting link will have the same Jones polynomial as the original link.

Examples: The following pairs of knots have identical Jones polynomials:

(1) 8_8 and 10_{129}:

(2) $4_1 \sharp 4_1$ and 8_9:

The above examples are in Jones' paper [J], along with five pairs of ten crossing knots whose duplicate Jones polynomials can be explained by this phenomenon. Notice that this type of generalized mutation can change the (minimum) crossing number and turn a composite knot into a prime one.

6 Wheel balancing and symmetries in $\mathcal{TL}_3$

The ideas in this section are due to Hoste and Przytycki [P], although I am adopting a somewhat different approach than they use.

We will call an element Z of the skein of the room R *algebraically symmetric* with respect to a geometric motion f of the room, if $Z = f(Z)$ in $\mathcal{S}(R)$.

Proposition 9 (Wheel balancing) *Let Y be any inhabitant of the room R_6, then there exists T in $\mathcal{S}(R_4)$ such that the following diagram is algebraically symmetric with respect to 180 degree rotation (about the z-axis) of R_6 in the plane.*

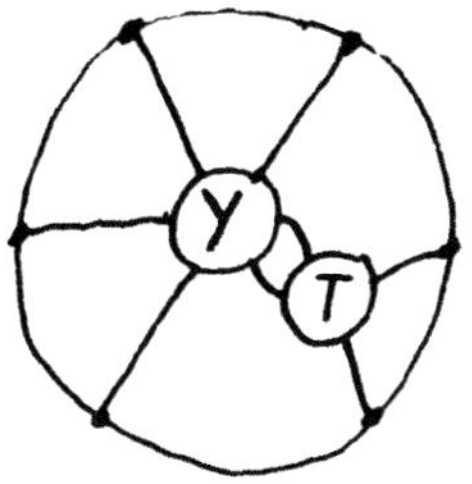

Corollary 1 *For any Y, there exists T such that:*

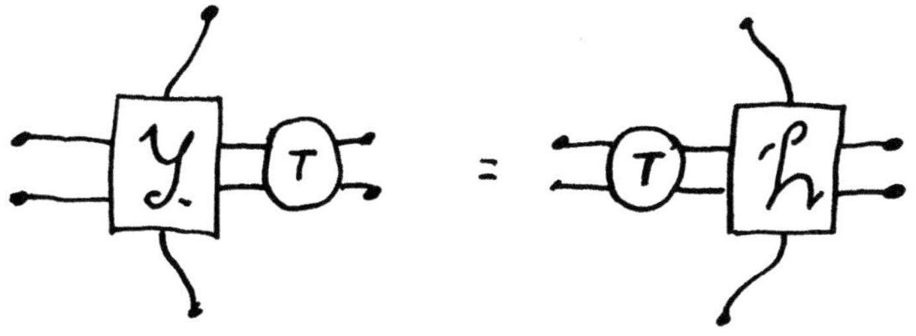

Proof: Athough the T gets turned over in the rotation, it can right itself by mutation.

You can see that Proposition 7 follows from 9 in the same way.

Let's prove proposition 9 using the Temperley-Lieb algebra. Although one could deal entirely in pictures, the algebra assists in calculation. Express an element Z of $\mathcal{TL}_3$ in terms of the standard basis:

$$Z = z_0\mathbf{1} + z_1\mathbf{e}_1 + z_2\mathbf{e}_2 + z_{12}\mathbf{e}_1\mathbf{e}_2 + z_{21}\mathbf{e}_2\mathbf{e}_1.$$

Since three of the basis elements, $\mathbf{1}, \mathbf{e}_1\mathbf{e}_2, \mathbf{e}_2\mathbf{e}_1$, are each symmetric under rotation, while $\mathbf{e}_1$ and $\mathbf{e}_2$ get interchanged, we see that Z is algebraically symmetric if and only if $z_1 = z_2$. Now write $Y = y_0\mathbf{1} + y_1\mathbf{e}_1 + y_2\mathbf{e}_1 + y_{12}\mathbf{e}_1\mathbf{e}_2 + y_{21}\mathbf{e}_2\mathbf{e}_1$ and $T = t_0\mathbf{1} + t_1\mathbf{e}_1$. If you then work out the multiplication table of $Z = YT$, you see that

$$z_1 = y_1t_0 + y_0t_1 + \delta y_1t_1 + y_{12}t_1$$

and

$$z_2 = y_2t_0.$$

Therefore Z is symmetric iff $(y_2 - y_1)t_0 = (y_0 + \delta y_1 + y_{12})t_1$. Clearly, given Y, we always have the solution

$$t_0 = y_0 + \delta y_1 + y_{12}, \qquad t_1 = y_2 - y_1$$

and all solutions (t_0, t_1) are scalar multiples of this one.

Problem: Verify that Proposition 9 also is true for rotation in the y-axis, and that the solution T is, in general different than for z-rotation. Show, on the other hand, solutions may not exist for x-rotation. Notice that the product of a y-flip and a z-flip is an x-flip.

The π rotation of an inhabitant V about the x- y- or z-coordinate axes induces three involutions in the Temperley-Lieb algebras, which we might call the x- y- and z-transpose V^x, V^y, V^z. Each is a linear isomorphism and x-transposition is an algebra automorphism as well, whereas y- or z-transposition (as with transpose of matrices) is antimultiplicative. In this terminology, we see that any element of $\mathcal{TL}_3$ can be y- or z-symmetrized by multiplication (on either side) by a suitable element of $\mathcal{TL}_2 \subset \mathcal{TL}_3$. Similar results hold more generally, which is work still in progress. Notice that *all* the algebra generators of $\mathcal{TL}_n$ are y-symmetric; in general, the y-transpose of a sum of products of the $\mathbf{e}_i$ is formed by reversing the order of the products.

Proposition 10 *Consider link diagrams as follows, where $Y, V_1, V_2, \ldots$ are arbitrary and the shading denotes a closed (possibly knotted and twisted) band. The other free ends are connected in any manner by curves which are disjoint from the band:*

Type 1:

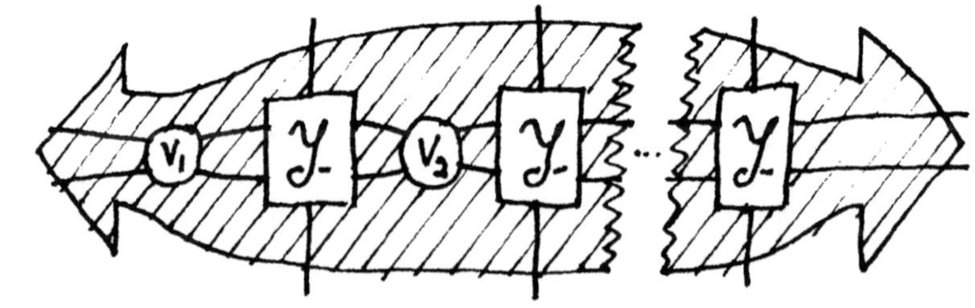

Type 2:

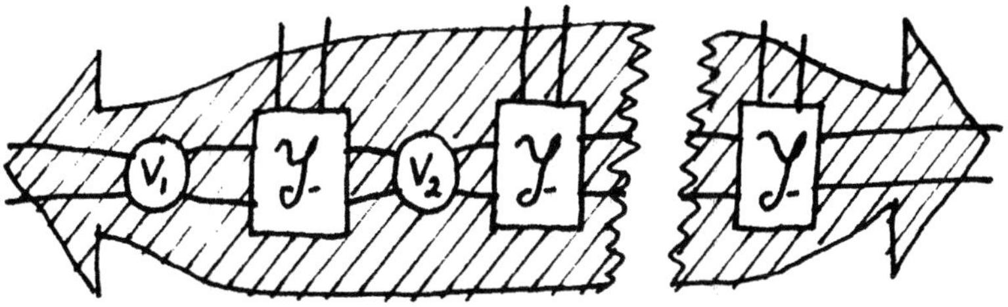

Then the Jones polynomial of the resulting link is the same as the original if one replaces all pictured occurences of Y as follows

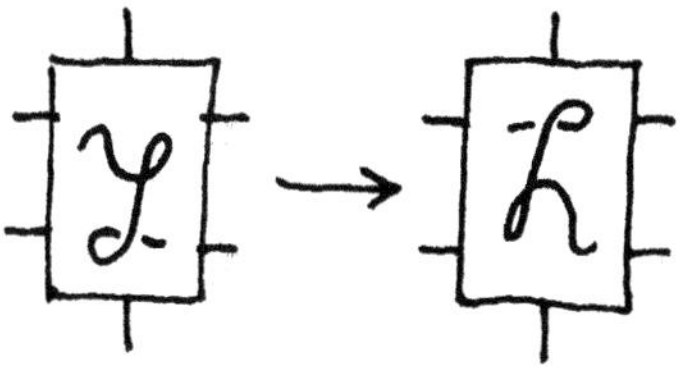 *for type 1, or* 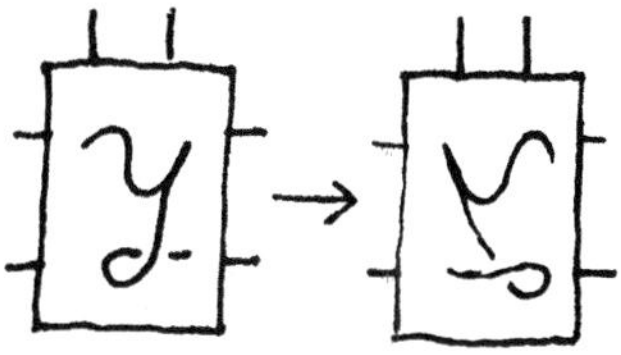 *for type 2.*

The proof is to use the Jones trick, introducing cancelling tangles T and $\overline{T}$ in the band and sending the T once around, flipping the Y's as it passes them, and finally annihilating with $\overline{T}$. The more skeptical among you might have noticed that this doesn't work if T happens to be noninvertible: t_0 might be 0. But it does work for a dense subset of Y in $\mathcal{S}(R_6)$, and that's enough, by the following reasoning. We are really trying to establish that two linear mappings:

$$F : \mathcal{S}(R_6) \to \mathcal{S}(\text{plane})$$

and

$$G : \mathcal{S}(R_6) \to \mathcal{S}(\text{plane})$$

are equal, where $G(Y) = F(Y\text{flipped})$. But this follows since we have demonstrated they agree on a dense set in the domain. Another triumph of abstract nonsense!

There are further tricks to be done, such as replacing some of the Y tangles in the original link by another Y' such that $Y'T$ and YT are simultaneously symmetric, illustrating the principle: generalized mutants are bountiful!

I'll close this lecture with a challenge. Get a nice long piece of string, fasten the ends together to make an unknot, and then try to lay it down on a table to make a complicated diagram of the unknot, of a type described in, say, Proposition 10, and with the further property that when you perform the appropriate generalized mutation, the resulting curve is really tied in a knot. It will be the desired knot with $V(t) = 1$ and will make you famous, at least among topologists!

7 References:

[APR] R. Anstee, J. Przytycki, D. Rolfsen, *Knot polynomials and generalized mutation*, Topology and its Applications 32 (1989), 237–249.

[J] V. F. R. Jones, *Commuting transfer matrices and link polynomials*, International Journal of Math. 3 (1992), 205–212.

[JR] G. T. Jin, D. Rolfsen, *Some remarks on rotors in link theory*, Canadian Math. Bull. 34 (1991), 480–484.

[K] L. Kauffman, *Knots and Physics*, Series on Knots and Everything, vol. 1, World Scientific, 1991.

[P] J. Przytycki, *Manuscript of lectures delivered at University of Tennessee*, Spring 1992.

TWISTED TOPOLOGICAL INVARIANTS ASSOCIATED WITH REPRESENTATIONS.

BOJU JIANG and SHICHENG WANG
Department of Mathematics
Peking University
Beijing 100871
China

ABSTRACT. The purpose of this note is to set up a framework for twisting the classical topological invariants via a matrix representation of the fundamental group, and to show how it works for two well known invariants — the Alexander polynomial and the Lefschetz number. As examples of knots with the same Alexander polynomial but different twisted Alexander polynomial have already been given by Lin, we supply some maps with zero Lefschetz number but non-zero twisted Lefschetz number.

0. Introduction.

The purpose of this note is to set up a framework for twisting the classical topological invariants via a matrix representation of the fundamental group, and to show how it works for two well known invariants — the Alexander polynomial and the Lefschetz number.

The Alexander polynomial $A(K)$ for a knot K in S^3 was introduced by Alexander (1928). Reidemeister (1934) introduced the Reidemeister torsion invariant for manifolds. Milnor [M1] noticed a relation between the Alexander polynomial and a certain Reidemeister torsion. A systematic study of this relation is carried out by Turaev [T].

The Lefschetz number $L(f)$ for a self-map f of a manifold M was introduced by Lefschetz (1923). Weil (1949) (cf. [B]) introduced the zeta function $\zeta(f) = \exp\sum_n L(f^n)t^n/n$ and proved that $\zeta(f) = \prod_q p_q(t)^{(-1)^{q+1}}$, where each $p_q(s)$ is a polynomial closely related to the characteristic polynomial of the linear transformation $f_{*q} : H_q(M, \mathbf{Q}) \to H_q(M, \mathbf{Q})$. Milnor [M2] noticed a relation between $\zeta(f)$ and the Reidemeister torsion of the mapping torus T_f of f. Generalizing in this direction, Fried [F] introduced a twisted Lefschetz zeta function.

Turaev and Fried define their invariants via abelian coverings, because the determinant is not well defined for square matrices in a non-abelian group ring.

Some attempts have been made to obtain stronger invariants by considering non-abelian coverings. In an unpublished note [J3] of 1987, for a selfmap $f : M \to M$ the first author defined the twisted Lefschetz zeta function $\zeta_\rho(f)$ associated with a representation $\rho : \pi_1(T_f) \to \mathrm{GL}_l(R)$ where T_f is the mapping torus of f and R is a commutative ring. In a preprint [L] of 1990, for a knot K in S^3 Lin defined the twisted Alexander polynomial $A(K, \rho)$ associated with a representation $\rho : \pi_1(S^3 \setminus K) \to \mathrm{GL}_l(R)$. Two examples of knots in S^3 were given with the same classical Alexander polynomial but different twisted Alexander polynomials. Lin's definition of $A(K, \rho)$ was based on the special fact that there is a

M.E. Bozhüyük (ed.), Topics in Knot Theory, 211–227.

Seifert surface cutting the knot complement into a handlebody, so Lin asked the question of how to generalize his definition to links in S^3 or to knots in homology spheres. We shall see his invariant fits into our framework.

The content of the paper is as below. In Section 1 we define the elementary factors associated with a representation. In Section 2 we define the Alexander invariant. In Section 3, we define the twisted Alexander polynomial for aspherical 3-manifolds which are closed or with boundary a union of tori. We show that in the case of knot complement in S^3, Lin's twisted Alexander polynomial coincides with that of ours. In Section 4, which is developed from a part of [J3], we study the twisted Lefschetz number $L_\rho(f)$ and Lefschetz zeta function $\zeta_\rho(f)$. In Section 5, we give examples of maps with $L(f) = 0$ but $L_\rho(f) \neq 0$ for some simple representation ρ.

The early version of this paper was prepared in the spring of 1991 in Beijing. It is revised and improved under the push of The Conference "Topic in knot theory", 1-12, Sep. 1992, in Erzrum and its Proceedings. We would like to thank the organizers of the Conference.

Both authors are partially supported by NSFC of China. The second author is also partially supported by Alexander von Humboldt Foundation. He would like to thank Prof. G. Burde and Prof. H. Zieschang for hospitality.

Section 1. Algebra.

In this section we discuss modules over a ring (associative but not necessarily commutative) with unity. The facts will be stated without proof. A good reference is §III.3 of [Z].

Let Λ be an associative ring with unity. Let

$$A = \langle g_1, \cdots, g_n \mid r_1, \cdots, r_m \rangle \tag{1.1}$$

be a finite presentation of a left Λ-module A, where $r_i = \sum_{j=1}^n b_{ij} g_j$, $b_{ij} \in \Lambda$. Two types of elementary transformations of presentations are defined:

(T_1) Add a relation r_{m+1} which is a consequence of $r_1, \cdots, r_m$.

(T_2) Add a generator g_{n+1} together with a relation $g_{n+1} - w(g_1, ..., g_n) = 0$ which defines g_{n+1} as a Λ-linear combination of the old generators.

The following is the analogue of the Tietze Theorem in group theory.

Proposition 1.1. *Any two finite presentations of a given Λ-module A are convertible into each other via a finite sequence of T_1, T_2 and their inverses.*

Let $\mathcal{M}_{m\times n}(\Lambda)$ denote the set of m by n matrices in Λ. We define the matrix $B = (b_{ij})_{m\times n} \in \mathcal{M}_{m\times n}(\Lambda)$ to be the *presentation matrix* for A corresponding to the above presentation.

We define three elementary operations on a matrix B:

(1) Add a new row which is a Λ-linear combination of rows of B.

(2) Replace B with the matrix

$$\begin{pmatrix} B & 0 \\ * & 1 \end{pmatrix}.$$

(3) Interchange two rows or two columns.

The operations (1) and (2) correspond to the transformations T_1 and T_2 respectively. The operation (3) is introduced for changing the ordering of the generators and relations.

We say that two matrices B and B' in Λ are *equivalent*, written $B \sim B'$, if B can be transformed to B' via a finite number of operations (1), (2), (3) and their inverses. By Theorem 15 on p.119 of [Z] we know

Proposition 1.2. *Two matrices in Λ present isomorphic Λ-modules if and only if they are equivalent.*

Let R be a *commutative* ring with unity. Suppose $\rho : \Lambda \to \mathcal{M}_{l\times l}(R)$ is a representation (i.e. a ring homomorphism). Then it gives a natural map

$$\rho : \mathcal{M}_{m\times n}(\Lambda) \to \mathcal{M}_{ml\times nl}(R), \qquad B = (b_{ij}) \mapsto B^\rho = (b_{ij}^\rho). \tag{1.2}$$

In words, ρ converts matrices in Λ into block matrices in R. This map preserves matrix multiplication.

For a square matrix B in Λ, we define its *ρ-determinant* to be $\det B^\rho$. It is evident that $\det(BC)^\rho = (\det B^\rho)(\det C^\rho)$.

Suppose $B \in \mathcal{M}_{m\times n}(\Lambda)$ and k is an integer. We define the k-th elementary ideal $E_k^\rho(B)$ of B associated with ρ as follows.

If $0 < n-k \le m$, then $E_k^\rho(B)$ is the ideal in R generated by the ρ-determinants of all $n-k$ by $n-k$ submatrices of B.

If $n-k > m$ then $E_k^\rho(B) = 0$.

If $n-k \le 0$ then $E_k^\rho(B) = R$.

Proposition 1.3. *Two equivalent matrices have the same sequence of elementary ideals. Hence the elementary ideals $E_k^\rho(B)$ associated to the representation ρ are invariants of the Λ-module A, independent of its presentation matrix B.*

The proof is similar to that of Theorem 16 on p.120 of [Z].

When R is a unique factorization domain, we define the elementary factor $A_k^\rho(B)$ to be the greatest common divisor of the elements of $E_k^\rho(B)$. It is well defined up to multiplication by a unit in R. Therefore, any equality involving $A_k^\rho(B)$ should be understood up to multiplication by a unit in R.

2. Alexander invariant.

Let X be a connected finite CW-complex and v be the base point of X. For any regular covering $p : \tilde{X}, \tilde{v} \to X, v$, let $\mathfrak{D} = \pi_1(X, v)/p_*\pi_1(\tilde{X}, \tilde{v})$ be the deck transformation group of the covering, acting on $\tilde{X}$ from the left.

For a subcomplex Y of X, we shall call the left $\mathbf{Z}\mathfrak{D}$-module $H_*(\tilde{X}, p^{-1}(Y))$ to be the *Alexander invariant* of the pair (X, Y) associated with the covering p. We are only concerned with the simplest cases when Y is either empty or consists of the single base point v. $H_i(\tilde{X})$ (respectively $H_i(\tilde{X}, p^{-1}(v))$) will be called the i-th Alexander module (resp. the i-th reduced Alexander module) of X associated with the covering p.

Proposition 2.1. *For any regular covering $p : \tilde{X} \to X$ we have*

(1) $H_i(\tilde{X}) = H_i(\tilde{X}, p^{-1}(v))$ *for* $i > 1$. *If X is aspherical and p is the universal covering, then* $H_*(\tilde{X}) = H_0(\tilde{X}) = \mathbf{Z}$, *and* $H_*(\tilde{X}, p^{-1}(v)) = H_1(\tilde{X}, p^{-1}(v))$.

(2) *There is an exact sequence of $\mathbf{Z}\mathfrak{D}$-modules*

$$0 \longrightarrow H_1(\tilde{X}) \longrightarrow H_1(\tilde{X}, p^{-1}(v)) \longrightarrow \mathbf{Z}\mathfrak{D} \longrightarrow \mathbf{Z} \longrightarrow 0$$

where $\mathbf{Z}$ is regarded as a trivial $\mathbf{Z}\mathfrak{D}$-module. If p is the universal covering, then $H_1(\tilde{X}, p^{-1}(v))$ is the kernel of the projection $\mathbf{Z}\mathfrak{D} \to \mathbf{Z}$ which is determined by the group $\mathfrak{D} = \pi_1(X)$.

Proof. Use the homology exact sequence of the pair $\left(\tilde{X}, p^{-1}(v)\right)$. □

Example 2.2. Suppose X is the complement of a knot K in S^3.

(1) If $p : \tilde{X} \to X$ is the trivial covering, then $H_i(\tilde{X})$ is $\mathbf{Z}$ for $i = 0, 1$, and is 0 for $i > 1$.

(2) If $p : \tilde{X} \to X$ is the infinite cyclic covering, then $H_1(\tilde{X})$ is the Alexander invariant in the sense of [Ro] and [T]. It is one of the most important invariants of classical knot theory.

(3) If $p : \tilde{X} \to X$ is the universal covering, then $\tilde{X}$ is contractible, so $H_i(\tilde{X}) = 0$ for $i > 0$.

Suppose X is a connected finite CW-complex with base point v, and we have a presentation

$$G := \pi_1(X, v) = \langle g_1, \cdots, g_n \mid r_1, \cdots, r_m \rangle \tag{2.1}$$

where each r_i is a word in $g_1^{\pm 1}, \cdots, g_n^{\pm 1}$. Let $p : \tilde{X} \to X$ be a regular covering, K be the normal subgroup $p_*\pi_1(\tilde{X})$ in G. Let $\phi : G \to \mathfrak{D} = G/K$ be the projection onto the deck transformation group. We shall calculate a presentation matrix for the $\mathbf{Z}\mathfrak{D}$-module $H_1(\tilde{X}, p^{-1}(v))$.

Construct a finite CW-complex X' with a single 0-cell v', the 1-cells labelled $g_1, \cdots, g_n$, the 2-cells $e_1, \cdots, e_m$ such that $\partial e_i = r_i$, and no other cells. It is clear that $\pi_1(X', v') = G$, and there is a map $h : X', v' \to X, v$ such that $h_* : \pi_1(X', v') \to \pi_1(X, v)$ is the identity automorphism of G. Let $p' : \tilde{X}' \to X'$ be the corresponding regular covering with $p'_*\pi_1(\tilde{X}') = K$. Then $\tilde{X}'$ has the same deck transformation group $\mathfrak{D}$. The map h lifts to a $\mathfrak{D}$-equivariant map $\tilde{h} : \tilde{X}', p'^{-1}(v') \to \tilde{X}, p^{-1}(v)$. We get a commutative diagram of $\mathbf{Z}\mathfrak{D}$-module homomorphisms.

$$\begin{array}{ccccccccc}
0 & \longrightarrow & H_1(\tilde{X}') & \longrightarrow & H_1(\tilde{X}', p'^{-1}(v')) & \longrightarrow & \mathbf{Z}\mathfrak{D} & \longrightarrow & \mathbf{Z} \\
 & & \downarrow{\scriptstyle \tilde{h}_*} & & \downarrow{\scriptstyle \tilde{h}_*} & & \| & & \| \\
0 & \longrightarrow & H_1(\tilde{X}) & \longrightarrow & H_1(\tilde{X}, p^{-1}(v)) & \longrightarrow & \mathbf{Z}\mathfrak{D} & \longrightarrow & \mathbf{Z}
\end{array}$$

Since $\tilde{h}$ induces an isomorphism $\pi_1(\tilde{X}') \to \pi_1(\tilde{X})$, it also induces an isomorphism $H_1(\tilde{X}') \to H_1(\tilde{X})$. It then follows from the 5-Lemma that $H_1(\tilde{X}', p'^{-1}(v'))$ and $H_1(\tilde{X}, p^{-1}(v))$ are isomorphic.

To find a presentation matrix for the $\mathbf{Z}\mathfrak{D}$-module $H_1(\widetilde{X}', p'^{-1}(v))$, we examine the cellular structure of $\widetilde{X}'$.

Pick a base point $\tilde{v}' \in p'^{-1}(v')$. Then for every path w in X' starting from v' there is a unique lift $\tilde{w}$ in $\widetilde{X}'$ starting from $\tilde{v}'$. For each $\tilde{r}_i$, there is a unique lift $\tilde{e}_i$ of e_i such that $\partial\tilde{e}_i = \tilde{r}_i$. As free $\mathbf{Z}\mathfrak{D}$-modules, we have

$$C_1(\widetilde{X}', p'^{-1}(v')) = \langle \tilde{g}_1, \cdots, \tilde{g}_n \rangle,$$
$$C_2(\widetilde{X}', p'^{-1}(v')) = \langle \tilde{e}_1, \cdots, \tilde{e}_m \rangle.$$

Since every element of $C_1(\widetilde{X}', p'^{-1}(v'))$ is a (relative) cycle, we have

$$(2.2) \qquad H_1(\widetilde{X}', p'^{-1}(v')) = \langle \tilde{g}_1, \cdots, \tilde{g}_n \mid \tilde{r}_1, \cdots, \tilde{r}_m \rangle,$$

where the right hand side is the $\mathbf{Z}\mathfrak{D}$-module with generators $\{\tilde{g}_1, \cdots, \tilde{g}_n\}$ and relators $\{\tilde{r}_1, \cdots, \tilde{r}_m\}$.

Suppose w_1, w_2 are loops in X' at v'. Then the lift of the product loop $w_1 w_2$, regarded as a 1-chain in $\widetilde{X}'$, is $\widetilde{w_1 w_2} = \tilde{w}_1 + w_1^{\phi}\tilde{w}_2$, where w_1^{ϕ} is the element of $\mathfrak{D}$ represented by the loop w_1. This leads to the observation that we can express $\tilde{r}_i$ in terms of the Fox free calculus

$$(2.3) \qquad \tilde{r}_i = \sum_{j=1}^{n} \left(\frac{\partial r_i}{\partial g_j}\right)^{\phi} \tilde{g}_j, \qquad i = 1, \cdots, m.$$

For details, see Chapter 9 of [BZ]. For simplicity, the ϕ in (2.3) will often be omitted in the notation.

Thus, from (2.3) we see

Theorem 2.3. *The $\mathbf{Z}\mathfrak{D}$-module $H_1(\widetilde{X}, p^{-1}(v))$ is completely determined by the group $\pi_1(X)$ and the projection $\phi : \pi_1(X) \to \mathfrak{D}$. The Fox calculus Jacobian*

$$(2.4) \qquad J = \left(\frac{\partial r_i}{\partial g_j}\right)_{m \times n}$$

(or more precisely J^{ϕ}) is a presentation matrix for it. □

The notion of mapping torus will play an important role in this paper. Let $f : X \to X$ be a map. The *mapping torus* T_f of f is the quotient space of $X \times [0, 1]$ with $(x, 1)$ identified to $(f(x), 0)$ for every $x \in X$. If $f \simeq g : X \to X$ are homotopic, their mapping tori T_f and T_g have the same homotopy type.

Now suppose X is a connected finite CW-complex with a single 0-cell v. Suppose $f : X \to X$ is a cellular map inducing a homomorphism $f_* : G \to G$. Suppose $G := \pi_1(X, v)$ is presented as (2.1). Regard X as $X \times \{0\}$ imbedded in T_f, and take $v \in X$ also as the base point in T_f. The loop formed by $v \times I$ represents an element z in $\Gamma := \pi_1(T_f, v)$. By the van Kampen Theorem, Γ is obtained from G by adding the new generator z and adding the relations $z^{-1}gz = f_*(g)$ for all $g \in G$. Thus

$$(2.5) \qquad \Gamma = \langle g_1, \cdots, g_n, z \mid g_1 z f_*(g_1)^{-1} z^{-1}, \cdots, g_n z f_*(g_n)^{-1} z^{-1}, r_1, \cdots, r_m \rangle.$$

Let $p : \tilde{T}_f \to T_f$ be a regular covering. Apply Theorem 2.3 and note that

$$\frac{\partial(g_i z f_*(g_i)^{-1} z^{-1})}{\partial g_j} = \delta_{ij} - g_i z f_*(g_i)^{-1} \frac{\partial f_*(g_i)}{\partial g_j} = \delta_{ij} - z \frac{\partial f_*(g_i)}{\partial g_j},$$

$$\frac{\partial(g_i z f_*(g_i)^{-1} z^{-1})}{\partial z} = g_i - g_i z f_*(g_i)^{-1} z^{-1} = g_i - 1,$$

we get a presentation matrix for the module $H_1(\tilde{T}_f, p^{-1}(v))$:

$$(2.6) \qquad \begin{pmatrix} (I - zD)_{n \times n} & (g_i - 1)_{n \times 1} \\ J_{m \times n} & 0_{m \times 1} \end{pmatrix}_{(m+n)\times(n+1)}$$

where J is the matrix of (2.4), I is the identity matrix, and D is the Fox Jacobian of the homomorphism f_*

$$(2.7) \qquad D = \left(\frac{\partial f_*(g_i)}{\partial g_j} \right)_{n \times n}$$

Example 2.4. Suppose X is a bouquet of n circles. Then

$$\pi_1(X) = \langle a_1, \cdots, a_n \rangle.$$

Hence a presentation matrix for $H_1(\tilde{T}_f, p^{-1}(v))$ is

$$(2.8) \qquad ((I - zD)_{n \times n}, (a_i - 1)_{n \times 1}).$$

Example 2.5. Suppose X is an orientable closed surface of genus g. Then

$$\pi_1(X) = \langle a_1, \cdots, a_{2g} \mid r \rangle$$

where $r = \prod_{i=1}^{g} [a_{2i-1}, a_{2i}]$. Hence a presentation matrix for $H_1(\tilde{T}_f, p^{-1}(v))$ is

$$(2.9) \qquad \begin{pmatrix} (I - zD)_{2g \times 2g} & (a_i - 1)_{2g \times 1} \\ \left(\frac{\partial r}{\partial a_j} \right)_{1 \times 2g} & 0_{1 \times 1} \end{pmatrix}.$$

3. Alexander polynomial twisted.

3.1. Twisted Alexander Polynomial of 3-manifolds.

Let X be a connected finite CW-complex and v be its base point. Let $G := \pi_1(X, v)$. Suppose F is a field and $\rho_1 : G \to \mathrm{GL}_l(F)$ is a matrix representation. Suppose H is a free commutative group with a given basis $\{t_1, \cdots, t_\beta\}$ and $\rho_2 : G \to H$ is a homomorphism. Then the homomorphism $\rho := (\rho_1, \rho_2) : G \to \mathrm{GL}_l(F) \times H$ induces a representation ρ :

$\mathbf{Z}G \to \mathbf{Z}[\mathrm{GL}_l(F) \times H] = \mathcal{M}_{l\times l}(FH)$. It is well known that the group algebra FH, i.e. the algebra of Laurent polynomials $F[t_1^{\pm 1}, \cdots, t_\beta^{\pm 1}]$, is a unique factorization domain.

Suppose given a presentation

$$G = \langle g_1, \cdots, g_n \mid r_1, ..., r_m \rangle. \tag{3.1}$$

According to Theorem 2.3, the Jacobian

$$B = \left(\frac{\partial r_i}{\partial g_j}\right)_{m\times n} \tag{3.2}$$

is a presentation matrix of the $\mathbf{Z}G$-module $H_1(\widetilde{X}, p^{-1}(v))$, where $p : \widetilde{X}, \tilde{v} \to X, v$ is the universal covering. We can study the elementary factors $A_k^\rho(H_1(\widetilde{X}, p^{-1}(v)))$.

Now we assume that X is a compact 3-manifold M. The fundamental group of a closed 3-manifold M always admits a presentation of deficiency 0, i.e., with n generators and n relators (pick a Heegaard splitting, then the n generators of one handlebody will serve as the n generators and the n meridian disc of another handlebody will serve as n relators). For a compact 3-manifold M with ∂M a non-empty union of tori, the fundamental group admits a presentation of deficiency 1, i.e., with n generators and $n-1$ relators (collapse M onto a CW 2-complex with one 0-cell, since $\chi(M) = 0$, if this 2-complex has n 1-cells, it must have $n-1$ 2-cells). Actually it turns out that when the 3-manifold X is aspherical, the deficiencies of the above presentations are the deficiencies of the groups; for details, see Chapter V of [Ja]. Many well-known presentations, say, the Wirtinger presentations for the fundamental group of the complement of a link in S^3, the HNN extension presentation of the fundamental group of a surface (closed or not) bundle over the circle, satisfy the above requirement on deficiency.

We now define the twisted Alexander polynomial $A^\rho(M)$ for these 3-manifolds M.

(1) Suppose ∂M is a non-empty union of tori. For any presentation of G with deficiency 1, let $B = (b_1, \cdots, b_n)$, b_j being the j-th column of B. Let B_j be the square submatrix obtained by deleting b_j from B. Define

$$A^\rho(M) = A_1^\rho(B) = \gcd\{\det B_j^\rho \mid 1 \le j \le n\}. \tag{3.3}$$

By tradition, when M is the complement of a knot K in S^3, $A^\rho(M)$ is also denoted by $A^\rho(K)$.

There is a relation between the determinants $\det B_j^\rho$ in (3.3).

By the fundamental formula of Fox calculus, we have

$$\sum_{j=1}^{n} \frac{\partial r_i}{\partial g_j}(1 - g_j) = 1 - r_i = 0 \qquad \text{for every } i. \tag{3.4}$$

Let $I^j(a)$ be the matrix obtained from the identity matrix by changing the j-th diagonal entry to a. Let $1 \le j < k \le n$. Then we have

$$\begin{aligned} B_j I^{k-1}(1-g_k) &= (\cdots, \widehat{b}_j, \cdots, b_k(1-g_k), \cdots) \\ &= (\cdots, \widehat{b}_j, \cdots, -\sum_{i\neq k} b_i(1-g_i), \cdots) \\ &\sim (\cdots, \widehat{b}_j, \cdots, -b_j(1-g_j), \cdots) \\ &\sim (\cdots, b_j(1-g_j), ..., \widehat{b}_k, ...) \\ &= B_k I^j(1-g_j), \end{aligned}$$

where $\sim$ is the equivalence relation of §1. Note that the operation of adding a right Λ-multiple of a column to another column is the composition of a sequence of elementary operations.

We arrive at a formula very useful in computing $A_1^\rho(B)$.

Proposition 3.1. *For $1 \le j < k \le n$, we have*

$$\det B_j^\rho \det(1-g_k)^\rho = \pm \det B_k^\rho \det(1-g_j)^\rho. \qquad \square$$

(2) Suppose M is closed. For any presentation of G with deficiency 0, B is a square matrix.

We define

$$A^\rho(M) = A_0^\rho(B) = \det B^\rho \tag{3.5}$$

if $\det(1-g_i)^\rho = 0$ for all $i = 1, \cdots, n$; and define

$$A^\rho(M) = A_1^\rho(B) \tag{3.5'}$$

otherwise.

For example, if ρ trivial, then B^ρ is a presentation matrix of $\mathbf{Z}$-module $H_1(M, v)$ by Theorem 2.3. By Proposition 2.2 (2), we have $H_1(M) = H_1(M, v)$. Then we know that $\det B^\rho \neq 0$ if and only if $H_1(M)$ is finite, and $\det B^\rho$ is the order of $H_1(M)$. That is the reason for the definition (3.5).

The reason for (3.5') is the following fact

Proposition 3.2. *Suppose $G := \langle g_1, \cdots, g_n \mid r_1, \cdots, r_n \rangle$, with deficiency 0, and B is the Fox Jacobian matrix of the presentation. Then $\det B^\rho = 0$ unless all $\det(1-g_i)^\rho = 0$.*

Proof. Let $\bar{G} := G * \mathbf{Z} = \langle g_1, \cdots, g_n, g' \mid r_1, \cdots, r_n \rangle$, with deficiency 1. Now any representation $\rho : G \to GL_l(F)$ can be extended to $\bar{\rho} : \bar{G} \to GL_l(F)$ by an arbitrary assignment of $\bar{\rho}(g')$.

Then the Fox Jacobian of $\bar{G}$ is

$$\bar{B} = (\,B \quad 0\,)$$

where B is the square Jacobian matrix $\frac{\partial r_i}{\partial g_j}$. Now apply Proposition 3.1 to $\bar{B}$. We get $\det \bar{B}_{n+1}^{\bar{\rho}} = \det B^\rho$ and $\det \bar{B}_i^{\bar{\rho}} = 0$ for all $1 \le i \le n$. So Proposition 3.1 tells us that $\det B^\rho \cdot \det(1-g_i^\rho) = 0$ for all $1 \le i \le n$. Thus, $\det B^\rho = 0$ unless all $\det(1-a_i^\rho) = 0$. $\square$

Definition (3.5') generalizes the Alexander polynomial $\Delta(M)$ of a compact 3-manifold M defined by Turaev in [T], pp.126–127. $\Delta(M)$ is defined there to be the order of the $\mathbf{Z}\mathfrak{D}$-module $H_1(\tilde{M})$, where $p : \tilde{M} \to M$ is the maximum free abelian cover and $\mathfrak{D}$ is the free part of $H_1(M)$. Turaev pointed out on p.127 that his $\Delta(M)$ is the first Alexander polynomial of $\pi_1(M)$ in the sense of Fox. In our language, this means $\Delta(M) = A_1^\rho(B)$, where B is the Fox Jacobian matrix of a presentation of $\pi_1(M)$, ρ_1 is trivial, $\beta = \beta_1(M)$ and ρ_2 is the projection $\pi_1(M) \to H_1(M)/\text{torsion}$. Thus $\Delta(M) = A^\rho(M)$ according to (3.5').

3.2. On the Twisted Alexander Polynomial of Lin.

Let us roughly describe Lin's presentation of the knot group and the definition of his twisted Alexander polynomial $A(K,\rho)$.

Suppose K is a knot in S^3. Then there is a so-called regular Seifert surface S of K such that the resulting 3-manifold by cutting $X := S^3 \setminus K$ along S is a $2g$-handlebody V, here g is the genus of S. Pick a certain wedge at the base point v of $2g$ circles $a_1, \cdots, a_{2g}$ such that $\pi_1(S) = \langle a_1, \cdots, a_{2g} \rangle$. Let z be the meridian of K. Let $\pi_1(V) = \langle x_1, \cdots, x_{2g} \rangle$. Let $\alpha_1, \cdots, \alpha_{2g}$ and $\beta_1, \cdots, \beta_{2g}$ be copies of $a_1, \cdots, a_{2g}$ in V on the positive side and negative side of S respectively. Then a presentation of the knot group $G = \pi_1(X)$ is

$$G = \langle x_1, ..., x_{2g}, z \mid z\alpha_i z^{-1} = \beta_i, \text{ for } i = 1, \cdots, 2g \rangle. \tag{3.6}$$

For each representation $\rho_1 : G \to \mathrm{GL}_l(\mathbf{C})$, Lin defines the $\mathbf{C}[t, t^{-1}]$-module presented by the matrix

$$B' = \left(tz^{\rho_1} (\frac{\partial \alpha_i}{\partial x_j})^{\rho_1} - (\frac{\partial \beta_i}{\partial x_j})^{\rho_1} \right)_{2gl \times 2gl} \tag{3.7}$$

to be the twisted Alexander module and $\det B'$ to be the twisted Alexander polynomial of K associated to ρ_1.

Now we are going to verify that our $A^\rho(K)$ and Lin's $A(K,\rho)$ are essentially the same.

Take $F = \mathbf{C}$. Take $H = H_1(X)$, the infinite cyclic group $\langle t \rangle$ where t is represented by the meridian z. The x_i's are all null homologous in X because they do not intersect the Seifert surface S. Take $\rho_2 : G \to H$ to be the abelianization, then $\rho_2(z) = t$ and $\rho_2(x_i) = 1$ for all $1 \leq i \leq 2g$. Hence $\det(1 - z)^\rho = \det(1 - tz^{\rho_1})$ is a polynomial in t of degree l, whereas $\det(1 - x_i)^\rho = \det(1 - x_i^{\rho_1})$ is in $\mathbf{C}$. It is easy to verify from the presentation (3.6) that

$$B^\rho_{2g+1} = \left(z\frac{\partial \alpha_i}{\partial x_j} - \frac{\partial \beta_i}{\partial x_j} \right)^\rho = B'. \tag{3.8}$$

The formula of Proposition 3.1 gives $\det B^\rho_{2g+1} \det(1 - x_i)^\rho = \det B^\rho_i \det(1 - z)^\rho$, or $A(K,\rho) \det(1 - x_i^{\rho_1}) = \det B^\rho_i \det(1 - tz^{\rho_1})$ for every $1 \leq i \leq n$. Thus,

Proposition 3.3. *Suppose K is a knot in S^3. Then $A(K,\rho)$ is divisible by $\det(1 - tz^{\rho_1})$ and*

$$A^\rho(K) = A^\rho_1(B) = \frac{A(K,\rho)}{\det(1 - tz^{\rho_1})}, \tag{3.9}$$

except in the degenerate case where all $\det(1 - x_i^{\rho_1}) = 0$ and $A^\rho(K) = A^\rho_1(B) = A(K,\rho)$. □

In any case, the invariant $A^\rho(K)$ is as powerful as $A(K,\rho)$.

Lin [L] has displayed a knot K which has the same classical Alexander polynomial as the trefoil knot, but can be distinguished from the latter by the twisted Alexander polynomial associated with metabelian representations of $\pi_1(S^3 \setminus K)$ into $\mathrm{SU}(2)$.

Remark. If ρ_1 is trivial, then all $\det(1 - x_i^{\rho_1}) = 0$, and both invariants coincide with the classical Alexander polynomial.

3.3. More Examples.

Example 3.4. Let K be the trefoil knot. A presentation of $G := \pi_1(S^3 \setminus K)$ is

$$G = \langle x, z \mid zxz^{-1} = xz^{-1}x^{-1}z \rangle. \tag{3.10}$$

Then

$$\left(\tfrac{\partial r}{\partial x},\ \tfrac{\partial r}{\partial z} \right) = \left(z - 1 + xz^{-1}x^{-1},\ \ 1 - zxz^{-1} + xz^{-1} - xz^{-1}x^{-1} \right). \tag{3.11}$$

Take $F = \mathbf{C}$. Take $H = H_1(S^3 \setminus K)$, the infinite cyclic group $\langle t \rangle$ where t is represented by the meridian z. Let $\rho_2(z) = t$ and $\rho_2(x) = 1$.

Let $\rho_1 : G \to \mathrm{SL}_2(\mathbf{C})$ be a representation that sends the meridian z to a hyperbolic element of $\mathrm{SL}_2(\mathbf{C})$. Then up to conjugacy we may assume

$$\rho_1(z) = \begin{pmatrix} \lambda & 0 \\ 0 & \lambda^{-1} \end{pmatrix}, \qquad \rho_1(x) = \begin{pmatrix} a & b \\ c & d \end{pmatrix} \text{ where } ad - bc = 1. \tag{3.12}$$

Substituting (3.12) into the relation in (3.10) to solve a, b, c, d in terms of λ, we find

$$\det \left(\frac{\partial r}{\partial x} \right)^{\rho} = (t + t^{-1})(t - \lambda - \lambda^{-1} + t^{-1}),$$

$$\det \left(\frac{\partial r}{\partial z} \right)^{\rho} = (1 - \lambda^2 - \lambda^{-2})(1 + t^{-2}).$$

Hence

$$A^{\rho}(K) = \begin{cases} t + t^{-1}, & \text{if } \lambda^2 + \lambda^{-2} \neq 1, \\ (t + t^{-1})(t - \lambda - \lambda^{-1} + t^{-1}) & \text{if } \lambda^2 + \lambda^{-2} = 1. \end{cases} \tag{3.13}$$

Thus $A^{\rho}(K)$ is a family of polynomials parametrized by $\lambda \in \mathbf{C} \setminus \{0\}$.

Example 3.5. Suppose S is a compact surface with boundary and M is a surface S bundle over the circle with gluing map f. Then M is the mapping torus of the map $f : S \to S$.

$$\pi_1(M) = \langle a_1, \cdots, a_n, z \mid a_i z = z f_*(a_i),\ i = 1, \cdots, n \rangle. \tag{3.14}$$

Now

$$B = \left(\delta_{ij} - z \tfrac{\partial f_*(a_i)}{\partial a_j},\ \ a_i - 1 \right). \tag{3.15}$$

Pick $\rho = (\rho_1, \rho_2) : G \to \mathrm{GL}_l(F) \times \langle t \rangle$ such that $\rho_2(z) = t$ and all $\rho_2(a_i) = 1$. So all $\det(1 - a_i)^{\rho} \in F$ and $B_{n+1} = 1 - zD$, where

$$D = \left(\frac{\partial f_*(a_i)}{\partial a_j} \right)_{n \times n}$$

and $B^{\rho}_{n+1} = 1 - t(zD)^{\rho_1}$.

If some $\det(1 - a_i^{\rho_1}) \neq 0$, then

$$A^{\rho}(M) = A^{\rho}_1(B) = \frac{\det\left(1 - t(zD)^{\rho_1}\right)}{\det(1 - tz^{\rho_1})}. \tag{3.16}$$

If all $\det(1 - a_i^{\rho_1}) = 0$, then

$$A^{\rho}(M) = A^{\rho}_1(B) = \det\left(1 - t(zD)^{\rho_1}\right). \tag{3.17}$$

Example 3.6. Suppose S is an orientable closed surface as in Example 2.5, M is an S-bundle over the circle with gluing map f. Then similarly we have

$$(3.18)\qquad B = \begin{pmatrix} I - zD & a_i - 1 \\ \frac{\partial r}{\partial a_j} & 0 \end{pmatrix}.$$

Pick $\rho = (\rho_1, \rho_2) : \pi_1(M) \to \mathrm{GL}_l(F) \times \langle t \rangle$ such that $\rho_2(z) = t$ and all $\rho_2(a_i) = 1$. Since $\det(I - zD)^\rho$ is a non-zero polynomial of t, so $A^\rho(M) = A_1^\rho(B) \neq 0$. If further ρ_1 is the trivial representation, the last row of B^ρ becomes 0, and D^{ρ_1} becomes the integer matrix F_1 of the homomorphism $f_* : H_1(M) \to H_1(M)$. Hence

$$(3.19)\qquad A^\rho(M) = A_1^\rho(B) = \det(1 - t(zD)^{\rho_1}) = \det(1 - tF_1)).$$

Example 3.7. Suppose M is the real projective 3-space. Then

$$\pi_1(M) = \langle a \mid a^2 = 1 \rangle$$

and

$$D = (1 + a).$$

Let ρ be the trivial representation. Then $\det D^\rho = 2$ which is the order of $H_1(M)$.

4. Lefschetz number and Lefschetz zeta function twisted.

4.1. Nielsen Fixed Point Theory via the Mapping Torus Approach.

Let X be a compact connected polyhedron, $f : X \to X$ be a map. The fixed point set $\mathrm{Fix}\, f := \{x \in X \mid x = f(x)\}$ splits into a disjoint union of *fixed point classes.* Two fixed points are in the same class if and only if they can be joined by a path which is homotopic (relative to end-points) to its own f-image. Each fixed point class $\mathfrak{x}$ is an isolated subset of $\mathrm{Fix}\, f$ hence its index $\mathrm{ind}(\mathfrak{x}, f) \in \mathbf{Z}$ is defined. The number of fixed point classes with non-zero index is called the Nielsen number $N(f)$ of f. It is a homotopy invariant of f, so that every map homotopic to f must have at least $N(f)$ fixed points. (Cf. p.19 of [J1].)

Equivalently, we can work on the mapping torus.

Here we describe the *mapping torus* T_f of $f : X \to X$ as the space obtained from $X \times \mathbf{R}_+$ by identifying $(x, s+1)$ with $(f(x), s)$ for all $x \in X$, $s \in \mathbf{R}_+$, where $\mathbf{R}_+$ stands for the real interval $[0, \infty)$. On T_f there is a natural semi-flow ("sliding along the rays")

$$\varphi : T_f \times \mathbf{R}_+ \to T_f, \qquad \varphi_t(x, s) = (x, s + t) \text{ for all } t \geq 0.$$

A point $x \in X$ and a positive number $\tau > 0$ determine an orbit curve $\varphi_{(x,\tau)} := \{\varphi_t(x)\}_{0 \leq t \leq \tau}$ in T_f. We may identify X with the cross-section $X \times 0 \subset T_f$, then the map $f : X \to X$ is just the return map of the semi-flow φ.

A point $x \in X$ is a fixed point of f if and only if the time-one orbit curve $\varphi_{(x,1)}$ is a closed curve. It turns out that $x, y \in \mathrm{Fix}\, f$ are in the same fixed point class if and only if the closed curves $\varphi_{(x,1)}$ and $\varphi_{(y,1)}$ are homotopic in T_f (see [J2]).

Now from this point of view the notion of fixed point class naturally generalizes to the notion of periodic orbit class.

Let $\mathrm{PP}f := \{(x,n) \in X \times \mathbf{N} \mid x = f^n(x)\}$ be the periodic point set of f, where $\mathbf{N}$ denotes the set of natural numbers. Suppose $x \in X$ and $n \in \mathbf{N}$. Then $(x,n) \in \mathrm{PP}f$, or equivalently $x \in \mathrm{Fix}\, f^n$, if and only if $\varphi_{(x,n)}$ is a closed curve in T_f. We define $x, y \in \mathrm{Fix}\, f^n$ to be in the same *periodic orbit class of order n* if and only if $\varphi_{(x,n)}$ and $\varphi_{(y,n)}$ are in the same free homotopy class of closed curves in T_f. (The term "free homotopy" means homotopy with no concern about base point). Thus a periodic orbit class of order 1 is nothing but a fixed point class.

Let $\mathrm{o}^{(n)}$ be a periodic orbit class of order n. It is easily seen that $\mathrm{o}^{(n)}$ is an isolated subset of $\mathrm{Fix}\, f^n$. So the fixed point index $\mathrm{ind}(\mathrm{o}^{(n)}, f^n)$ is well defined.

Remark. If $x \in \mathrm{o}^{(n)}$, then the whole f-orbit $\{x, f(x), \cdots, f^{n-1}(x)\} \subset \mathrm{o}^{(n)}$. Thus $\mathrm{o}^{(n)}$ is a union of f-orbits, hence the name "periodic orbit class".

Since for all $x \in \mathrm{o}^{(n)}$ the closed curves $\varphi_{(x,n)}$ are freely homotopic in T_f, they represent a well defined conjugacy class $[\varphi_{(x,n)}]$ in the fundamental group $\Gamma := \pi_1(T_f)$. This conjugacy class will be called the *coordinate* of $\mathrm{o}^{(n)}$ in Γ, written

$$\mathrm{cd}_\Gamma(\mathrm{o}^{(n)}) = [\varphi_{(x,n)}] \in \Gamma_c$$

where Γ_c denotes the set of conjugacy classes in Γ.

Let $\mathbf{Z}\Gamma_c$ denote the free abelian group with basis Γ_c. For each natural number n, we define the *Lefschetz number of order n* as

$$L_\Gamma^{(n)}(f) := \sum_{\mathrm{o}^{(n)}} \mathrm{ind}(\mathrm{o}^{(n)}, f^n) \cdot \mathrm{cd}_\Gamma(\mathrm{o}^{(n)}) \quad \in \mathbf{Z}\Gamma_c \tag{4.1}$$

the summation being over all n-th order periodic orbit classes $\mathrm{o}^{(n)}$ of f. When every fixed point of f^n is isolated, we also have

$$L_\Gamma^{(n)}(f) = \sum_{(x,n)\in \mathrm{PP}f} \mathrm{ind}(x, f^n) \cdot [\varphi_{(x,n)}] \quad \in \mathbf{Z}\Gamma_c. \tag{4.1'}$$

We shall write $L_\Gamma(f)$ for $L_\Gamma^{(1)}(f)$.

So far $L_\Gamma^{(n)}(f)$ is defined as a formal sum organizing the index and coordinate information of the periodic orbit classes. Its importance lies with its computability.

4.2. The Trace Formula for the Lefschetz Numbers.

Pick a base point $v \in X$ and a path w from v to $f(v)$. Let $G := \pi_1(X, v)$ and let $f_G : G \to G$ be the composition

$$\pi_1(X,v) \xrightarrow{f_*} \pi_1(X, f(v)) \xrightarrow{w_*} \pi_1(X,v).$$

Let $p : \tilde{X}, \tilde{v} \to X, v$ be the universal covering. The deck transformation group is identified with G. Let $\tilde{f} : \tilde{X} \to \tilde{X}$ be the lift of f such that the reference path w lifts to a path from $\tilde{v}$ to $\tilde{f}(\tilde{v})$. Then for every $g \in G$ we have $\tilde{f} \circ g = f_G(g) \circ \tilde{f}$ (cf. pp.24–25 of [J1]).

Pick a cellular decomposition $\{e_j^d\}$ of X, the base point v being a 0-cell. It lifts to a G-invariant cellular structure on the universal covering $\tilde{X}$. Choose an arbitrary lift $\tilde{e}_j^d$ for each e_j^d. They constitute a free $\mathbf{Z}G$-basis for the cellular chain complex of $\tilde{X}$. Without loss we assume f to be a cellular map. In every dimension d, the cellular chain map $\tilde{f}$ gives rise to a $\mathbf{Z}G$-matrix $\tilde{F}_d$ with respect to the above basis, i.e. $\tilde{F}_d = (a_{ij})$ if $\tilde{f}(\tilde{e}_i^d) = \sum_i a_{ij}\tilde{e}_j^d$, $a_{ij} \in \mathbf{Z}G$.

For the mapping torus, take the base point v of X as the base point of T_f (recall that X is regarded as imbedded in T_f). Let $\Gamma = \pi_1(T_f, v)$. By the van Kampen Theorem, Γ is obtained from G by adding a new generator z represented by the loop $\varphi_{(v,1)}w^{-1}$, and adding the relations $z^{-1}gz = f_G(g)$ for all $g \in G$:

$$\Gamma = \langle G, z \mid z^{-1}gz = f_G(g) \text{ for all } g \in G \rangle. \tag{4.2}$$

Note that the homomorphism $G \to \Gamma$ induced by the inclusion $X \subset T_f$ is not necessarily injective.

In this notation, we can adapt the Reidemeister trace formula ([R], [We], see §1 of [HJ] for an exposition) to our mapping torus setting, and get a simple trace formula

$$L_\Gamma(f) = \sum_d (-1)^d \left[\operatorname{tr}(z\tilde{F}_d)\right] \quad \in \mathbf{Z}\Gamma_c, \tag{4.3}$$

where $z\tilde{F}_d$ is regarded as a matrix in $\mathbf{Z}\Gamma$. Similarly, for higher order Lefschetz numbers we have

$$L_\Gamma^{(n)}(f) = \sum_d (-1)^d \left[\operatorname{tr}(z\tilde{F}_d)^n\right] \quad \in \mathbf{Z}\Gamma_c. \tag{4.4}$$

4.3. Twisted Lefschetz numbers and Lefschetz zeta function.

Suppose a group representation $\rho : \Gamma \to \mathrm{GL}_l(F)$ is given, where F is a field of characteristic 0. Then ρ extends to an algebra representation $\rho : \mathbf{Q}\Gamma \to \mathcal{M}_{l\times l}(F)$.

Define the twisted Lefschetz numbers

$$L_\rho^{(n)}(f) := \operatorname{tr}\left(L_\Gamma^{(n)}(f)\right)^\rho = \sum_{\mathrm{o}^{(n)}} \operatorname{ind}(\mathrm{o}^{(n)}, f^n)\cdot \operatorname{tr}\left(\mathrm{cd}_\Gamma(\mathrm{o}^{(n)})\right)^\rho \quad \in F \tag{4.5}$$

for all $n \in \mathbf{N}$, the summation being over all periodic orbit classes $\mathrm{o}^{(n)}$ of order n. When every fixed point of f^n is isolated, we have

$$L_\rho^{(n)}(f) = \sum_{(x,n)\in \mathrm{PP}f} \operatorname{ind}(x, f^n)\cdot \operatorname{tr}\left(\varphi_{(x,n)}^\rho\right) \quad \in F. \tag{4.5'}$$

It has the trace formula

$$\begin{aligned} L_\rho^{(n)}(f) &= \sum_d (-1)^d \operatorname{tr}\left((z\tilde{F}_d)^n\right)^\rho \\ &= \sum_d (-1)^d \operatorname{tr}\left((z\tilde{F}_d)^\rho\right)^n \quad \in F. \end{aligned} \tag{4.6}$$

We now define the formal power series

$$\zeta_\rho(f) := \exp \sum_n L_\rho^{(n)}(f)\frac{t^n}{n}. \tag{4.7}$$

It has constant term 1, so it is in the multiplicative subgroup $1 + tF[[t]]$ of the formal power series ring $F[[t]]$.

It follows from (4.6) that

Theorem 4.1. *$\zeta_\rho(f)$ is a rational function in F.*

$$\begin{aligned}\zeta_\rho(f) &= \exp \sum_d (-1)^d \sum_n tr\left((z\tilde{F}_d)^\rho\right)^n \frac{t^n}{n} \\ &= \prod_d \det\left(1 - t(z\tilde{F}_d)^\rho\right)^{(-1)^{d+1}} \qquad \in F(t),\end{aligned} \tag{4.8}$$

where 1 *stands for suitable identity matrices.* □

The following invariance can be proved in a similar way as the basic homotopy invariance (cf. Theorem I.4.5 of [J1]) of Nielsen fixed point theory.

Theorem 4.2 (Homotopy invariance). *Suppose $f \simeq f' : X \to X$ via a homotopy $\{f_t\}_{0\leq t\leq 1}$. The homotopy gives rise to a homotopy equivalence $T_f, v \simeq T_{f'}, v$ in a standard way. If we identify $\Gamma' = \pi_1(T_{f'}, v)$ with $\Gamma = \pi_1(T_f, v)$ via this homotopy equivalence, then $L_\Gamma^{(n)}(f') = L_\Gamma^{(n)}(f)$, $L_\rho^{(n)}(f') = L_\rho^{(n)}(f)$ and $\zeta_\rho(f') = \zeta_\rho(f)$.* □

By (4.5') and the homotopy invariance, we have

Theorem 4.3 (Twisted version of the Lefschetz fixed point theorem). *If a map $f : X \to X$ is homotopic to a fixed point free map $g : X \to X$, then for any representation $\rho : \pi_1(T_f) \to \mathrm{GL}_l(F)$ we have $L_\rho(f) = 0$.*

If $f : X \to X$ is homotopic to a periodic point free map $g : X \to X$, then for any representation $\rho : \pi_1(T_f) \to \mathrm{GL}_l(F)$ we have $\zeta_\rho(f) = 1$. □

Corollary 4.4. *For a map $f : X \to X$, if $L_\rho(f) \neq 0$ for some representation $\rho : \pi_1(T_f) \to \mathrm{GL}_l(F)$, then the Nielsen number $N(f) > 0$.* □

Remark. When $F = \mathbf{Q}$ and $\rho : \Gamma \to \mathrm{GL}_1(\mathbf{Q}) = \mathbf{Q}$ is trivial (sending everything to 1), then $L_\rho(f) \in \mathbf{Z}$ is the ordinary Lefschetz number, and $\zeta_\rho(f)$ is the classical Lefschetz zeta function defined by Weil.

Example 4.5. Let S be a surface with boundary, and $f : S \to S$ be a map. Suppose $\{a_1, \cdots, a_n\}$ is a free basis for $G = \pi_1(S)$. Then f has the homotopy type of a self-map of a bouquet of n circles which can be decomposed into one 0-cell and n 1-cells corresponding to the a_i's. As pointed out in [FH],

$$\tilde{F}_0 = (1),$$

$$\tilde{F}_1 := D = \left(\frac{\partial f_G(a_i)}{\partial a_j}\right).$$

Then

$$L_\Gamma(f) = [z] - \sum_{i=1}^{n} \left[z \frac{\partial f_G(a_i)}{\partial a_i} \right] \quad \in \mathbf{Z}\Gamma_c, \tag{4.9}$$

$$L_\rho(f) = \mathrm{tr}(z^\rho) - \sum_{i=1}^{n} \mathrm{tr} \left(z \frac{\partial f_G(a_i)}{\partial a_i} \right)^\rho \tag{4.10}$$

$$\zeta_\rho(f) = \frac{\det(1 - t(zD)^\rho)}{\det(1 - tz^\rho)} \quad \in F[[t]]. \tag{4.11}$$

When f is a homeomorphism, its mapping torus T_f is the S-bundle over S^1 with gluing map f. If some $\det(1 - a_i)^\rho \neq 0$, then by (3.16),

$$\zeta_\rho(f) = A^\rho(T_f). \tag{4.12}$$

That is, the twisted Lefschetz zeta function of f coincides with the twisted Alexander polynomial of T_f.

Section 5. Examples with $L(f) = 0$ but $L_\rho(f) \neq 0$.

Example 5.1. Let X be a bouquet of two oriented circles a and b, and $f : X \to X$ be a map such that

$$f_*(a) = a^{-1}, \qquad f_*(b) = a^{2m}b^2.$$

Then $L(f) = 1 - (-1 + 2) = 0$. Now we have

$$\pi_1(T_f) = \langle a, b, z \mid z^{-1}az = a^{-1},\ z^{-1}bz = a^{2m}b^2 \rangle$$

and the Fox Jacobian matrix

$$D = \begin{pmatrix} -a^{-1} & 0 \\ 1 + a + \ldots + a^{2m-1} & a^{2m}(1 + b) \end{pmatrix}$$

Let $\rho : \pi_1(T_f) \to U(1)$ be the 1-dimensional unitary representation determined by

$$a^\rho = -1, \qquad b^\rho = 1, \qquad z^\rho = 1.$$

By (4.10) we have

$$L_\rho(f) = z^\rho - (-za^{-1} + za^{2m}(1 + b))^\rho = 1 - (1 + 2) = -2.$$

It follows that $N(f) > 0$. When $m = 0$, this is the example discussed in §II.B and §VII.E of [Br], where $N(f) > 0$ is proved by combinatorial arguments.

Finally, for the classical and the ρ-twisted Lefschetz zeta functions, we have

$$\zeta(f) = \frac{(1 + t)(1 - 2t)}{1 - t}, \qquad \zeta_\rho(f) = 1 - 2t.$$

Example 5.2. Let X be the closed orientable surface of genus two. Then

$$\pi_1(X) = \langle a_1, a_2, a_3, a_4 \mid [a_1, a_2][a_3, a_4] = 1\rangle.$$

Let $f : X \to X$ be an orientation preserving homeomorphism such that

$$f_*(a_1) = a_1^{-1}a_2^{-1}, \quad f_*(a_2) = a_2 a_1^2, \quad f_*(a_3) = a_3, \quad f_*(a_4) = a_4.$$

It is easy to see that $L(f) = 0$. McCord first showed $N(f) > 0$ by proving that the mod H Nielsen number $N(f, H) > 0$ for a certain normal subgroup $H \subset \pi_1(X)$ [Mc]. It was reproved in [FH] using a different approach. They first calculated the Reidemeister trace, then claimed that there are different Reidemeister classes in the trace. Usually it is very difficult to distinguish Reidemeister classes. The fact $N(f) > 0$ can be shown easily by our approach.

As calculated in p.66 of [FH], we have

$$\tilde{F}_1 = D = \begin{pmatrix} -a^{-1} & -a_1^{-1}a_2^{-1} & 0 & 0 \\ a_2(1+a_1) & 1 & 0 & 0 \\ 0 & 0 & 1 & 0 \\ 0 & 0 & 0 & 1 \end{pmatrix}$$

and

$$\tilde{F}_0 = \tilde{F}_2 = (1).$$

Now

$$\pi_1(T_f) = \langle a_1, a_2, a_3, a_4, z \mid [a_1, a_2][a_3, a_4] = 1, \ z^{-1}a_i z = f_*(a_i) \text{ for } i = 1, 2, 3, 4\rangle.$$

Let the representation $\rho : \pi_1(T_f) \to U(1)$ be chosen so that

$$a_1^\rho = -1, \qquad a_2^\rho = a_3^\rho = a_4^\rho = z^\rho = 1.$$

Then

$$L_\rho(f) = z^\rho - (-za_1^{-1} + 3z)^\rho + z^\rho = 1 - 4 + 1 = -2.$$

So $L_\rho(f) \neq 0$, and it follows that $N(f) > 0$.

For the zeta functions, it is easy to see

$$\zeta(f) = 1 + t^2, \qquad \zeta_\rho(f) = (1-t)^2.$$

Added to the paper. Just before the deadline for submitting the paper, we received M. Wada's paper "Twisted Alexander polynomial for finitely presentable groups" which contains an example of two knots with respect to same HOMFLY polynomial but different twisted Alexander polynomails with representations into $SL_2(\mathbf{Z}_7)$.

References.

[B] Bott, R., (1975) On the shape of a curve, Advances in Math. 16 23–38.
[Br] Brown, R.F., (1971) The Lefschetz Fixed Point Theorem, Scott-Foresman, Chicago.
[BZ] Burde, G., Zieschang, H., (1985) Knots, de Gruyter, Berlin.
[FH] Fadell, E., Husseini, S., (1983) The Nielsen number on surfaces Topological Methods in Nonlinear Functional Analysis (eds S.P. Singh et al.) Contemp. Math. vol. 21 Amer. Math. Soc., Providence, 59-98.
[F] Fried, D., (1983) Periodic points and twisted coefficients Geometric Dynamics (ed J. Palis Jr.) Lecture Notes in Math. vol. 1007 Springer-Verlag, Berlin, Heidelberg, New York 261–293.
[HJ] Huang, H.-H., Jiang, B.-J., (1989) Braids and periodic solutions Topological Fixed Point Theory and Applications (ed B. Jiang) Lecture Notes in Math. vol. 1411 Springer-Verlag, Berlin, Heidelberg, New York 107–123.
[Ja] Jaco, W., (1980) Lectures on Three-Manifold Topology Amer. Math. Soc. Providence.
[J1] Jiang, B., (1983) Lectures on Nielsen Fixed Point Theory Contemp. Math. vol. 14 Amer. Math. Soc., Providence.
[J2] Jiang, B., (1988) A characterization of fixed point classes, Fixed Point Theory and its Applications ed R.F. Brown Contemp. Math. vol. 72 Amer. Math. Soc. Providence, 157–160
[J3] Jiang, B., (1987) Reidemeister trace and Reidemeister torsion unpublished notes.
[L] Lin, X.-S., (1990) Representations of knot groups and twisted Alexander polynomial, preprint.
[Mc] McCord, D., (1976) An estimate of the Nielsen number and an example concerning the Lefschetz fixed point theorem Pacific J. Math. 66, 195-203
[M1] Milnor, J.W., (1962) A duality theorem for Reidemeister torsion Annals of Math. 137–147.
[M2] Milnor, J.W., (1968) Infinite cyclic coverings Conference on the Topology of Manifolds (ed J.G. Hocking) Prindle, Weber and Schmidt, 115–133.
[R] Reidemeister, K., (1936) Automorphismen von Homotopiekettenringen Math. Ann. 112 586–593.
[Ro] Rolfsen, D., (1976) Knots and Links Publish or Perish, Berkeley.
[T] Turaev, V.G., (1986) Reidemeister torsion in knot theory Russian Math. Surveys 41 1 119–182.
[Wa] Wada, M., (1992) Twisted Alexander polynomial for finitely presentable groups, preprint.
[We] Wecken, F., (1942) Fixpunktklassen, II Math. Ann. 118 216–234.
[Z] Zassenhauss, H.J., (1958) The Theory of Groups Chelsea, New York.

On the Alexander and Jones Polynomial

Heiner Zieschang (Ruhr-Universität Bochum)

The most used knot invariant is the Alexander polynomial because it is easily calculated and has a straight forward theory. However it is by far not strong enough to classify knots. Recently a new strong polynomial invariant has been introduced by Jones which has been in the centre of interest since 1985. We will introduce the Alexander polynomial in a form which shows directly that it is an invariant of the knot while the definition of the Jones polynomial seems to depend on the special presentation of the knot and we have to show its invariance.

1. The Infinite Cyclic Covering and the Alexander Module of a Knot

Let $\mathfrak{k} \subset S^n$ be a knot, let $C = \overline{S^3 - N(\mathfrak{k})}$ denote the *knot complement* of $\mathfrak{k}$, and $\mathfrak{G} = \pi_1 C$ the *knot group*. The abelianization of $\mathfrak{G}$ is an infinite cyclic group $\mathfrak{Z}$ and the kernel of the projection is the commutator subgroup $\mathfrak{G}'$. Let $p_\infty : C_\infty \to C$ be the covering adjoint to $\mathfrak{G}'$, that is $p_\#(\pi_1 C_\infty) = \mathfrak{G}' \triangleleft \mathfrak{G}$ and the quotient group $\mathfrak{Z} = \mathfrak{G}/\mathfrak{G}' \cong \mathbb{Z}$ operates on C_∞ as the group of covering transformations. Therefore the covering $p_\infty : C_\infty \to C$ is called *the infinite cyclic covering* of the complement of $\mathfrak{k}$. We will now determine the homology of C_∞ which is obviously an invariant of the knot. The infinite cyclic covering can be obtained as follows: we "cut C along some Seifert surface S" and obtain a space C^* with two copies of S (an *upper* and *lower side*) in its boundary. Take for every integer n a copy C_n^* of this space and glue the upper side of C_n^* along the lower side of C_{n+1}^*. We describe first the homology of C^*.

Since a Seifert surface S of genus h has one boundary curve it can isotopically be deformed to the regular neighbourhood of canonical system of curves $(a_1, a_2, \dots, a_{2h-1}, a_{2h})$ that is into a disk with $2h$ bands, compare Fig. 1.1 (a). These bands may have drills, but one

M.E. Bozhüyük (ed.), Topics in Knot Theory, 229–257.

can get rid of them as is shown in Fig. 1.2 and one obtains a form as in Fig. 1.1 (b).

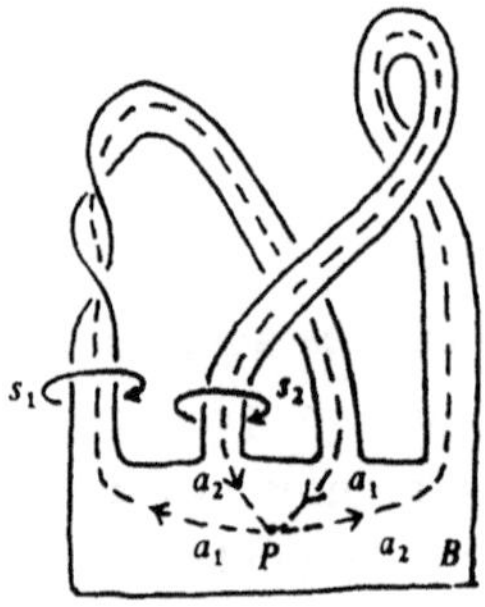

Fig. 1.1 (a)

Fig. 1.1 (b)

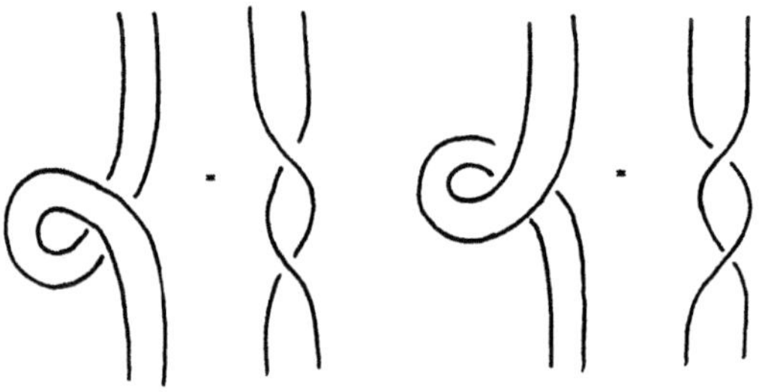

Fig. 1.2

1.1 Proposition (band projection of a knot). *Every knot $\mathfrak{k}$ can be represented as the boundary of an orientable surface S embedded in 3-space with the following properties:*

(a) $S = D^2 \cup B_1 \cup \ldots \cup B_{2k}$ *where* D^2 *and each* B_j *is a disk.*

(b) $B_i \cap B_j = \infty$ *for* $i \neq j$, $\partial B_i = \alpha_i \nu_i \beta_i {\gamma_i'}^{-1}$, $D^2 \cap B_i = \alpha_i \cup \beta_i$, $\partial D^2 = \alpha_1\delta_1\beta_2^{-1}\delta_2\beta_1^{-1}\delta_3\alpha_2\delta_4 \ldots \alpha_{2h-1}\delta_{4h-3}\beta_{2h}^{-1}\delta_{4h-2}\beta_{4h-2}^{-1}\beta_{2h-1}^{-1}\delta_{4h-1}\alpha_{2h}\delta_{4h}$.

(c) *There is a projection which is locally homeomorphic on S (there are no twists in the bands B_i.)*

A projection of this kind is called a band projection of S *or* of $\mathfrak{k}$ (see Fig. 1.1 (b)). □

Next we "thicken" this disk with bands a bit and obtain a handlebody W of genus $2h$ with the following properties:

(a) $S \subset W$,

(b) $\partial W = S^+ \cup S^-$, $S^+ \cap S^- = \partial S^+ = \partial S^- = S \cap \partial W = K$, $S^+ \approx S^- \approx S$,

(c) S is a deformation retract of W.

We call S^+ the *upside* and S^- the *downside of* W. The curves $a_1, \ldots, a_{2h}$ are projected to curves $a_1^+, \ldots, a_{2h}^+$ on S^+ and $a_1^-, \ldots, a_{2h}^-$ on S^-, respectively. After connecting the basepoints of S^+ and S^- by an arc (and a small deformation) they form a canonical system of curves of the closed surface ∂W of genus $2h$ and, thus, define a basis of $H_1(\partial W) \cong \mathbb{Z}^{4h}$. Clearly,

$$a_i^+ \sim a_i^- \quad \text{in } W.$$

Moreover we choose a curve s_i on the boundary of the neighbourhood of the i-th band such that s_i bounds a disk in W; the orientation of the disk and of s_i are chosen such that the intersection number is +1 (right-hand-rule), see Fig. 1.3.

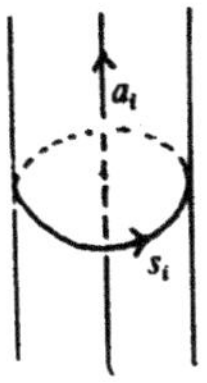

Fig. 1.3

1.2 Lemma. (a) *$\{a_1^+, \ldots, a_{2h}^+, a_1^-, \ldots, a_{2h}^-\}$ and $\{s_1, \ldots, s_{2h}, a_1^\varepsilon, \ldots, a_{2h}^\varepsilon\}$, ($\varepsilon =$ +or −) are bases of $H_1(\partial W) \cong \mathbb{Z}^{4h}$.*
(b) *$\{a_1^\varepsilon, \ldots, a_{2h}^\varepsilon\}$, $\varepsilon \in \{+, -\}$, is a basis of $H_1(W)$, and $\{s_1, \ldots, s_{2h}\}$ is a basis of $H_1(\overline{S^3 - W}) \cong \mathbb{Z}^{2h}$.*

Proof. We will give details only for the second statement of (b). The Mayer-Vietoris sequence to $(W, \overline{S^3 - W})$ gives

$$0 = H_2(S^3) \to H_1(\partial W) \xrightarrow{\varphi} H_1(W) \oplus H_1(\overline{S^3 - W}) \to H_1(S^3) = 0,$$

where $\varphi(s_i) = (0, s_i)$. From $H_1(\partial W) \cong \mathbb{Z}^{4h}$ and $H_1(\overline{S^3 - W}) \cong \mathbb{Z}^{2h}$ we obtain $H_1(\overline{S^3 - W}) \cong \mathbb{Z}^{2h}$. Now it follows from (a) that $\{s_1, \ldots, s_{2h}\}$ is a basis of $H_1(\overline{S^3 - W})$. □

Let v_{jk} be the linking number of the curves a_j^- and a_k: $v_{jk} = lk(a_j^-, a_k) \in \mathbb{Z}$. The matrix $V = (v_{jk})$ is called a *Seifert matrix* of $\mathfrak{k}$. Moreover define $f_{jk} = lk(a_j^- - a_j^+, a_k)$ and $F = (f_{jk})$.
A Seifert matrix (v_{jk}) can be read off a band projection in the following way: Consider the j-th band B_j directed as its core a_j. Denote by l_{jk} (resp. r_{jk}) the number of times B_j overcrosses B_k from left to right (resp. from right to left). Then $v_{jk} = l_{jk} - r_{jk}$.

1.3 Lemma.
(a) *Let i^ε: $S^\varepsilon \to \overline{S^3 - W}$ denote the inclusion. Then*

$$i_*^+(a_j^+) = \sum_{k=1}^{2h} v_{kj} s_k \text{ and } i_*^-(a_j^-) = \sum_{k=1}^{2h} v_{jk} s_k.$$

$$
\text{(b)} \qquad F = \begin{pmatrix} 0 & 1 & & & & & \\ -1 & 0 & & & & & \\ & & 0 & 1 & & & \\ & & -1 & 0 & & & \\ & & & & \ddots & & \\ & & & & & 0 & 1 \\ & & & & & -1 & 0 \end{pmatrix}.
$$

Proof. (a) Let Z_j^- be a projecting cylinder of the curve a_j^-, and close Z_j^- by a point at infinity. Now $Z_j^- \cap (S^3 - W)$ represents a 2-chain realizing $a_j^- \sim \sum_{k=1}^{2h} v_{jk} s_k$, Figure 1.4. The same construction applied to a_j^-, using a projecting cylinder Z_j^- directed upward, yields $a_j^+ \sim \sum_k v_{kj} s_k$. We write these equations frequently in matrix form, $a^- = Vs, a^+ = V^T s$, where a^+, a^-, s denote the $2h$-columns of the elements a_j^+, a_j^-, s_j, and V^T is the transposed matrix of V.

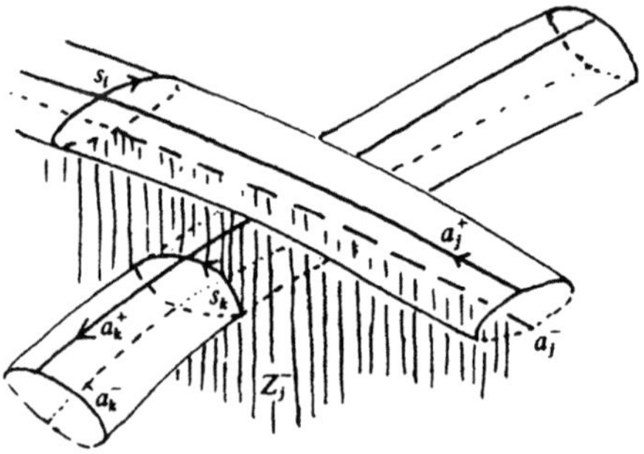

Fig. 1.4

(b) There is an annulus bounded by $a_i^- - a_i^+$. It follows from the definition of the canonical system $\{a_j\}$ that

$$
\begin{aligned}
f_{2n-1,2n} &= lk(a_{2n-1}^- - a_{2n-1}^+, a_{2n}) &&= int(a_{2n-1}, a_{2n}) &&= +1, \\
f_{2n,2n-1} &= lk(a_{2n}^- - a_{2n}^+, a_{2n-1}) &&= int(a_{2n}, a_{2n-1}) &&= -1,
\end{aligned}
$$

$f_{ik} = 0$ otherwise (Fig. 1.5). (A compatible convention concerning the sign of the intersection number is supposed to have been agreed on.) The matrix $F = (f_{jk})$ is the intersection matrix of the canonical curves $\{a_j\}$ (Fig. 1.5). □

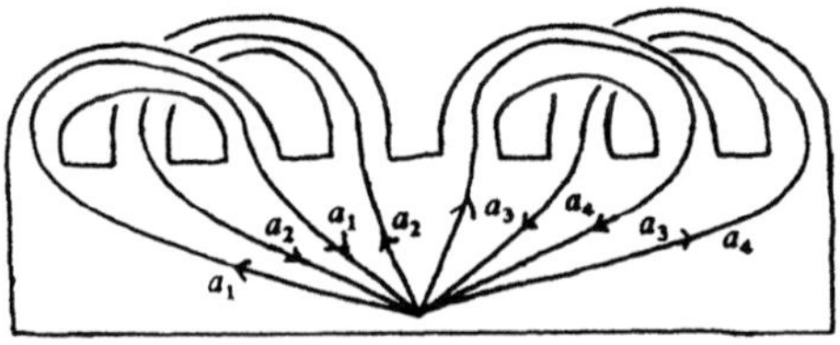

Fig. 1.5

1.3 implies certain properties of Seifert matrizes. The following proposition uses these properties to characterize Seifert matrices:

1.4 Proposition (characterization of Seifert matrices). *A Seifert matrix V of a knot $\mathfrak{k}$ satisfies the equation $V - V^T = F$. (V^T is the transposed matrix of V and F is the intersections matrix defines in* 1.3 (b)).

Every square matrix V of even order satisfying $V - V^T = F$ is a Seifert matrix of a knot.

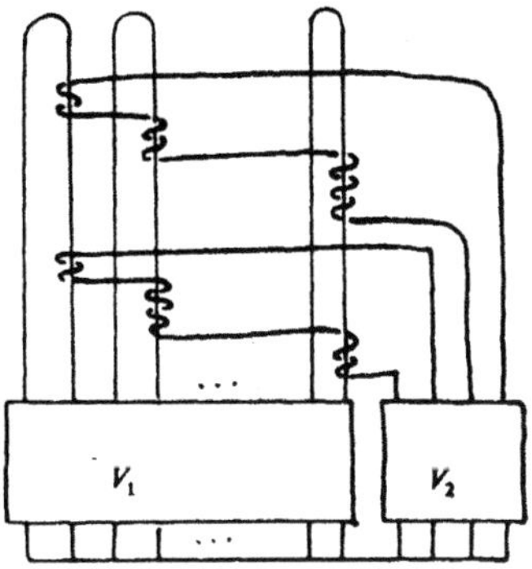

Fig. 1.6

Proof. Figure 1.5 shows a realization of the matrix

$$V_0 = \begin{pmatrix} 0 & 1 & & & & & \\ 0 & 0 & & & & & \\ & & 0 & 1 & & & \\ & & 0 & 0 & & & \\ & & & & \ddots & & \\ & & & & & 0 & 1 \\ & & & & & 0 & 0 \end{pmatrix}.$$

Any $2h \times 2h$ matrix V satisfying $V - V^T = F$ is of the form $V = V_0 + Q, Q = Q^T$. A realization of V is easily obtained by an inductive argument on h as shown in Figure 1.6. (Here a $(2h-2) \times (2h-2)$ matrix V_1 and a 2×2 matrix V_2 are assumed to be already realized; the band are represented just by lines.) The last two bands can be given arbitrary linking numbers with the first $2h - 2$ bands. □

The total space C_∞ of the infinite cyclic covering is the union of the "pieces" C_n^* and a generator τ of the group $\mathfrak{Z}$ maps C_n^* to C_{n+1}^*. Now we identify C_0^* and C^* and put $C_n^* = \tau^n C^*$. For the homology we denote the generators of $H_1(C^*) \cong H_1(\overline{S^3 - W})$ by the same symbols as the curves, that is by a_j^+, s_j etc. The generators for the other piece $\tau^n C^*$ will be denoted in a convenient way by $t^n a_j^*$, $t^n s_j$ etc. Moreover we introduce formal sums of the powers of t and calculate with them as with polynomials. Then we obtain the *group* *ring* $\mathbb{Z}\mathfrak{Z} = \{\sum_i n_i t^i \mid n_i \in \mathbb{Z}\}$ where $n_i \neq 0$ only for a finite number of coefficients. The addition and multiplication are given by

$$\left(\sum_i n_i t^i\right) + \left(\sum_j m_j t^j\right) = \sum_i (n_i + m_i) t^i,$$

$$\left(\sum_i n_i t^i\right) \cdot \left(\sum_j m_j t^j\right) = \sum_i \left(\sum_k n_k \cdot m_{i-k}\right) t^i;$$

the multiplication is well defined since the sums are finite. With these operations $\mathbb{Z}\mathfrak{Z}$ becomes a commutative ring. By 1.2, $\{t^i s_j \mid i \in \mathbb{Z},\ 1 \leq j \leq 2h\}$ forms a generating system for $H_1(C_\infty)$ and the relations are due to the identification of the upside of $\tau^i C^*$ and the downside of $\tau^{i+1} C^*$ (a consequence of the Mayer-Vietoris sequence or the Seifert-van Kampen theorem):

$$t^i a_j^+ = \sum_{k=1}^{2h} t^i v_{kj} s_k = \sum_{k=1}^{2h} t^{i+1} v_{jk} s_k = t^{i+1} a_j^-.$$

Now $H_1(C_\infty)$ becomes a module over $\mathbb{Z}\mathfrak{Z} = \mathbb{Z}(t)$, the *Alexander module* $M(t)$ of the knot group $\mathfrak{G} = \pi_1 C$ where t denotes either a generator of $\mathfrak{Z} = \mathfrak{G}/\mathfrak{G}'$ or a representative of its coset in $\mathfrak{G}$. One says that $M(t)$ is a $\mathbb{Z}\mathfrak{Z}$*-module* or, shorter, a $\mathfrak{Z}$*-Module*. Hence:

1.5 Theorem. *Let V be a Seifert matrix of a knot. Then $A(t) = V^T - tV$ is a presentation matrix of the Alexander module $H_1(C_\infty) = M(t)$. (We call a presentation matrix of the Alexander module an* Alexander matrix.*) More explicitly: $H_1(C_\infty)$ is generated by the elements*

$$t^i s_j, i \in \mathbb{Z}, 1 \leqq j \leqq 2h, \text{ and}$$

$$t^i a_j^+ = \sum_{k=1}^{2h} t^i v_{kj} s_k = \sum_{k=1}^{2h} t^{i+1} v_{jk} s_k = t^{i+1} a_j^-,\ 1 \leq j \leq 2h,$$

are defining relations. □

Writing the first homology group of C_∞ like a vector space using a basis and coefficients from $\mathbb{Z}\mathfrak{Z}$ gives a finite description of the group which in general has an infinite number of generators and relators. The group is interpreted as a *module over the ring* $\mathbb{Z}\mathfrak{Z}$. Since $\mathbb{Z}\mathfrak{Z}$ is not a field it is much more difficult to classify the modules than vector spaces (which are classified by their dimension). We will deal with this problem in the next section.

As a good exercise in homology theory one can determine the other homology groups of C_∞. The result is as follows.

1.6 Proposition.

$$H_m(C_\infty) = 0 \quad \text{for } m > 1,$$
$$H_1(C_\infty, \partial C_\infty) \cong H_1(C_\infty),$$
$$H_2(C_\infty, \partial C_\infty) \cong \mathbb{Z},$$
$$H_m(C_\infty, \partial C_\infty) = 0 \quad \text{for } m > 2. \quad \square$$

2. Alexander Polynomials

The Alexander module $M(t)$ of a knot is a finitely presented $\mathfrak{Z}$-module. In the preceding section we have described a method of obtaining a presentation matrix $A(t)$ (an Alexander matrix) of $M(t)$ from a knot projection. An algebraic classification of Alexander modules is not known, since the group ring $\mathbb{Z}(t)$ is not a principal ideal domain. But the theory of finitely generated modules over principal ideal domains can nevertheless be applied to obtain algebraic invariants of $M(t)$.

We call *Alexander matrixes* $A(t)$, $A'(t)$ *equivalent*, $A(t) \sim A'(t)$, if they present isomorphic modules.

Let R be a commutative ring with a unity element 1, and A an $m \times n$-matrix over R. We define *elementary ideals* $E_k(A) \subset R$, for $k \in \mathbb{Z}$

$$E_k(A) = \begin{cases} 0, \text{ if } n-k > m \text{ or } k < 0 \\ R, \text{ if } n-k \leqq 0 \\ \text{ideal, generated by the } (n-k) \times (n-k) \text{ minors of } A, \\ \text{if } 0 < n-k \leqq m. \end{cases}$$

It follows from the Laplace expansion theorem that the elementary ideals form an ascending chain

$$0 = E_{-1}(A) \subset E_0(A) \subset E_1(A) \subset \ldots \subset E_n(A) = E_{n+1}(A) = \ldots = R.$$

Given a knot $\mathfrak{k}$, its Alexander module $M(t)$ and an Alexander matrix $A(t)$, we call $E_k(t) = E_{k-1}(A(t))$ the k-th elementary ideal of $\mathfrak{k}$. The proper ideals $E_k(t)$ are invariants of $M(t)$, and hence, of $\mathfrak{k}$. Compare [BZ, A6] or [CF, VII].

2.1 Definition (Alexander polynomials). The greatest common divisor $\Delta_k(t)$ of the elements of $E_k(t)$ is called the *k-th Alexander polynomial of* $M(t)$, resp. of the knot. The first Alexander polynomial $\Delta_1(t)$ is usually called simply the *Alexander polynomial* and is denoted by $\Delta(t)$ (without an index). If there are no proper elementary ideals, we say that the Alexander polynomials are trivial, $\Delta_k(t) = 1$.

Remark: $\mathbb{Z}(t)$ is a unique factorization ring. So $\Delta_k(t)$ exists, and it is determined up to a factor $\pm t^\nu$, a unit of $\mathbb{Z}(t)$. It will be convenient to introduce the following notation:

$$f(t) \doteq g(t) \text{ for } f(t), g(t) \in \mathbb{Z}(t),\ f(t) = \pm t^\nu g(t), \nu \in \mathbb{Z}.$$

2.2 Proposition. *The (first) Alexander polynomial* $\Delta(t)$ *is obtained from the Seifert matrix* V *of a knot by*

$$|V^T - tV| = \det(V^T - tV) = \Delta(t).$$

The first elementary ideal $E_1(t)$ is a principal ideal.

Proof. $V^T - tV = A(t)$ is a $2h \times 2h$-matrix. The determinante
$|A(t)|$ generates the elementary ideal $E_0(A(t)) = E_1(t)$. Since det $(A(1)) = 1$, the ideal does not vanish, $E_1(t) \neq 0$. □

2.3 Proposition. *The Alexander matrix $A(t)$ of a knot $\mathfrak{k}$ satisfies*
(a) $A(t) \sim A^T(t^{-1})$ *(duality).*
The Alexander polynomials $\Delta_k(t)$ are polynomials of even degree with integral coefficients subject to the following conditions:
(b) $\Delta_k(t)|\Delta_{k-1}(t)$,
(c) $\Delta_k(t) \doteq \Delta_k(t^{-1})$ *(symmetry),*
(d) $\Delta_k(1) = \pm 1$.

Remark: The symmetry (c) implies, together with deg $\Delta_k(t) \equiv 0 \bmod 2$, that $\Delta_k(t)$ is a symmetric polynomial:

$$\Delta_k(t) = \sum_{i=0}^{2r} a_i t^i, \quad a_{2r-i} = a_i.$$

Proof. Duality follows from the fact that $A(t) = V^T - tV$ is an Alexander matrix, by 1.5, $(V^T - t^{-1}V)^T = -t^{-1}(V^T - tV)$. This implies $E_k(t) = E_k(t^{-1})$ and (c). For $t = 1$ we get: $A(1) = F^T$, and since det $F = 1$, we have $E_k(1) = \mathbf{Z}(1) = \mathbf{Z}$, which proves (d). The fact that $\Delta_k(t)$ is of even degree is a consequence of (c) and (d). Property (b) follows from the definition. □

By simple geometric constructions one can show that every polynom $\Delta(t)$ with the above properties is the Alexander polynom of a knot, see [BZ, 8.13]. Even more, any system of polynomials $\Delta_k(t)$ as above is the system of Alexander polynomials of some knot, see [L].

2.4 Examples. The Alexander polynomials of the trivial knot are trivial: $\Delta_k(t) = 1$. Figure 2.1 show band projection of the trefoil knot 3_1 and the four-knot 4_1. The Seifert matrices are

$$V_{3_1} = \begin{pmatrix} -1 & -1 \\ 0 & -1 \end{pmatrix}, \quad V_{4_1} = \begin{pmatrix} 1 & -1 \\ 0 & -1 \end{pmatrix},$$

and the Alexander polynomials are

$$\det(V_{3_1}^T - tV_{3_1}) \doteq t^2 - t + 1, \quad \det(V_{4_1}^T - tV_{4_1}) \doteq t^2 - 3t + 1.$$

Next we describe the homology of a covering space considered as module over the group ring of the group of covering transformations. As a result we obtain an algebraic calculation of the Alexander matrices.

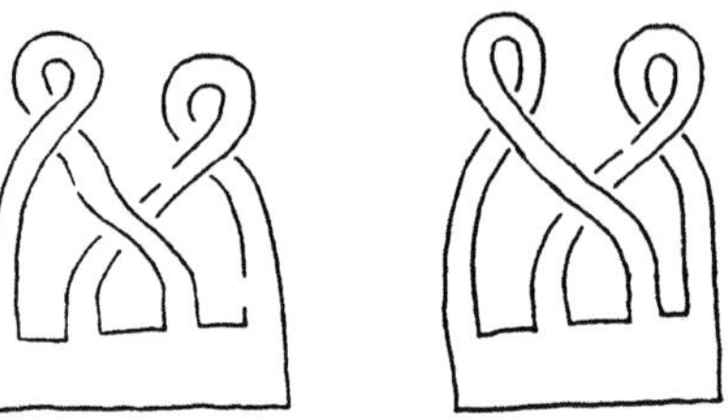

3_1 4_1

Fig. 2.1

2.5 On the homologie of a covering space. Let $p\colon \tilde{X} \to X$ be a regular covering of a connected 2-complex. We assume X to be a finite CW-complex with one 0-cell P. Then a presentation

$$\mathfrak{G} = \pi_1(X,P) = \langle s_1,\ldots,s_n | R_1,\ldots,R_m\rangle$$

of the fundamental group of X is obtained by assigning a generator s_i to each (oriented) 1-cell (also denoted by s_i), and a defining relation to (the boundary of) each 2-cell e_j of X. Choose a base point $\tilde{P} \subset \tilde{X}$ over P, $p_\#(\pi_1(\tilde{X},\tilde{P})) = \mathfrak{N} \triangleleft \mathfrak{G}$, and let $\mathfrak{D} \cong \mathfrak{G}/\mathfrak{N}$ denote the group of covering transformations. Let $\varphi\colon \mathfrak{G} \to \mathfrak{D}$, $w \mapsto w^\varphi$ be the canonical homomorphism. The linear extension to the group ring is also denoted by $\varphi\colon \mathbf{Z}\mathfrak{G} \to \mathbf{Z}\mathfrak{D}$. Observe: $(w_1w_2)^\varphi = w_1^\varphi w_2^\varphi$. Our aim is to present $H_1(\tilde{X},\tilde{X}^0)$ as a $\mathbf{Z}\,\mathfrak{D}$-module ($\tilde{X}^0$ denotes the 0-skeleton of $\tilde{X}$).

The (oriented) edges s_i lift to edges $\tilde{s}_i$ with initial point $\tilde{P}$. By w we denote a closed path in the 1-skeleton X^1 of X, and, at the same time, the element it represents in the free group $\mathfrak{S} = \pi_1(X^1,P) = \langle s_1,\ldots,s_n|-\rangle$. There is a unique lift $\tilde{w}$ of w starting at $\tilde{P}$. Clearly $\tilde{w}$ is a special element of the relative cycles $Z_1(\tilde{X},\tilde{X}^0)$ which are called *homotopy* 1-*chains*. Every 1-chain can be written in the form $\sum_{i=1}^n \xi_j\tilde{s}_j$, $\xi_j \in \mathbf{Z}\mathfrak{D}$.

There is a rule

(1) $$\widetilde{w_1w_2} = \tilde{w}_1 + w_1^\varphi \cdot \tilde{w}_2.$$

To understand it, first lift w_1 to $\tilde{w}_1$. Its endpoint is $w_1^\varphi \cdot \tilde{P}$. The covering transformation w_1^φ maps $\tilde{w}_2$ onto a chain $w_1^\varphi\tilde{w}_2$ over w_2 which starts at $w_1^\varphi\tilde{P}$. If $\tilde{w}_k = \sum_{j=1}^n \xi_{kj}\tilde{s}_j$ with $\xi_{kj} \in \mathbf{Z}\mathfrak{D}$, $k = 1,2$, then $\widetilde{w_1w_2} = \sum_{j=1}^n \xi_j\tilde{s}_j$ with

(2) $$\xi_j = \xi_{1j} + w_1^\varphi \cdot \xi_{2j} \text{ for } 1 \le j \le n.$$

(The coefficient ξ_{kj} is the algebraic intersection number of the path $\tilde{w}_k$ with the cover of s_j.) This defines mappings

(3) $$\left(\frac{\partial}{\partial s_j}\right)^\varphi : \mathfrak{G} = \pi_1(X,P) \to \mathbf{Z}\mathfrak{D},\ w \mapsto \xi_j,\ \text{with } \tilde{w} = \sum_{j=1}^n \xi_j\tilde{s}_j,$$

satisfying the rule

$$(4)\qquad \left(\frac{\partial}{\partial s_j}(w_1w_2)\right)^{\varphi} = \left(\frac{\partial}{\partial s_j}w_1\right)^{\varphi} + w_1^{\varphi}\cdot\left(\frac{\partial}{\partial s_j}w_2\right)^{\varphi}.$$

There is a linear extension to the group ring $\mathbb{Z}\mathfrak{G}$:

$$(5)\qquad \left(\frac{\partial}{\partial s_j}(\eta+\chi)\right)^{\varphi} = \left(\frac{\partial}{\partial s_j}\eta\right)^{\varphi} + \left(\frac{\partial}{\partial s_j}\chi\right)^{\varphi} \quad \text{for } \eta,\chi\in\mathbb{Z}\mathfrak{G}.$$

From the definition it follows immediately that

$$(6)\qquad \left(\frac{\partial}{\partial s_j}s_k\right)^{\varphi} = \delta_{jk},\ \tilde{w} = \sum\left(\frac{\partial w}{\partial s_j}\right)^{\varphi}\tilde{s}_j,\ \delta_{jk} = \begin{cases}1 & j=k\\ 0 & j\neq k.\end{cases}$$

We may now use this terminology to present $H_1(\tilde{X},\tilde{X}^0)$ as a $\mathfrak{D}$-module: The 1-chains $\tilde{s}_i, 1\leq i\leq n$, are generators, and $\tilde{R}_j$, the lifts of the boundaries $R_j = \partial e_j$ are defining relations. (The boundary of an arbitrary 2-cell of $\tilde{X}$ is of the form $\delta(\tilde{R}_j)$, $\delta\in\mathfrak{D}$. Hence, for a presentation of $H_1(\tilde{X},\tilde{X}^0)$ as a $\mathfrak{D}$-module it suffices to include the $\tilde{R}_j$, $1\leq j\leq m$, as defining relations.)

2.6 Proposition. $H_1(\tilde{X},\tilde{X}^0) = \langle \tilde{s}_1,\ldots,\tilde{s}_n|\tilde{R}_1,\ldots,\tilde{R}_m\rangle$, $0=\tilde{R}_j = \sum\left(\frac{\partial R_j}{\partial s_i}\right)^{\varphi}\tilde{s}_i$, $1\leq i\leq m$, *is a presentation of* $H_1(\tilde{X},\tilde{X}^0)$ *as a* $\mathfrak{D}$*-module.* □

These geometric arguments lead to the so-called *Fox or free differential calculus*, a purely algebraic approach to the above construction.

2.7 Definition. (a) There is a homomorphism ε: $\mathbb{Z}\mathcal{G}\to\mathbb{Z}$, $\tau = \sum n_ig_i \mapsto \sum n_i = \tau^{\varepsilon}$, called the *augmentation homomorphism*. Its kernel is denoted by $I\mathfrak{G} = \varepsilon^{-1}$ (a) and is called the *augmentation ideal*.
(b) A mapping Δ: $\mathbb{Z}\mathfrak{G}\to\mathbb{Z}\mathfrak{G}$ is called a *derivation* (of $\mathbb{Z}\mathfrak{G}$) if

$$\Delta(\xi+\eta) = \Delta(\xi)+\Delta(\eta) \quad \text{(linearity)},$$

and

$$\Delta(\xi\cdot\eta) = \Delta(\xi)\cdot\eta^{\varepsilon} + \xi\cdot\Delta(\eta) \quad \text{(product rule)},$$

for $\xi,\eta\in\mathbb{Z}\mathfrak{G}$.

From the definition it follows by simple calculations:

2.8 Lemma. (a) *The derivations of* $\mathbb{Z}\mathfrak{G}$ *form a (right)* $\mathfrak{G}$*-module under the operations defined by:*

$$(\Delta_1+\Delta_2)(\tau) = \Delta_1(\tau)+\Delta_2(\tau),$$
$$(\Delta\gamma)(\tau) = \Delta(\tau)\cdot\gamma.$$

(b) *Let Δ be a derivation. Then:*

$$\begin{aligned}
&\Delta(m) = 0 \text{ for } m \in \mathbf{Z},\\
&\Delta(g^{-1}) = -g^{-1}\cdot\Delta(g),\\
&\Delta(g^{n}) = (1+g+\ldots+g^{n-1})\cdot\Delta(g),\\
&\Delta(g^{-n}) = -(g^{-1}+g^{-2}+\ldots+g^{-n})\cdot\Delta(g) \text{ for } n \leq 1.
\end{aligned}$$

2.9 Examples (a) Δ_ε: $\mathbf{Z}\mathfrak{G} \to \mathbf{Z}\mathfrak{G}$, $\tau \mapsto \tau - \tau^\varepsilon$, is a derivation.

(b) If $a, b \in \mathfrak{G}$ commute, $ab = ba$, then $(a-1)\Delta b = (b-1)\Delta a$. (We write Δa instead of $\Delta(a)$ when no confusion can arise.) It follows that a derivation Δ: $\mathbf{Z}\mathfrak{Z}^n \to \mathbf{Z}\mathfrak{Z}^n$ of the group ring of a free abelian group $\mathfrak{Z}^n = \langle S_1\rangle \times \ldots \times \langle S_n\rangle$, $n \geq 2$, with $\Delta S_i \neq 0$, $1 \leq i \leq n$, is a multiple of Δ_ε, in the module of derivations.

Contrary to the situation in group rings of abelian groups the group ring of a free group admits many derivations.

2.10 Proposition. *Let $\mathfrak{F} = \langle\{S_i | i \in J\}|\rangle$ be a free group. There is a uniquely determined derivation Δ: $\mathbf{Z}\mathfrak{F} \to \mathbf{Z}\mathfrak{F}$, with $\Delta S_i = w_i$, for arbitrary elements $w_i \in \mathbf{Z}\mathfrak{F}$.*

Proof. $\Delta(S_i^{-1}) = -S_i^{-1}w_i$ follows from $\Delta(1) = 0$ and the product rule. Linearity and product rule imply uniqueness. Define $\Delta(S_{i_1}^{\eta_1}\ldots S_{i_k}^{\eta_k})$ using the product rule:

$$\Delta(S_{i_1}^{\eta_1}\ldots S_{i_k}^{\eta_k}) = \Delta S_{i_1}^{\eta_1} + S_{i_1}\Delta S_{i_2}^{\eta_2} + \ldots + S_{i_1}^{\eta_1}\ldots S_{i_{k-1}}^{\eta_{k-1}}\Delta S_{i_k}^{\eta_k}.$$

The product rule then follows for combined words $w = uv$: $\Delta w = \Delta u + u\Delta v$. The equation

$$\Delta(uS_i^\eta S_i^{-\eta}v) = \Delta u + u\Delta S_i^\eta + uS_i^\eta\Delta S_i^{-\eta} + u\Delta v = \Delta u + u\Delta v = \Delta(uv),$$

for $\eta = \pm 1$ shows that Δ is well defined on $\mathfrak{F}$. □

2.11 Definition The derivations

$$\frac{\partial}{\partial S_i}: \mathbf{Z}\mathfrak{F} \to \mathbf{Z}\mathfrak{F},\ S_j \mapsto \begin{cases} 1 & \text{for } i = j \\ 0 & \text{for } i \neq j, \end{cases}$$

of the group ring of a free group $\mathfrak{F} = \langle S_i|\rangle$ are called partial derivations.

The partial derivations form a basis of the module of derivations:

(a) $\Delta = \sum_{i\in J} \frac{\partial}{\partial S_i}\cdot\Delta(S_i)$ for every derivation Δ: $\mathbf{Z}\mathfrak{F} \to \mathbf{Z}\mathfrak{F}$. (The sum in (a) may be infinite, however, for any $\tau \in \mathbf{Z}\mathcal{F}$ there are only finitely many $\frac{\partial\tau}{\partial S_i} \neq 0$.)

(b) $\sum_{i\in J}\frac{\partial}{\partial S_i}\cdot\tau_i = 0 \iff \tau_i = 0$, $i \in J$.

(c) $\Delta_\varepsilon(\tau) = \tau - \tau^\varepsilon = \sum_{i\in J}\frac{\partial\tau}{\partial S_i}(S_i - 1)$ (fundamental formula).

(d) $\tau - \tau^\varepsilon = \sum_{i\in J} v_i(S_i - 1) \iff v_i = \frac{\partial \tau}{\partial S_i}, i \in J.$

We return now to the cyclic covering $p\colon \tilde{X} \to X$ of a knot space. Let

$$\psi\colon \mathfrak{F} = \langle S_1, \dots, S_n | - \rangle \to \langle S_1, \dots, S_n | R_1, \dots, R_m \rangle = \mathfrak{G}$$

denote the canonical homomorphism of the groups and, at the same time, its extension to the group rings:

$$\psi\colon \mathbb{Z}\mathfrak{F} \to \mathbb{Z}\mathfrak{G},\ (\Sigma n_i f_i)^\varphi = \Sigma n_i f_i^\varphi,\ \text{for } f_i \in \mathfrak{F}, n_i \in \mathbb{Z}.$$

Combining ψ with the map $\varphi\colon \mathbb{Z}\mathfrak{G} \to \mathbb{Z}\mathfrak{D}$ of 2.5 (we use the notation $(\xi)^{\varphi\psi} = (\xi^\psi)^\varphi, \xi \in \mathbb{Z}\mathfrak{F}$), we may state Proposition 2.6 in terms of the differential calculus.

2.12 Proposition. $\left(\left(\frac{\partial R_k}{\partial S_j}\right)^{\varphi\psi}\right)$, $1 \le k \le m, 1 \le j \le n$, *is a presentation matrix of* $H_1(\tilde{X}, \tilde{X}^0)$ *as a* $\mathfrak{D}$*-module. (*k *= row index,* j *= column index.)*

Proof. Comparing the linearity and the product rule of the Fox derivations 2.7 with (4) and (5) of 2.5, we deduce from 2.10 that the mappings $\left(\frac{\partial}{\partial S_i}\right)^\varphi$ in 2.5 (6) coincide with those defined by $\left(\frac{\partial}{\partial S_i}\right)^{\varphi\psi}$ in 2.11. □

Remark. The fact that the partial derivation of 2.5 (6) and 2.11 are the same lends a geometric interpretation also to the fundamental formular: For $w \in \mathfrak{G}$, and $\tilde{w}$ its lift,

$$\partial\tilde{w} = (w^{\varphi\psi} - 1)\tilde{P} = \Sigma\left(\frac{\partial W}{\partial S_i}\right)^{\varphi\psi}(S_i^{\varphi\psi} - 1)\tilde{P} = \Sigma\left(\frac{\partial W}{\partial S_i}\right)^{\varphi\psi}\partial\tilde{S}_i.$$

To obtain information about $H_1(\tilde{X})$ we consider the exact homology sequence

$$(7)\qquad 0 = H_1(\tilde{X}^0) \to H_1(\tilde{X}) \to H_1(\tilde{X}, \tilde{X}^0) \xrightarrow{\partial} H_0(\tilde{X}^0) \xrightarrow{i_*} H_0(\tilde{X}) \to 0.$$

$H_0(\tilde{X}^0)$ is generated by $\{w^{\varphi\psi} \cdot \tilde{P} \mid w \in \mathfrak{F}\}$ as an abelian group. The kernel of i_* is the image $(IF)^{\varphi\psi}$ of the augmentation ideal $I\mathfrak{F} \subset \mathbb{Z}\mathfrak{F}$ (see 2.7). The fundamental formula shows that ker i_* is generated by $\{(S_j^{\varphi\psi} - 1)\tilde{P} \mid 1 \le j \le n\}$ as a $\mathfrak{D}$-module.
Thus we obtain from (7) a short exact sequence:

$$(8)\qquad 0 \to H_1(\tilde{X}) \to H_1(\tilde{X}, \tilde{X}^0) \xrightarrow{\partial} \ker i_* \to 0.$$

In the case of a knot group $\mathfrak{G}$, and its infinite cyclic covering C_∞ now ($\mathfrak{U} = \mathfrak{G}'$) the group of covering transformations is cyclic: $\mathfrak{D} = \mathfrak{Z} = \langle t \rangle$, and ker i_* is a free $\mathfrak{Z}$-module generated by $(t-1)\tilde{P}$. The sequence (8) splits, and

$$(9)\qquad H_1(\tilde{X}, \tilde{X}^0) \cong H_1(\tilde{X}) \oplus \sigma(\mathbb{Z}\mathfrak{Z} \cdot (t-1)\tilde{P}),$$

where σ is a homomorphism $\sigma\colon \ker i_* \to H_1(\tilde{X}, \tilde{X}^0)$, $\partial\sigma = id$. This yields the following

2.13 Theorem. *Let $\mathfrak{G} = \langle S_1, \ldots, S_n | R_1, \ldots, R_n\rangle$ and $\left(\left(\frac{\partial R_j}{\partial S_i}\right)^{\varphi\psi}\right)$, its Jacobian, $\varphi\colon \mathfrak{G} \to \mathfrak{G}/\mathfrak{G}' = \mathfrak{Z} = \langle t\rangle$. A presentation matrix (Alexander matrix) of $H_1(\tilde{X}) \cong H_1(C_\infty)$ as a $\mathfrak{Z}$-module is obtained from the Jacobian by omitting its i-th column, if $S_i^{\varphi\psi} = t^{\pm 1}$. (In the case of a Jacobian derived from a Wirtinger presentation any column may be omitted.)*

Proof. It remains to show that the homomorphism $\sigma\colon \ker i_* \to H_1(\tilde{X}, \tilde{X}^0)$ can be chosen in such a way that $\sigma(\ker i_*) = \mathbb{Z}\mathfrak{Z}\tilde{s}_i$. Put $\sigma(t-1)\tilde{P} = \pm t^\mu \tilde{s}_i, S_i^{\varphi\psi} = t^\nu$, $\partial\sigma = \text{id}$. Then

$$(t-1)\tilde{P} = \partial\sigma(1-t)\tilde{P} = \partial(\pm t^\mu \tilde{s}_i) = \pm t^\mu (S_i^{\varphi\psi} - 1)\tilde{P} = \pm t^\mu (t^\nu - 1)\tilde{P}$$

that is, $(t-1) = \pm t^\mu(t^\nu - 1)$. It follows $\nu = \pm 1$, and in these cases σ can be chosen as desired. □

If $\mathfrak{D}$ is not free cyclic, the sequence (8) does not necessarily split, and $H_1(\tilde{X})$ cannot be identified as a direct summand of $H_1(\tilde{X}, \tilde{X}^0)$. This is interesting for the cases $\mathfrak{D} \cong \mathbb{Z}n$ (finite cyclic coverings) and $\mathfrak{D} \cong \mathbb{Z}^\mu$ (for coverings related to links), see [BZ, 9D].

There is a useful corollary to Theorem 2.13:

2.14 Corollary. *Every $(n-1) \times (n-1)$ minor Δ_{ij} of the $n \times n$ Jacobian of a Wirtinger presentation $\langle S_i | R_j\rangle = \mathfrak{G}$ of a knot group $\mathfrak{G}$ is a presentation matrix of $H_1(C_\infty)$. Furthermore, det $\Delta_{ij} \doteq \Delta(t)$. The elementary ideals of the Jacobian are the elementary ideals of the knot.*

Proof. Every Wirtinger relator R_k is a consequence of remaining ones. Thus, by 2.13, a presentation matrix of $H_1(C_\infty) = M(t)$ is obtained from the Jacobian by leaving out an arbitrary row and arbitrary column. □

Corollary 2.14 shows that a Jacobian of a Wirtinger presentation has nullity one. The following lemma explicitely describes the linear dependence of the rows and columns of the Jacobian of a Wirtinger presentation:

2.15 Lemma. (a) $\sum_{i=1}^n \left(\frac{\partial R_j}{\partial S_i}\right)^{\varphi\psi} = 0.$

(b) $\sum_{j=1}^n \eta_j \left(\frac{\partial R_j}{\partial S_i}\right)^{\varphi\psi} = 0$, $\eta_j = t^{\nu_j}$ *for suitable $\nu_j \in \mathbb{Z}$ for a Wirtinger presentation $\langle S_1, \ldots, S_n | R_1, \ldots, R_n\rangle$ of a knot group.*

Proof. Equation (a) follows from the fundamental formula 2.12 (c) applied to R_j:

$$0 = (R_j - 1)^{\varphi\psi} = \left[\sum_{i=1}^n \left(\frac{\partial R_j}{\partial S_i}\right)(S_i - 1)\right]^{\varphi\psi} = \sum_{i=1}^n \left(\frac{\partial R_j}{\partial S_i}\right)^{\varphi\psi}(t-1).$$

Since $\mathbb{Z}\mathfrak{Z}$ has no divisors of zero equation (a) is proved. To prove (b) we use the dependence of Wirtinger relators by the equation $\prod_{j=1}^{n} L_j R_j L_j^{-1} = 1$ in the free group $\langle S_1, \ldots, S_n | - \rangle$ (see, e.g., [BZ, 3.6]). Now

$$\left(\frac{\partial}{\partial S_i} L_j R_j L_j^{-1}\right)^{\varphi\psi} = \left(\frac{\partial L_j}{\partial S_i}\right)^{\varphi\psi} + L_j^{\varphi\psi}\left(\frac{\partial R_j}{\partial S_i}\right)^{\varphi\psi} - (L_j R_j L_j^{-1})^{\varphi\psi}\left(\frac{\partial L_j}{\partial S_i}\right)^{\varphi\psi}$$
$$= L_j^{\varphi\psi}\left(\frac{\partial R_j}{\partial S_i}\right)^{\varphi\psi}, \text{ as } (L_j R_j L_j^{-1})^{\varphi\psi} = 1.$$

By the product rule

$$0 = \frac{\partial}{\partial S_i}\left(\prod_{j=1}^{n} L_j R_j L_j^{-1}\right)^{\varphi\psi} = \sum_{j=1}^{n}\left(\prod_{k=1}^{j-1}(L_k R_k L_k^{-1})\right)^{\varphi\psi} L_j^{\varphi\psi}\left(\frac{\partial R_j}{\partial S_i}\right)^{\varphi\psi}$$
$$= \sum_{j=1}^{n} L_j^{\varphi\psi}\left(\frac{\partial R_j}{\partial S_i}\right)^{\varphi\psi},$$

which proves (b) with $L_j^{\varphi\psi} = t^{\nu_j} = \eta_j$. □

2.16 Example. A Wirtinger presentation of the group of the trefoil is:

$$\langle S_1, S_2, S_3 | S_1 S_2 S_3^{-1} S_2^{-1}, S_2 S_3 S_1^{-1} S_3^{-1}, S_3 S_1 S_2^{-1} S_1^{-1} \rangle.$$

If $R = S_1 S_2 S_3^{-1} S_2^{-1}$ then

$$\frac{\partial R}{\partial S_1} = 1, \quad \frac{\partial R}{\partial S_2} = S_1 - S_1 S_2 S_3^{-1} S_2^{-1}, \quad \frac{\partial R}{\partial S_3} = -S_1 S_2 S_3^{-1} \text{ and}$$
$$\left(\frac{\partial R}{\partial S_1}\right)^{\varphi\psi} = 1, \quad \left(\frac{\partial R}{\partial S_1}\right)^{\varphi\psi} = t - 1, \quad \left(\frac{\partial R}{\partial S_3}\right)^{\varphi\psi} = -t.$$

By similar calculations we obtain the matrix of derivatives and apply $\varphi\psi$ to get

$$\begin{pmatrix} 1 & t-1 & -t \\ -t & 1 & t-1 \\ t-1 & -t & 1 \end{pmatrix}.$$

It is easy to verify 2.15 (a) and (b). The 2×2 minor $\Delta_{11} = \begin{pmatrix} 1 & t-1 \\ -t & 1 \end{pmatrix}$, for instance, is a presentation matrix. $|\Delta_{11}| = 1 - t + t^2 = \Delta(t)$, $E_1(t) = (1 - t + t^2)$. For $k > 1$: $E_k(t) = (1) = \mathbb{Z}(t)$, $\Delta_k(t) = 1$.

2.17 Proposition. *Let $\langle S_1, \ldots, S_n | R_1, \ldots, R_m \rangle = \mathfrak{G} = \langle S_1', \ldots, S_{n'}' | R_1', \ldots, R_{m'}' \rangle$ be two finite presentations of a knot group. The elementary ideals of the respective Jacobian $\left(\left(\frac{\partial R_j}{\partial S_i}\right)^{\varphi\psi}\right)$ and $\left(\left(\frac{\partial R_j'}{\partial S_i'}\right)^{\varphi\psi}\right)$ coincide, and are those of the knot.*

Proof. This follows from 2.14, and from the fact [BZ, Appendix A 6], that the elementary ideals are invariant under Tietze processes. □

The theory of Alexander modules and polynomials can be extended to the case of links with more than one leaf; unfortunately, the algebra becomes more difficult, see [BZ], [H]. The methods may also be used for the study of finite branched coverings of a knot or link complement.

3. On link polynomials

Next we will desribe other polynomials to classify knots which are much stronger then the Alexander polynomial. The definitions and studies of these polynomials were initiated by V.F.R. Jones. One can define them by adjoining a polynom to a projection using a "skein relation" like for the Conway polynomial and to verifiy that the obtained quantity is an invariant of the knot (or better link). We will construct a polynomial in two variables which generalizes the Jones polynomial and the Conway polynomial (an absolute variant of the Alexander polynomial).

It can be shown that the Alexander polynomial $\Delta(t)$ of a knot may be written as a polynomial with integral coefficients in $u = t + t^{-1} - 2$, $\Delta(t) = f(u)$, see [BZ, 8.13]. Hence, $\Delta(t^2)$ is a polynomial in $z = (t - t^{-1})$. (It is even a polynomial in z^2.) J.H. Conway (1970) defined a polynomial $\nabla_\ell(z)$ with integral coefficients for (oriented) links which can be inductively computed from a regular projection of a link k in the following way:

3.1 Definition. To each link ℓ is adjoined a so-called *Conway potential function* which has the following properties:

(1) $\nabla_\ell(z) = 1$, if ℓ is trivial.

(2) $\nabla_\ell(z) = 0$, if ℓ is a split link.

(3) $\nabla_{\ell_+} - \nabla_{\ell_-} = z \cdot \nabla_{\ell_0}$ *(skein relation)*, if ℓ_+, ℓ_-, and ℓ_0 differ by a local operation of the kind depicted in Figure 3.1

ℓ_+ ℓ_- ℓ_0

Fig. 3.1

Changing overcrossings into undercrossings eventually transforms any regular projection into that of a trivial knot or splittable link. Equation (3) may therefore be used as an algorithm (Conway algorithm) to compute $\nabla_k(z)$ with initial conditions (1) and (2). Thus, if there is a function $\nabla_\ell(z)$ satisfying conditions (1), (2), (3) which is an invariant of the link, it must be unique.

3.2 Proposition. (a) *There is a unique integral polynomial* $\nabla_\ell(z)$ *satisfying* (1), (2), (3); *it is called the* Conway potential function *and is an invariant of the link.*

(b) $\nabla_\ell(t - t^{-1}) = \Delta(t^2)$ *for* $\mu = 1$,

$\nabla_{\mathfrak{k}}(t - t^{-1}) = (t^2 - 1)^{\mu-1}\nabla(t^2)$

for $\mu > 1$.

(Here $\Delta(t)$ *denotes the Alexander polynomial, and* $\nabla(t)$ *the so-called* Hosokawa polynomial *of* $\mathfrak{k}$.*)*

We will not give a construction of a Conway potential function; we will do this later for a some more complicated polonomial invariant. Here we show with an example, namely the trefoil, how one can calculate the polynomial for a given knot projection.

Fig. 3.2

3.3 Example. The following notation is from Fig. 3.2. We want to calculate the Conway potential function of the trefoil $\mathfrak{k}_+$ and apply the rule 3.1 (3) at the encircled crossing. The result is the equation

$$(i) \qquad \nabla_{\mathfrak{k}_+} - \underbrace{\nabla_{\mathfrak{k}_-}}_{1} = z \cdot \nabla_{\mathfrak{k}_0}.$$

The value of $\mathfrak{k}_-$ ist obtained from 3.1 (1). Next we put $\mathfrak{h}_+ = \mathfrak{k}_0$, apply 3.1 (3) to the encircled crossing in $\mathfrak{h}_+$ in Fig. 3.2, and obtain

$$(ii) \qquad \nabla_{\mathfrak{h}_+} - \underbrace{\nabla_{\mathfrak{h}_-}}_{0} = z \cdot \underbrace{\nabla_{\mathfrak{h}_0}}_{1}.$$

Hence, with $z = t - t^{-1}$ we "verify" 3.2 (b):

$$(iii) \qquad \nabla_{\mathfrak{k}_+}(z) = 1 + z^2 = 1 + (t - t^{-1})^2 = t^{-2}(1 - t^2 + t^4) \doteq \Delta(t^2)$$

The famous Jones polynomial admits a similar "skein relation" as the Conway function. We start with a more general polynomial.

3.4 Theorem. *There is a uniquely determined function, the* Homfly polynomial

$$P : \{\text{oriented links in } S^3\} \to \mathbb{Z}[\ell^{\pm 1}, m^{\pm 1}]$$

with the following properties:

(1) P *is well defined on the equivalence classes of oriented links;*

(2) $P_{\bigcirc} = 1$, *that is P of the trivial knot* $\bigcirc$ *is 1;*

(3) *for links* $\mathfrak{l}_+, \mathfrak{l}_-, \mathfrak{l}_0$ *which projections differ only at one double point in the form as described in Fig.* 3.1, *the following* skein relation *is valid:*

$$\ell P_{\mathfrak{l}_+} + \ell^{-1} P_{\mathfrak{l}_-} + m P_{\mathfrak{l}_0} = 0.$$

Clearly, the skein relation allows to calculate the Homfly polynomial for a link given by a projection (of course, it is a question how many steps (= time) are needed and straight forword approaches have exponential time complexity).

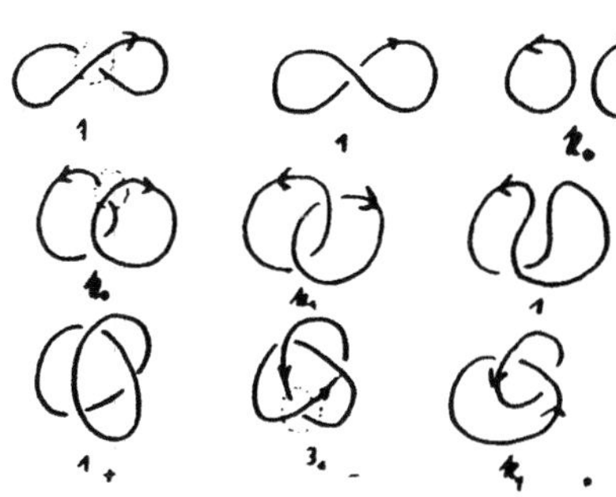

Fig. 3.3

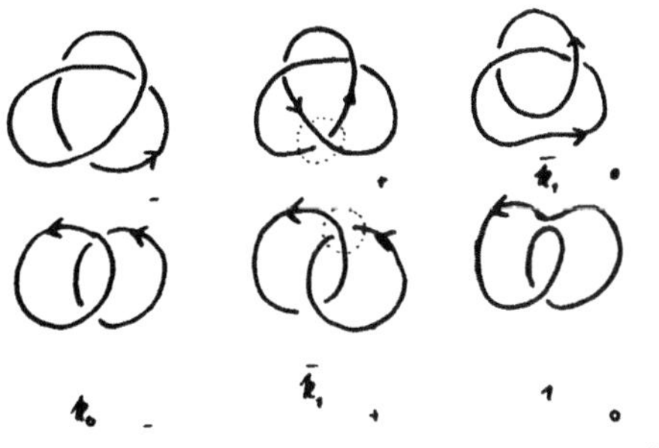

Fig. 3.4

3.5 Example. We calculate the polynomial for the trefoil and its mirror image. The used links are defined in Fig. 3.3; at trivial places the polynomial is written there. By (3), from the first line

$$\ell \cdot 1 + \ell^{-1} \cdot 1 + m \cdot P_{\mathfrak{l}_0} = 0 \quad \Longrightarrow \quad P_{\mathfrak{l}_0} = -(\ell^{-1} + \ell) \cdot m^{-1},$$

from the second

$$\ell \cdot P_{\mathfrak{l}_0} + \ell^{-1} \cdot P_{\mathfrak{l}_1} + m \cdot 1 = 0 \quad \Longrightarrow \quad P_{\mathfrak{l}_1} = (\ell + \ell^3)m^{-1} - \ell m,$$

and finally from the third

$$\ell \cdot 1 + \ell^{-1} \cdot P_{3_1} + m \cdot P_{\mathfrak{l}_1} = 0 \quad \Longrightarrow \quad P_{3_1} = -2\ell^2 - \ell^4 + \ell^2 m^2.$$

Similarly, for the mirror image $\bar{3}_1$ one has, according to the lines in Fig 3.4:

$$\ell \cdot P_{\bar{\mathfrak{l}}_1} + \ell^{-1} \cdot P_{\mathfrak{l}_0} + m \cdot 1 = 0 \quad \Longrightarrow \quad P_{\bar{\mathfrak{l}}_1} = -(\ell^{-1} + \ell^{-3}) \cdot m^{-1} - \ell^{-1} m$$
$$\ell \cdot P_{\bar{3}_1} + \ell^{-1} \cdot 1 + m \cdot P_{\bar{\mathfrak{l}}_1} = 0 \quad \Longrightarrow \quad P_{\bar{3}_1} = -2\ell^{-2} - \ell^{-4} + \ell^{-2} m^2.$$

These calculations give the impression that only the conditions (2) and (3) have been used; however we have used several times the invariance condition (1) in the form that the trivial knot given by a projection whatsoever has the polynomial 1. One can obtain the invariance checking that the polynomial is invariant under Reidemeister moves. We will present another proof in section 4.

In particular, this example shows the result of Dehn that the trefoil and its mirror image are not ambient isotopic, that is they are different knots. The relationship between the polynomials of the two knots is of general nature:

3.6 Theorem. *Let $\mathfrak{k}, \mathfrak{k}_1, \mathfrak{k}_2$ be oriented links in S^3. Let $\mathfrak{k}_1 + \mathfrak{k}_2$ denote a connected sum of the two links (if the links have several components it is not uniquely determined) and $\mathfrak{k}_1 \cup \mathfrak{k}_2$ the disjoint union of $\mathfrak{k}_1, \mathfrak{k}_2$ which are represented within two disjoint balls $\subset S^3$. Let $\rho\mathfrak{k}$ denote the link obtained from $\mathfrak{k}$ by inversing the directions of all components and $\bar{\mathfrak{k}}$ the mirror image. Then:*

(a) $$P_{\mathfrak{k}_1+\mathfrak{k}_2} = P_{\mathfrak{k}_1} \cdot P_{\mathfrak{k}_2},$$

(b) $$P_{\mathfrak{k}_1 \cup \mathfrak{k}_2} = -(\ell + \ell^{-1})m^{-1} \cdot P_{\mathfrak{k}_1} \cdot P_{\mathfrak{k}_2},$$

(c) $$P_{\rho\mathfrak{k}} = P_{\mathfrak{k}},$$

(d) $$P_{\bar{\mathfrak{k}}} = \overline{P_{\mathfrak{k}}} \quad \textit{defined by} \quad \bar{\ell} = \ell^{-1}, \bar{m} = m.$$

Proof. (b) is true for the union of two trivial circles, as we have seen for $\mathfrak{k}_0$ in 3.5. If we handle a double point like in 3.4 (3) the formula has to be applied for the union as well as for one component; hence, the product rule remains true. Similar arguments prove (c) and (d); (a) is a direct consequence of (b). □

3.7 Theorem *The Homfly polynomial $P_{\mathfrak{k}}$ is related to the Alexander polynomial $\Delta_{\mathfrak{k}}$ and the* Jones polynomial *$V_{\mathfrak{k}}$ by*

$$P_{\mathfrak{k}}(i, i(t^{1/2} - t^{-1/2})) = \Delta_{\mathfrak{k}}(t),$$
$$P_{\mathfrak{k}}(it^{-1}, -i(t^{1/2} - t^{-1/2})) = V_{\mathfrak{k}}(t).$$

□

Check the following skein relations for the Alexander and Jones polynomial:

$$\Delta_{\mathfrak{k}_+}(t) - \Delta_{\mathfrak{k}_-}(t) + (t^{1/2} - t^{-1/2})\Delta_{\mathfrak{k}_0}(t) = 0,$$
$$t^{-1}V_{\mathfrak{k}_+}(t) - tV_{\mathfrak{k}_-}(t) + (t^{-1/2} - t^{1/2})V_{\mathfrak{k}_0}(t) = 0.$$

4. On the Homfly polynomial

To construct the Homfly polynomial we follow the original route of Jones using the presentation of knots and links by closed braids.

4.1 On the symmetric group. The symmetric group S_n has a presentation

$$S_n = \langle \tau_1, \dots, \tau_{n-1} \mid \tau_i^2 = 1, \tau_i\tau_j = \tau_j\tau_i \text{ if } |i-j| \geq 2, \tau_i\tau_{i+1}\tau_i = \tau_{i+1}\tau_i\tau_{i+1} \rangle$$

where τ_i is the transposition $(i, i+1)$. To every permutation π we construct below a special word $b_\pi \equiv \prod_{j=1}^{r} \tau_{i_j}$ such that $\pi = b_\pi(\tau_i)$ in S_n and the word b_π has minimal length. Moreover the set $W = \{b_\pi \mid \pi \in S_n\}$ fulfills the Schreier condition, that is every subword $\prod_{i=1}^{k} \tau_{i_j}$, $1 \leq k \leq r$ of b_π belongs also to W.
For the construction we consider (projections of) n−braids. A presentation of a permutation π as word in the generators τ_i corresponds to a regular projection of a braid with the permutation π of the strings. Let the word $b_\pi(\tau_i)$ begin with the subword $\tau_k\tau_{k+1}\dots\tau_{n-1}$ where $k = \pi^{-1}(n)$; it corresponds to the braid of Fig. 4.1 which brings the k-th string to n. By induction one handles the remaining $n-1$ strings, see Fig 4.2 for $\pi = (1325)$. This construction gives us a uniquely determined word of minimal length, which is equal to the number of fail positions in π. The Schreier condition is valid. Remark that the generator τ_{n-1} appears at most once in $b_\pi(\tau_i)$ for $\pi \in S_n$.

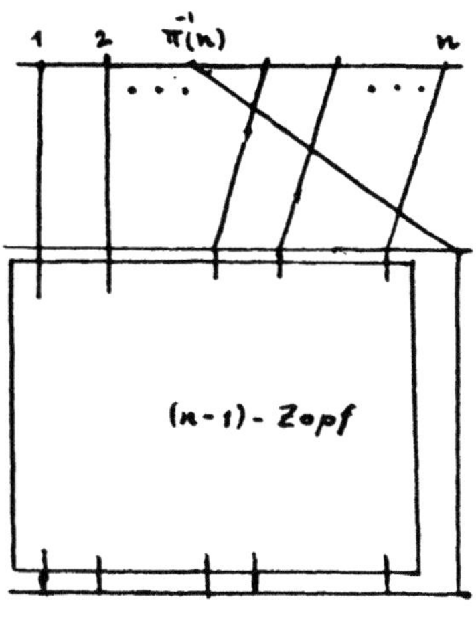

Fig. 4.1

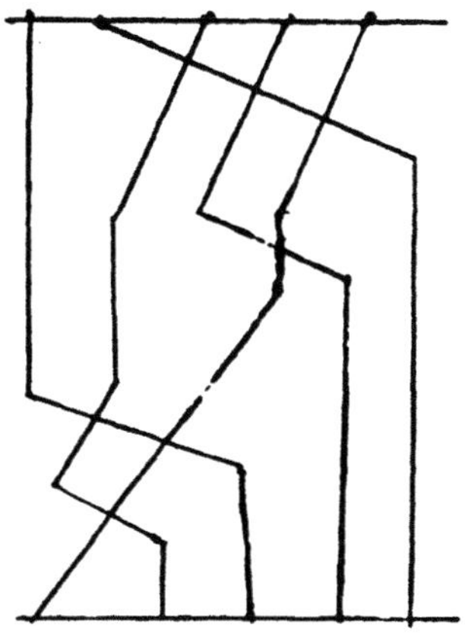

Fig. 4.2

We write $w(\tau_i) \triangleq w'(\tau_i)$ if one word is changed to the other by the "braid relations", that is by replacing $\tau_i\tau_{i+1}\tau_i$ by $\tau_{i+1}\tau_i\tau_{i+1}$ or $\tau_i\tau_j$ by $\tau_j\tau_i$ if $|i-j| \geq 2$. These procedures do not change the length.

4.2 Lemma.

$$b_\pi \tau_i \mathrel{\hat{=}} \begin{cases} b_\rho & \text{if } \rho = \pi\tau_i \in W, \\ b_\rho \cdot \tau_i^2 & \text{if } \rho\tau_i = \pi. \end{cases}$$

The first case arises iff $\pi^{-1}(i) < \pi^{-1}(i+1)$.

Proof. The method described above to construct the normal form gives the same result for π and $\rho = \pi\tau_i$ for the strings which end in a number bigger than $i+1$. Needed changes to bring $b_\pi\tau_i$ to normal form only effect the first $i+1$ strings and one obtains, by using braid relations, a situation like on the left side in Fig. 4.3 (as in Fig. 4.2). The construction gives the right side and all necessary steps only use braid relations, that is they deal with words of minimal length. This proves the first statement. The second is illustrated in Fig. 4.4. □

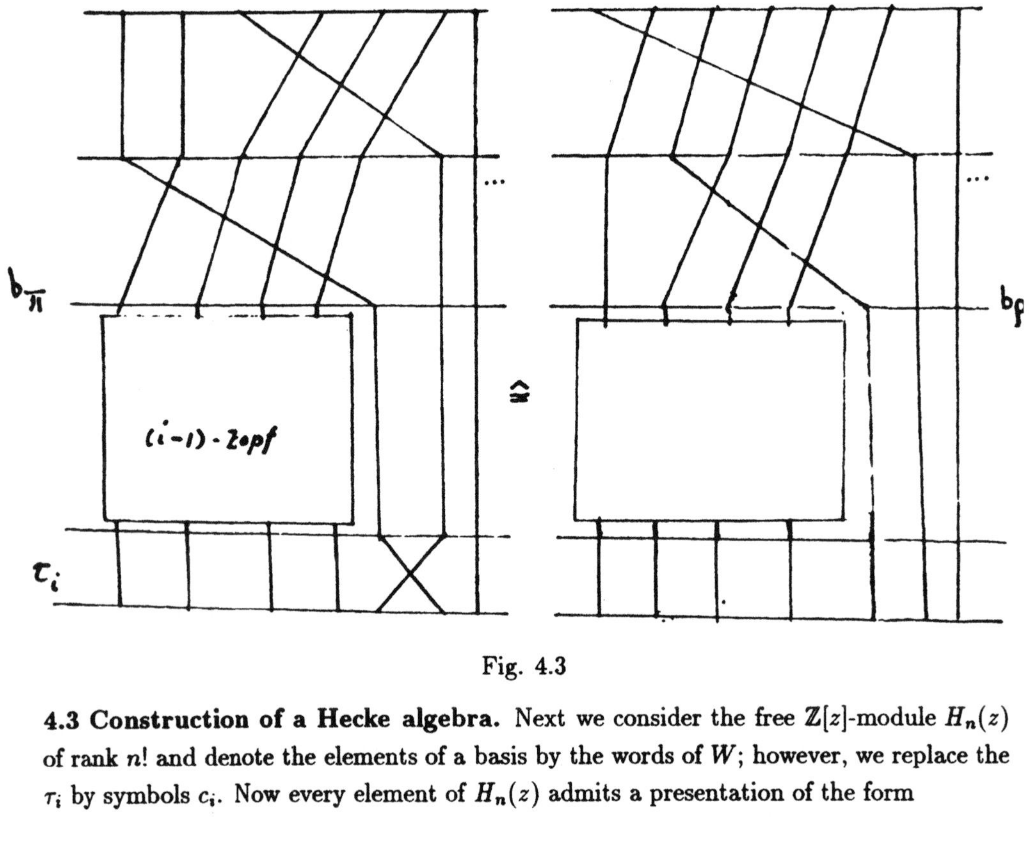

Fig. 4.3

4.3 Construction of a Hecke algebra. Next we consider the free $\mathbb{Z}[z]$-module $H_n(z)$ of rank $n!$ and denote the elements of a basis by the words of W; however, we replace the τ_i by symbols c_i. Now every element of $H_n(z)$ admits a presentation of the form

$$\sum_{\pi \in S_n} \alpha_\pi b_\pi(c_i) \quad \text{where } \alpha_\pi \in \mathbb{Z}[z].$$

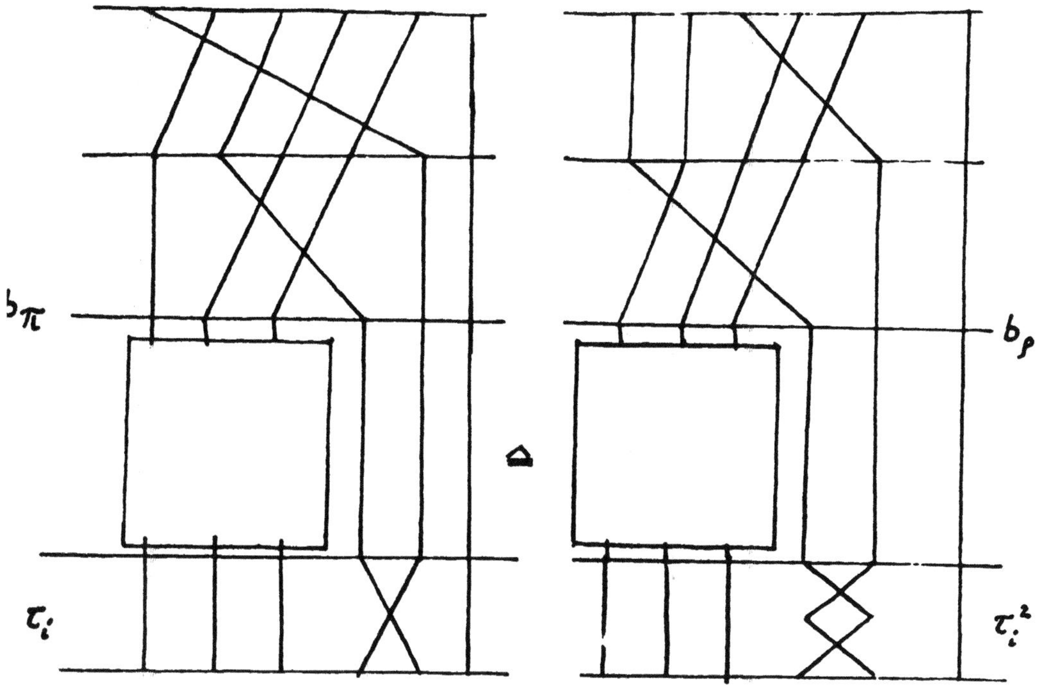

Fig. 4.4

Next we make $H_n(z)$ to an associative algebra by defining an associative multiplication for the elements of the basis. Clearly, the c_i and the empty word $b_{id} = 1$ are basic elements. We postulate:

(1) $$c_i^2 = zc_i + 1,$$

(2) $$c_i c_j = c_j c_i \text{ for } |i-j| \geq 2,$$

(3) $$c_i c_{i+1} c_i = c_{i+1} c_i c_{i+1}.$$

These equations give rise to a multiplication. Using Lemma 4.2 we put

$$b_\pi(c_i) \cdot c_k = b_\rho(c_i) \quad \text{if } \rho = \pi\tau_k \in W_n,$$
$$b_\pi(c_i) \cdot c_k = b_\rho(c_i) \cdot (zc_k + 1) = z \cdot b_k(c_i) + b_\rho(c_i) \quad \text{if } \rho\tau_k = \pi,\ \rho \in W_n.$$

This defines a multiplication for the basic elements by successive multiplication from the right and extends by linearity to the $\mathbf{Z}$-module $H_n[z]$.

We claim that *this multiplication is associative*. For products of three elements of the form c_i it is a direct consequence of the definition; for example:

$$(c_1c_2)c_2 = c_1(zc_2 + 1), \text{by the induction definition}$$
$$c_1(c_2c_2) = c_1(zc_2 + 1). \text{by (1)}$$

The proof of associativity for the general case is done by induction on the sum of the lengths of the basic words. We show

$$(b_1 b_2)(b_3 c_i) = b_1(b_2(b_3 c_i)) \quad \text{for } b_1, b_2, b_3 c_i \in W_n.$$

We assume that $b_1 \neq 1$ and prove the following equations:

$$(b_1 b_2)(b_3 c_i) \overset{(a)}{=} ((b_1 b_2) b_3) c_i \overset{(b)}{=} (b_1(b_2 b_3)) c_i \overset{(c)}{=} b_1((b_2 b_3) c_i) \overset{(d)}{=} b_1(b_2(b_3 c_i));$$

together they give the associative law. The equation (a) is valid by the definition of the multiplication for basic words since $b_3 c_i$ is a basic word. By induction hypothesis, (b) and (d) are true.

To prove (c), assume first that $b_2 b_3$ is not a basic word. A consequence of Lemma 4.2 is that

$$b_2 b_3 = \sum_\pi \alpha_\pi b_\pi$$

where the length of b_π if $\alpha_\pi \neq 0$ is smaller than the length of $b_2 b_3$. Now the equation follows from the distributive law and the induction hypothesis.

If $b_4 = b_2 b_3$ is basic and also $b_4 c_i$ then we have $(b_1 b_4) c_i = b_1(b_4 c_i)$ by the iterative definition of the multiplication. If $b_4 c_i \notin W$ then $b_4 c_i = z b_4 + b_4'$, $b_4' c_i = b_4$ and

$$(b_1 b_4) c_i = (b_1(b_4' c_i)) c_i \overset{(e)}{=} ((b_1 b_4') c_i) c_i \overset{(f)}{=} z b_1 b_4 + b_1 b_4' = b_1(z b_4 + b_4') = b_1(b_4 c_i);$$

here (e) follows from the induction hypothesis and (f) from the iterative definition of the multiplication.

For the generating elements c_i we have by (1):

$$(c_i - z) c_i = c_i^2 - z c_i = 1.$$

4.4 Proposition and Definition.

(a) *The construction above gives an associative algebra $H_n(z)$ of rank $n!$ over $\mathbb{Z}(z)$. It is called a* Hecke algebra *with generators $\{c_1, \ldots, c_n\}$ and defining relations (1)-(3). The generators c_i have inverses in $H_n(z)$, namely $c_i^{-1} = c_i - z$.*

(b) *Therefore the braid group B_n admits a representation in $H_n(z)$:*

$$\rho : B_n \to H_n(z), \quad \sigma_i \mapsto c_i.$$

(c) *By extending the ring of coefficients $\mathbb{Z}[z] \subset \mathbb{Z}[z^{\pm 1}, v^{\pm 1}]$, one obtains a Hecke algebra $H_n(z, v) = H_n$. The representation ρ can be extended to a representation*

$$\rho_v : B_n \to H_n, \quad \sigma_i \mapsto v c_i.$$

The extenison in (c) is possible since the braid relations $\sigma_i\sigma_{i+1}\sigma_i = \sigma_{i+1}\sigma_i\sigma_{i+1}$ and $\sigma_i\sigma_j = \sigma_j\sigma_i$ are homogeneous in the σ_k. □

4.5 Definition. (a) We use the obvious embedding $H_{n-1} \subset H_n$ and define $H = \cup_{n=1}^{\infty} H_n$ where $H_1 = \mathbb{Z}[v^{\pm 1}, z^{\pm 1}]$. Moreover we denote by W the set of the basic elements of the different H_n; clearly, the basic elements of H_{n-1} belong to the basis of H_n. We add a new variable T to the coefficient domain and get $\mathbb{Z}[v^{\pm 1}, z^{\pm 1}, T]$.

(b) A function $Tr : H \to \mathbb{Z}[v^{\pm 1}, z^{\pm 1}, T]$ is called *trace* if it fulfills the following conditions:

$$Tr(\sum \alpha_\pi b_\pi) = \sum \alpha_\pi Tr(b_\pi) \quad \text{where } \alpha_\pi \in \mathbb{Z}[z^{\pm 1}, v^{\pm 1}] \quad \text{(linearity)}, \tag{4}$$

$$Tr(xy) = Tr(yx) \quad \text{if } x, y \in H_n, \tag{5}$$

$$Tr(1) = 1, \tag{6}$$

$$Tr(xc_{n-1}) = T \cdot Tr(x) \quad \text{for } x \in H_{n-1}. \tag{7}$$

4.6 Lemma. *There is a uniquely determined trace on H.*

Proof. It suffices to show that a trace defined on H_n has a unique extension to H_{n+1}. By (5) and (7),

$$Tr(xc_{n-1}y) = Tr(yxc_{n-1}) = T \cdot Tr(yx) = T \cdot Tr(xy).$$

The basic elements of H_{n+1} which do not belong to H_n have the form xc_ny where $x, y \in W_n$. For the only possible extension we have to define

$$Tr(xc_ny) = T \cdot Tr(xy) \quad \text{for } xc_ny \in W_n$$

and to extend it to H_{n+1} linearly. Condition (7) is a consequence of the linearity, also $Tr(xc_ny) = T \cdot Tr(xy)$ for arbitrary $x, y \in H_n$. Now, for $a, x, y \in H_n$,

$$Tr(a \cdot xc_ny) = T \cdot Tr(axy) = T \cdot Tr(xya) = Tr(xc_ny \cdot a)$$

and, for to get (5), it suffices to prove

$$Tr(c_n \cdot xc_ny) = Tr(xc_ny \cdot c_n).$$

a) If $x, y \in H_{n-1}$ then $xc_n = c_nx$, $c_ny = yc_n$.

b) Let $x = ac_{n-1}b$, $a, b, y \in H_{n-1}$. Then

$$\begin{aligned}
Tr(c_n \cdot ac_{n-1}bc_ny) &= Tr(ac_nc_{n-1}c_nby) = Tr(ac_{n-1}c_nc_{n-1}by) \\
&= T \cdot Tr(ac_{n-1}^2by) = T \cdot Tr(a(zc_{n-1} + 1)by) \\
&= z \cdot T \cdot Tr(ac_{n-1}by) + T \cdot Tr(aby) = (zT^2 + T)Tr(aby); \\
Tr(ac_{n-1}bc_ny \cdot c_n) &= Tr(ac_{n-1}bc_n^2y) = Tr(ac_{n-1}b(zc_n + 1)y) \\
&= zTr(ac_{n-1}bc_ny) + Tr(ac_{n-1}by) \\
&= z \cdot T \cdot Tr(ac_{n-1}by) + T \cdot Tr(aby) = (zT^2 + T)Tr(aby).
\end{aligned}$$

c) The case $y = ac_{n-1}b,\ a, b, x \in H_{n-1}$ is analogous to b).

d) Let $x = ac_{n-1}b,\ y = dc_{n-1}e$ with $a, b, d, e \in H_{n-1}$. Then

$$\begin{aligned} Tr(c_n \cdot ac_{n-1}b \cdot c_n \cdot dc_{n-1}e) &= T \cdot Tr(ac_{n-1}^2 b \cdot dc_{n-1}e) \\ &= T \cdot z \cdot Tr(ac_{n-1}bdc_{n-1}e) + T^2 \cdot Tr(abde); \\ Tr(ac_{n-1}b \cdot c_n \cdot dc_{n-1}e \cdot c_n) &= T \cdot Tr(ac_{n-1}bdc_{n-1}^2 e) \\ &= T \cdot z \cdot Tr(ac_{n-1}bdc_{n-1}e) + T^2 \cdot Tr(abde). \quad \square \end{aligned}$$

Next we can introduce a new link invariant trying, for $\mathfrak{z} \in B_n$, a formula

$$P_{\mathfrak{z}_n} = k_n \cdot Tr\rho_v(\mathfrak{z}_n) \quad \text{with } k_n \in \mathbf{Z}[z^{\pm 1}, v^{\pm 1}, T].$$

Replacing $\mathfrak{z}_n$ by a conjugate braid does not change $P_{\mathfrak{z}_n}$ since (5) implies

$$Tr(ab \cdot a^{-1}) \ = \ Tr(a^{-1} \cdot ab) = Tr(b).$$

For to get $P_{\mathfrak{z}_n}$ as an invariant of the link represented by the closed braid $\hat{\mathfrak{z}}_n$ we need $P_{\mathfrak{z}_n} = P_{\mathfrak{z}_n \sigma_n^{\pm 1}}$, i.e. the invariance under the second Markov procedure: .

$$\begin{aligned} k_n \cdot Tr(\rho_v(\mathfrak{z}_n)) = P_{\mathfrak{z}_n} = P_{\mathfrak{z}_n \sigma_n^{\pm 1}} = k_{n+1} \cdot Tr(\rho_v(\mathfrak{z}_n \sigma_n)) = \\ k_{n+1} \cdot v \cdot Tr(\rho_v(\mathfrak{z}_n) \cdot c_n) = k_{n+1} v \cdot T \cdot Tr(\rho_v(\mathfrak{z}_n)) \quad \Longrightarrow \end{aligned}$$

$$(\alpha) \qquad k_n = k_{n+1} \cdot vT.$$

Similarly,

$$k_{n+1} Tr(\rho_v(\mathfrak{z}_n \sigma^{-1})) = k_{n+1} v^{-1} Tr(\rho_v(\mathfrak{z}_n) \cdot c_n^{-1}) = k_{n+1} v^{-1}(T - z) \cdot Tr(\rho_v(\mathfrak{z})) \quad \Longrightarrow$$

$$(\beta) \qquad k_n = k_{n+1} \cdot v^{-1}(T - z).$$

Go to the quotient field of $\mathbf{Z}[z^{\pm 1}, v^{\pm 1}, T]$ and put

$$(\gamma) \qquad T = \frac{zv^{-1}}{v^{-1} - v}, \quad k_{n+1} = k_n \cdot \frac{1}{v \cdot T} = k_n \cdot \frac{v^{-1} - v}{z}.$$

Then $P_{\mathfrak{z}_n} = k_n \cdot Tr(\rho_v(\mathfrak{z}_n)) \in \mathbf{Z}[z^{\pm 1}, v^{\pm 1}]$ is a link invariant. Namely, for $\mathfrak{z}_n \in H_n$ by (7):

$$\begin{aligned} P_{\mathfrak{z}_n \sigma_n} &= k_{n+1} \cdot Tr(\rho_v(\mathfrak{z}_n \sigma_n)) \\ &= (k_n \cdot z^{-1}(v^{-1} - v)) \cdot Tr(\rho_v(z_n) \cdot v c_n) \\ &= (k_n \cdot z^{-1}(v^{-1} - v)) \cdot vT \cdot Tr(\rho_v(z_n)) \\ &= P_{\mathfrak{z}_n}. \end{aligned}$$

If we put $k_1 = 1$ then we obtain $k_n = \frac{(v^{-1} - v)^{n-1}}{z^{n-1}}$. Now we obtain the following theorem; the second statement is a consequence of 4.5 (6).

4.7 Theorem. *Let $\mathfrak{z}_n \in B_n$ be a braid. Then the polynom*

$$P_{\mathfrak{z}_n}(z,v) = \frac{(v^{-1}-v)^{n-1}}{z^{n-1}} \cdot Tr(\rho_v(\mathfrak{z}_n))$$

is an invariant of the link $\mathfrak{k} = \hat{\mathfrak{z}}_n$. In the following we also denote the polynom by $P_{\mathfrak{k}}$. The trivial braid with n strings represents the trivial link with n strings; its polynomial is $\frac{(v^{-1}-v)^{n-1}}{z^{n-1}}$. □

A simple consequence is a lower bound for the braid index of a link:

4.8 Corollary. *The* braid index $\beta(\mathfrak{k})$ *of a link $\mathfrak{k}$ is the smallest number n such that $\mathfrak{k}$ is represented by a closed braid with n strings. For a polynomial $P = a_m(z)v^m + \ldots + a(z)v$, $a_i(z) \in \mathbf{Z}[z, z^{-1}]$ with $a_m(z) \neq 0 \neq a(z)$ define $Sp_v(P) = n - m$, the difference of the highest and smallest degree in v. Then, if the braid $\mathfrak{z}_n$ represents the link $\mathfrak{k}$,*

$$Sp_v(P_{\mathfrak{z}_n}) \leq 2\cdot(n-1), \qquad \textit{hence}$$

$$\beta(\mathfrak{k}) \geq 1 + \frac{1}{2} Sp_v(P_{\mathfrak{z}_n}).$$

Proof. From the definition 4.5 it follows by induction that trace of an element of H_n is polynomial in T of degree at most $n-1$. Hence, for $\mathfrak{z}_n = \prod \sigma_{i_j}^{\varepsilon_j}$ we obtain

$$\rho_v(\mathfrak{z}_n) = v^k \cdot \sum_{i=0}^{n-1} c_{i_j}^{\varepsilon_j} \quad \text{with } k = \sum \varepsilon_j \quad \Longrightarrow$$

$$Tr(\rho_v(\mathfrak{z}_n)) = v^k \cdot \sum_{i=0}^{n-1} a_i(z) T^i \quad \text{where } T^i = \frac{(v^{-1}-v)^{n-1}}{z^{n-1}} \quad \Longrightarrow$$

$$P_{\mathfrak{z}_n}(z,v) = \frac{(v^{-1}-v)^{n-1}}{z^{n-1}} \cdot Tr(\rho_v(\mathfrak{z}_n)) = v^k \cdot \sum_{i=0}^{n-1} a_i(z) \cdot z^{-n+i-1}(v^{-1}-v)^{n-2i-1}. \quad □$$

Using the method which will be developed next one can see that the knot $\mathfrak{k} = 6_1$ is not representable by a 3-braid since $Sp_v(P_{\mathfrak{k}}) = 6$; prove the same for the knots $7_2, 7_4$.

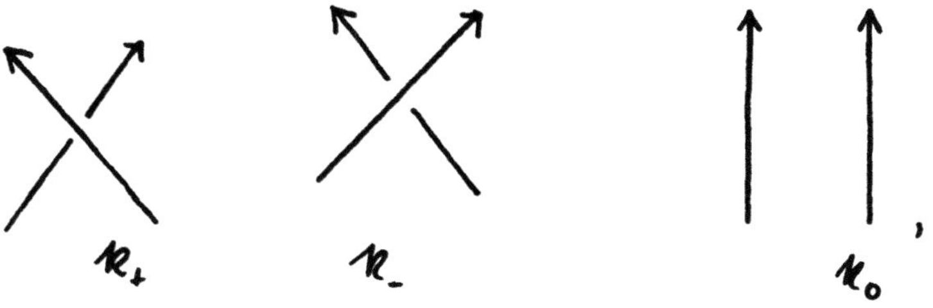

Fig. 4.5

4.9 Theorem. *Let $\mathfrak{k}_+, \mathfrak{k}_-$, and $\mathfrak{k}_0$ be link projections related as in Fig.* 4.5. *Then there is the* skein relation

$$v^{-1}P_{\mathfrak{k}_+} - vP_{\mathfrak{k}_-} = zP_{\mathfrak{k}_0}.$$

It gives an algorithm to calculate $P_{\mathfrak{k}}(z,v)$ for an arbitrary link $\mathfrak{k}$ given by a projection.

Proof. In the change of a given projection of a link to another of the form of a closed braid one can preserve a fixed neighbourhood of one double point. Therefore we may assume that the links $\mathfrak{k}_+, \mathfrak{k}_-, \mathfrak{k}_0$ are changed to braid form in the exterior of the neighbourhood, that is the braids there are the same, and we can take an appropriate braid within its conjugacy class such that the expressions $P(z,v)$ are calculated using the braids $\mathfrak{z}_n\sigma_i$, $\mathfrak{z}_n\sigma_i^{-1}$ and $\mathfrak{z}_0$. Now,

$$v^{-1}P_{\mathfrak{k}_+} - vP_{\mathfrak{k}_-} = v^{-1}k_n Tr(\rho_v(\mathfrak{z}_n\sigma_i)) - vk_n Tr(\rho_v(\mathfrak{z}_n\sigma_i^{-1}))$$

$$= k_n z Tr(\rho_v(\mathfrak{z}_n)) = zP_{\mathfrak{k}_0} \quad \text{since}$$

$$v^{-1}\rho_v(\mathfrak{z}_n\sigma_i) - v\rho_v(\mathfrak{z}_n\sigma_i^{-1}) = \rho_v(\mathfrak{z}_n)(v^{-1}\rho_v(\sigma_i) - v\rho_v(\sigma_i^{-1})) \quad \text{and}$$

$$\rho_v(\mathfrak{z}_n)(c_-c_i^{-1}) = \rho_v(\mathfrak{z}_n)\cdot z. \quad \square$$

Using the polynomial from above we can recover the "simpler" ones which have been considered above, the Jones and the Conway polynomials. We cannot give a proof here.

4.10 Theorem. *The polynomial $V_{\mathfrak{k}}(x) = P_{\mathfrak{k}}(x - x^{-1}, x^2)$ is the Jones polynomial and $\nabla_{\mathfrak{k}}(t) = P_{\mathfrak{k}}(x - x^{-1}, 1)$ with $t = x^2$ the Conway polynomial of the link $\mathfrak{k}$.*

The variable t is also often used for the Jones polynomial; however for links one obtains polynomials in $\sqrt{t}$.

Bibliography

The bibliography is rather short. Good references to the literature on knots till 1985 can be found in [BZ], [CF] or [R]. The number of recent articles on knot theory related with the Jones polynomial and succesors is enormous. Many of them can be found in [K].

[BZ] Burde, G.; Zieschang, H.: *Knots*. Berlin-New York: de Gruyter 1985

[CF] Crowell, R.H.; Fox, R.H.: *Introduction to Knot Theory*. New York: Ginn and Co. 1963

[H] Hillman, J.A.: *Alexander Ideals of Links*. Lecture Notes Math. 895. Berlin-Heidelberg-New York: Springer 1981

[K] Kauffman, L.H.: *On Knots*. Ann. Math. Studies 115. Princeton, N.J.: Princeton Univ. Press 1987

— *Knots and Physics*. Singapore-London: World Scientific 1991

[L] Levine, J.: *A Characterization of Knot Polynomials*. Topology 4 (1965), 135-141

[R] Rolfsen, D.: Knots and links. Berkeley, CA: Publish or Perish, Inc. 1976

HYPERBOLIC 3-DIMENSIONAL ORBIFOLDS

MARCEL A. HAGELBERG
Fakultät und Institut für Mathematik
Ruhr-Universität Bochum
Universitätsstraße 150
D-4630 *Bochum* 1
Germany

ABSTRACT. For the Hopf link with a tunnel in S^3 and certain branching indices it is shown that they are hyperbolic orbifolds and for this purpose that their fundamental group is discrete in $PSL_2(\mathbb{C})$.

Geometric methods turned out to be important to knot theory, since for example the complement of a knot which is neither a satellite nor a torus knot has a hyperbolic structure or more exactly is a hyperbolic three dimensional orbifold. Our concern now is to consider certain graphs in S^3 with branching indices which are connected to tunnel number one knots and show that they are hyperbolic orbifolds as well.

1. Foundations

The concept of manifolds may be generalized in a certain way by the concept of orbifolds. If a group Γ acts discontinuously on a manifold M, then the quotient space M/Γ is again a manifold if Γ acts fixed point free. If on the other hand Γ acts with fixed points, then the quotient space is no longer a manifold but what we an orbifold.
Since the orbifolds considered in this paper are only so called good orbifolds, we will introduce a definition of an orbifold here which is weaker than the one introduced by Thurston [13] but completely sufficient for our purposes:

Definition: Let M be a manifold and Γ a group acting discontinuously on M possibly having fixed points, then $\Gamma = M/\Gamma$ is called an *orbifold* and the former fixed points in the orbifold form the set of *singularities* and the order of the stabilizer of the fixed point in Γ is called the *branching index* of the singularity.
In this case M is called a *covering* of the orbifold M/Γ. If M is furthermore simply connected then M is called a *universal* covering and we have that $\Gamma = \pi_1(M/\Gamma)$.

This concept really is a generalization, since any manifold is an orbifold and coverings in the usual sense are also orbifold coverings.
Examples: Considering $\mathbb{R}^2/\mathbb{Z}\times\mathbb{Z}$ we get the torus and so the torus is an orbifold which has $\mathbb{R}^2$

M.E. Bozhüyük (ed.), Topics in Knot Theory, 259–268.

as its universal cover.
Considering now $\mathbb{R}^2 / _{<r>}$ where r is a rotation of $\frac{2\pi}{n}$ we get an orbifold which has the shape of a cone with the vertex as a point of singularity with the branching index n. This orbifold has again the universal covering $\mathbb{R}^2$.

Definition: We say an orbifold $\mathcal{O}$ admits a *geometry modelled* on a space X, if there exists a discrete subgroup Γ of , the group of Isometries of X, such that $\mathcal{O} = {}^{X} /_{\Gamma}$.
We also say $\mathcal{O}$ has the geometric structure of X.

We remark that a group acting discontinuously on X is also a discrete subgroup of all continuous functions of X in itself. If X is a complete Riemannian manifold and Γ a subgroup of Iso(X) then the converse is also true.

Consequently the two examples already considered have a Euclidian structure. By he uniformization theorem we know that any two-dimensionsal manifold carries either a Euclidian, spherical or hyperbolic structure and so this also holds for two-dimensional orbifolds (with our restricted definition), see for example Scott [12].

Thurston [14] conjectured that any compact three-manifold has a canonical decomposition into pieces which have a geometric structure. In fact he shows, that if one of these pieces allows a geometry essentially only the following eight geometries can occur: $E^3, \mathbb{H}^3, S^3, S^2 \times \mathbb{R}, \mathbb{H}^2 \times \mathbb{R}, \widetilde{\mathrm{SL}_2(\mathbb{R})}$, Nil and Sol. For more details about this topic see for example Apanasov [1] or Scott [12].

For later use we note that in view of Haeflieger and Quach [6] for an orbifold $\mathcal{O}$ with underlying space S^3 and the singular set given by a graph G with branching indices we have

$$\pi_1(\mathcal{O}) = \pi_1(S^3 \setminus G)/< x_1^{k_1}, x_2^{k_2}, ..., x_t^{k_t} >$$

where x_i is the path around the line of the graph G with the branching index k_i.

2. The Hyperbolic Space

For our purposes the most convenient model of the hyperbolic three dimensional space is the Poincaré model. It consists of the upper half space

$$\begin{aligned} \mathbb{H}^3 &= \mathbb{C} \times]0, \infty] \\ &= \{(z, r) \mid z \in \mathbb{C}, r \in \mathbb{R}, r > 0\} \\ &= \{(x, y, r) \mid x, y, r \in \mathbb{R}, r > 0\} \end{aligned}$$

equipped with the hyperbolic metric

$$ds^2 = \frac{dx^2 + dy^2 + dr^2}{r^2}.$$

The geodesics with respect to this metric, i.e. the hyperbolic lines, are the half circles or half lines orthogonal to the boundary $\mathbb{C}$ of $\mathbb{H}^3$ in the Euclidian sense. Consequently the hyperbolic planes are the Euclidian half spheres or Euclidian half planes orthogonal to the boundary.

On this model of hyperbolic geometry the group $\mathrm{PSL}_2(\mathbb{C})$ acts as the group of orientation preserving isometries by

$$\begin{pmatrix} \alpha & \beta \\ \gamma & \delta \end{pmatrix} (z, r) = \frac{(\alpha z + \beta)\overline{(\gamma z + \delta)} + \alpha\overline{\gamma} r^2 , r)}{|\gamma z + \delta|^2 + |\gamma|^2 r^2}.$$

According to the remark in the last section any discrete subgroup of $PSL_2(\mathbb{C})$ acts also discontinuously on $\mathbb{H}^3$. Later in section 3.3 we will prove the discreteness of a subgroup by the construction of its fundamental domain.

Thurston [14] proved that the complement of any knot which is neither a satellite knot nor a torus knot has a hyperbolic structure. Even an explicit construction for the figure eight knot is given in [13].

Furthermore it was proved in [11] that the Fibonacci manifold (for a definition see [9]) is an n-sheeted covering of the figure eight knot in the 3-sphere. Therefore the figure eight knot with the branching index n in S^3 is also an hyperbolic orbifold.

3. The Hyperbolic Structure of the Hopf Link with a Tunnel

Helling, Mennicke and Vinberg showed in [8] that the graph corresponding to the cloverleaf with a tunnel is a hyperbolic orbifold using essentially the following method.
First Γ is mapped in $PSL_2(\mathbb{C})$ by a homomorphism. Then looking at a certain extension of this group the fixed points of some finite subgroups become the vertices of a fundamental polyhedron. Poincaré's theorem then yields the faithfulness of the constructed homomorphism.
In this section using essentially the same method we will prove the following:

> *The Hopf link with a tunnel and branching indices k, l and m is a hyperbolic orbifold exactly in the cases $(3,3,3)$, $(3,4,2)$ and $(4,4,2)$ such that there exists a fundamental domain which is a tetrahedron.*

As A.D. Mednikh pointed out to me this result is essentially contained in a preprint of Zhuk [15], where he classified the fundamental tetrahedrons.

As the graph G we will consider the Hopf link where a tunnel joining the two circles is added. Let the branching indices be k at one of the circles, say the one at A, and l at the one at B and finally m at the tunnel AB (see figure 1). Here and throughout the rest of the work we assume of course the branching indices k, l and m to be greater than one.

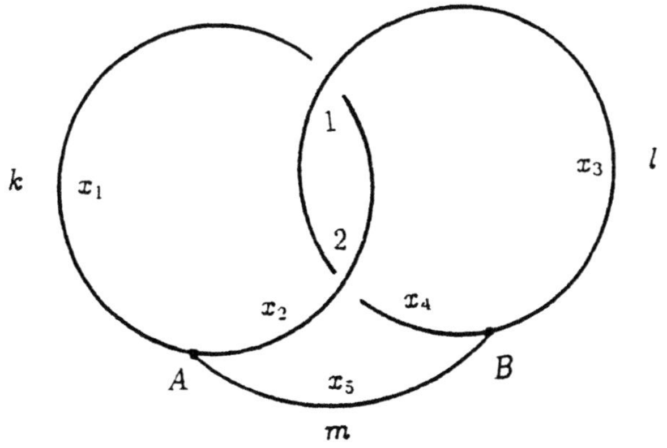

Figure 1: The Hopf link with a tunnel

3.1 THE FUNDAMENTAL GROUP

In this section it will turn out that the fundamental group of this object as an orbifold is a generalized triangle group (for a definition see Baumslag, Morgan and Shalen [2]). Calculating first the fundamental group of $S^3 \setminus G$ by the standard Wirtinger algorithm, which is exhibited e.g. in Burde-Zieschang [3] and then factorizing this group by a certain normal subgroup we get the fundamental group of the orbifold, according to the remark in section 1.

We remark that the fundamental group of many orbifolds obtained in a similar way, i.e. we take a tunnel number one knot and the tunnel and add the branching indices, is a generalized triangle group [7].

We will show that the fundamental group Γ of this graph as an orbifold with underlying topological space S^3 is

$$\Gamma = < x, y \mid x^k = y^l = (xyx^{-1}y^{-1})^m = 1 > .$$

First using the Wirtinger algorithm we get a representation of the fundamental group $\pi_1(S^3 \setminus G)$ of the complement of this graph in S^3, namely

$$\begin{array}{rll} \pi_1(S^3 \setminus G) = < & x_1,\, x_2,\, x_3,\, x_4,\, x_5 \mid & \\ A: & x_1{}^{-1} x_5 x_2 & = 1, \\ B: & x_4 x_5{}^{-1} x_3{}^{-1} & = 1, \\ 1: & x_3 x_1 x_3{}^{-1} x_2{}^{-1} & = 1, \\ 2: & x_2{}^{-1} x_4{}^{-1} x_2 x_3 & = 1 > . \end{array}$$

This is a free group of rank 2 as it becomes clear from the following.

$$\begin{array}{rcll} x_1 & = & x, & \\ x_2 & = & y x y^{-1} & \text{by } (1), \\ x_3 & = & y, & \\ x_4 & = & x_2 x_3 x_2{}^{-1} = (y x y^{-1}) y (y x y^{-1})^{-1} & \text{by } (2), \\ x_5 & = & x y x^{-1} y^{-1} & \text{by } (A). \end{array}$$

Adding now the branching relations

$$x_1{}^k = x_2{}^k = 1,\ x_3{}^l = x_4{}^l = 1,\ x_5{}^m = 1,$$

we obtain the following presentation of the fundamental group Γ_1 of the Hopf link graph with the branching indices

$$\Gamma = < x, y \mid x^k = y^l = (xyx^{-1}y^{-1})^m = 1 > .$$

For simplicity we denote by $w := [x, y] = xyx^{-1}y^{-1}$.

3.2 THE HOMOMORPHISM $\Gamma \longrightarrow \mathrm{PSL}_2(\mathbb{C})$

First we construct a homomorphism $\varphi : \Gamma \longrightarrow \mathrm{PSL}_2(\mathbb{C})$ such that the elements x, y and $w = xyx^{-1}y^{-1}$ are mapped to rotations of $\mathbb{H}^3$ through $\frac{2\pi}{k}, \frac{2\pi}{l}$ and $\frac{2\pi}{m}$ respectively, since $\mathrm{PSL}_2(\mathbb{C})$ can be regarded as the group of orientation preserving motions of $\mathbb{H}^3$ (see section 2).
For $A \in \mathrm{SL}_2(\mathbb{C})$ we denote by $\overline{A}$ the pair $\{A, -A\}$. Moreover set

$$\overline{X} := \varphi(x)\,, \overline{Y} := \varphi(y) \text{ and } \overline{W} := \varphi(w).$$

Now, we postulate

$$\operatorname{tr} X = \pm 2 \cos \frac{\pi}{k}$$

$$\operatorname{tr} Y = \pm 2 \cos \frac{\pi}{l}$$
$$\operatorname{tr} W = \pm 2 \cos \frac{\pi}{m}$$

in order to let φ fulfil the conditions formulated above.

Without loss of generality we may fix X and Y such that the trace is nonnegative, i.e.

$$\operatorname{tr} X = 2 \cos \frac{\pi}{k}, \operatorname{tr} Y = 2 \cos \frac{\pi}{l}.$$

Since $\det X = 1$ and $\det Y = 1$, the following equation holds

$$\operatorname{tr} W = \operatorname{tr}(XYX^{-1}Y^{-1}) = -2 - \operatorname{tr} X \operatorname{tr} Y \operatorname{tr} XY + (\operatorname{tr} X)^2 + (\operatorname{tr} Y)^2 + (\operatorname{tr} XY)^2.$$

If we set $\xi := \operatorname{tr} X, \eta := \operatorname{tr} Y, \zeta := \operatorname{tr} W$ and $\alpha := \operatorname{tr} XY$ this means

$$\zeta = -2 + \xi^2 + \eta^2 - \alpha\eta\xi + \alpha^2.$$

If we had that $\operatorname{tr} XY = \alpha \notin \mathbb{R}$, this would guarantee that $\varphi(\Gamma)$ is infinite (for later use note that this guarantees at the same time that the axes of the corresponding rotations of X and Y are skew), see for example Appendix 3 of [8]. Therefore we want α to be a complex root of the following polynomial

$$p(\alpha) = \alpha^2 - \eta\xi\,\alpha + \xi^2 + \eta^2 - 2 - \zeta.$$

α is a complex root of p, if and only if

$$0 > D := \frac{1}{4}\eta^2\xi^2 - \xi^2 - \eta^2 + 2 + \zeta$$

where now $\xi = 2 \cos \frac{\pi}{k}, \eta = 2 \cos \frac{\pi}{l}$ and $\zeta = \pm 2 \cos \frac{\pi}{m}$.
Since $\zeta \geq 0$ does not yield more cases for $D < 0$ we may restrict in the following to $\zeta \leq 0$.
We now consider the following cases separately, where without loss of generality we may assume $k \leq l$.

(i) If $k = 2$ then $\xi^2 = 0$ and hence $D = -\eta^2 + \zeta + 2$ and so we have $D < 0$ in the cases where either $l = 3, m > 3$ or $l = 4, m > 2$ or $l \geq 5$.

(ii) If $k = 3$ then $\xi^2 = 1$ and hence $D = -\frac{3}{4}\eta^2 + \zeta + 1$ and so we have $D < 0$ in the cases where either $l = 3, m > 2$ or $l \geq 4$.

(iii) If $k \geq 4$ then $2 \leq \xi^2 < 4$ and hence $D < \zeta \leq 0$ in all other cases.

So the required homomorphism exists except for $(2,2,m), (2,3,3), (2,3,2), (2,4,2)$ and $(3,3,2)$. These remaining cases are partially considered in a paper by Dunbar [5], i.e. $(2,3,2)$ and $(3,3,2)$. Especially the case $(2,3,3)$ can be found in Coxeter [4]. Turning now to $(2,2,m)$ and $(2,3,2)$ we find: **a)** In the case $(2,2,m)$ we have that

$$\begin{aligned}\Gamma &= \langle x, y \,|\, x^2 = y^2 = (xyx^{-1}y^{-1})^m = \mathbf{1} \rangle \\ &= \langle x, y \,|\, x^2 = y^2 = (xy)^{2m} = \mathbf{1} \rangle .\end{aligned}$$

this is a dihedral group D_{4m} of order $4m$ and hence finite. **b)** In the case $(2,4,2)$ we have that

$$\begin{aligned}\Gamma &= \langle x, y \,|\, x^2 = y^4 = (xyx^{-1}y^{-1})^2 = \mathbf{1} \rangle \\ &= \langle x, y \,|\, x^2 = y^4 = (xyxy^3)^2 = \mathbf{1} \rangle .\end{aligned}$$

The kernel N of the homomorphism

$$\begin{aligned} f: \Gamma &\to \Gamma \\ x &\mapsto 1 \\ y &\mapsto y \end{aligned}$$

is a normal divisor of Γ of index 4 with the following structure

$$N \cong (\mathbb{Z}_2 * \mathbb{Z}_2) \times (\mathbb{Z}_2 * \mathbb{Z}_2).$$

3.3 THE CONSTRUCTION OF THE FUNDAMENTAL POLYHEDRON

For the construction of the fundamental polyhedron we first consider some subgroups of $\varphi(\Gamma)$ which are triangle groups and their fixed points. Namely

$$\begin{array}{rcll} \Gamma_x & := & < \varphi(x), \varphi(yx^{-1}y^{-1}) > & \text{with the fixed point} \quad O_x \\ \varphi(y^{-1})\Gamma_x\varphi(y) & = & < \varphi(y^{-1}xy), \varphi(x^{-1}) > & \text{with the fixed point} \quad \varphi(y^{-1})O_x \\ \Gamma_y & := & < \varphi(y), \varphi(xy^{-1}x^{-1}) > & \text{with the fixed point} \quad O_y \\ \varphi(x^{-1})\Gamma_y\varphi(x) & = & < \varphi(x^{-1}yx), \varphi(y^{-1}) > & \text{with the fixed point} \quad \varphi(x^{-1})O_y. \end{array}$$

While the first two are triangle groups of type $T(k,k,m)$, the latter two are of type $T(l,l,m)$. Since they are finite, whenever $\frac{1}{k}+\frac{1}{k}+\frac{1}{m} > 1$ or $\frac{1}{l}+\frac{1}{l}+\frac{1}{m} > 1$, respectively, their corresponding fixed points exist as a usual point. In the case of equality this an infinite point.

We draw here attention to the fact that for a usual triangle group $\mathrm{T}(k,l,m)$ the following holds.

$$|\mathrm{T}(k,l,m)| < \infty \iff \frac{1}{k}+\frac{1}{l}+\frac{1}{m} > 1.$$

Now we verify some elementary facts concerning these fixed points using the following result which is proved for example in the Appendix 3 of [8]:

Let $X, Y \in \mathrm{SL}_2(\mathbb{C})$ and $\operatorname{tr} X, \operatorname{tr} Y \in [-2,2]$ then the following holds

$$\text{The axes of the rotations } X \text{ and } Y \text{ are coplanar} \iff \operatorname{tr} XY \in \mathbb{R}.$$

Since we have ensured that the $\operatorname{tr} XY \notin \mathbb{R}$ we know that the axes of $\varphi(x)$ and $\varphi(y)$ are skew. On the other hand $O_x, \varphi(y^{-1})O_x$ and $O_y, \varphi(x^{-1})O_y$ are on these axes respectively and so

$$O_x \neq O_y, \qquad O_x \neq \varphi(x^{-1})O_y$$

as well as

$$\varphi(y^{-1})O_x \neq O_y, \qquad \varphi(y^{-1})O_x \neq \varphi(x^{-1})O_y.$$

It still remains to be proved that

$$O_x \neq \varphi(y^{-1})O_x \text{ and } O_y \neq \varphi(x^{-1})O_y.$$

For this purpose suppose that $O_x = \varphi(y^{-1})O_x$, then we have

$$\varphi(y)\varphi(y^{-1})O_x = O_x = \varphi(y^{-1})O_x,$$

which means that $\varphi(y^{-1})O_x$ is on the rotation axis of $\varphi(y)$. This is a contradiction, since the axes of $\varphi(x)$ and $\varphi(y)$ are skew.
$O_y \neq \varphi(x^{-1})O_y$ can be verified analogously.

Therefore these four points are all different and we have the following tetrahedron T

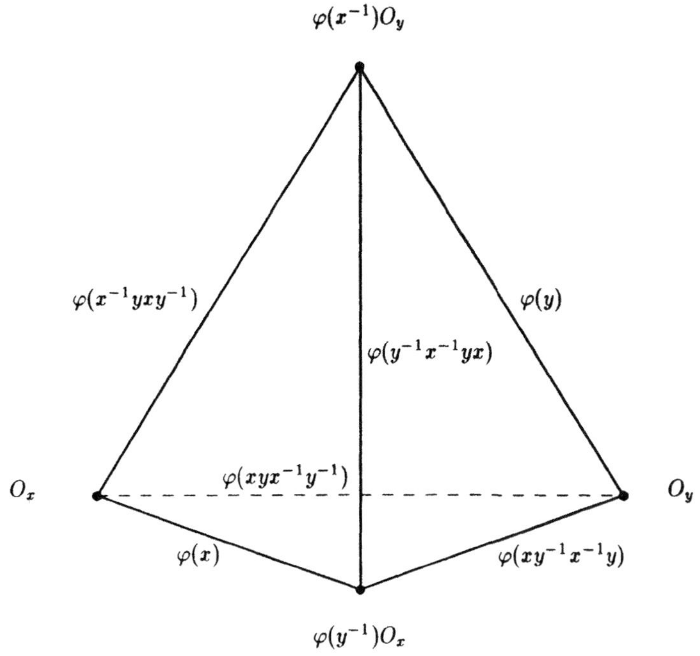

Figure 2: The tetrahedron T

Where we have the face identifications as follows

$$O_x\,\varphi(y^{-1})O_x\,\varphi(x^{-1})O_y \xrightarrow{\varphi(x)} O_x\,\varphi(y^{-1})O_x\,O_y$$

and

$$O_y\,\varphi(x^{-1})O_y\,\varphi(y^{-1})O_x \xrightarrow{\varphi(y)} O_y\,\varphi(x^{-1})O_y\,O_x.$$

To prove that this is really a fundamental polyhedron for $\varphi(\Gamma)$ we have to check on the Poincaré relations of this polyhedron. For this we have to check, that the sum of the angles of the edge cycles are equal to $\frac{2\pi}{n}$, furthermore that if the identifications at this edge are given by $h_1, h_2, ..., h_s$ n is the order of $h_s h_{s-1} \cdots h_1$. For the theorem of Poincaré see Maskit [10].

a) $O_x\,\varphi(y^{-1})O_x \xrightarrow{\varphi(x)} O_x\,\varphi(y^{-1})O_x$
The edge $O_x\,\varphi(y^{-1})O_x$ is equivalent only to itself. It is the axis of a rotation $\varphi(x)$ of order k. Hence precisely k copies of the tetrahedron meet in this edge, and hence the Poincaré condition holds true.

b) $O_y\,\varphi(x^{-1})O_y \xrightarrow{\varphi(y)} O_y\,\varphi(x^{-1})O_y$
The edge $O_y\,\varphi(x^{-1})O_y$ is equivalent only to itself. It is the axis of a rotation $\varphi(y)$ of order l. Hence precisely l copies of the tetrahedron meet in this edge, and hence the Poincaré condition

holds true.

c) $\varphi(y^{-1})O_x\,\varphi(x^{-1})O_y \xrightarrow{\varphi(x)} \varphi(y^{-1})O_x\,O_y \xrightarrow{\varphi(y)} O_x\,O_y \xrightarrow{\varphi(x^{-1})}$

$$O_x\,\varphi(x^{-1})O_y \xrightarrow{\varphi(y^{-1})} \varphi(y^{-1})O_x\,\varphi(x^{-1})O_y$$

The remaining four edges are all equivalent under the identification of faces. Join four copies of T as shown in figure 3, obtaining a polyhedron. Observe that $\varphi(xyx^{-1}y^{-1})$ is a rotation of order m with the axis $O_x\,O_y$ which maps $\varphi(yx)O_y$ to $\varphi(x)O_y$. Hence there are precisely m copies of the tetrahedron which meet in the edge $O_x\,O_y$, and hence the Poincaré condition holds true for this edge.

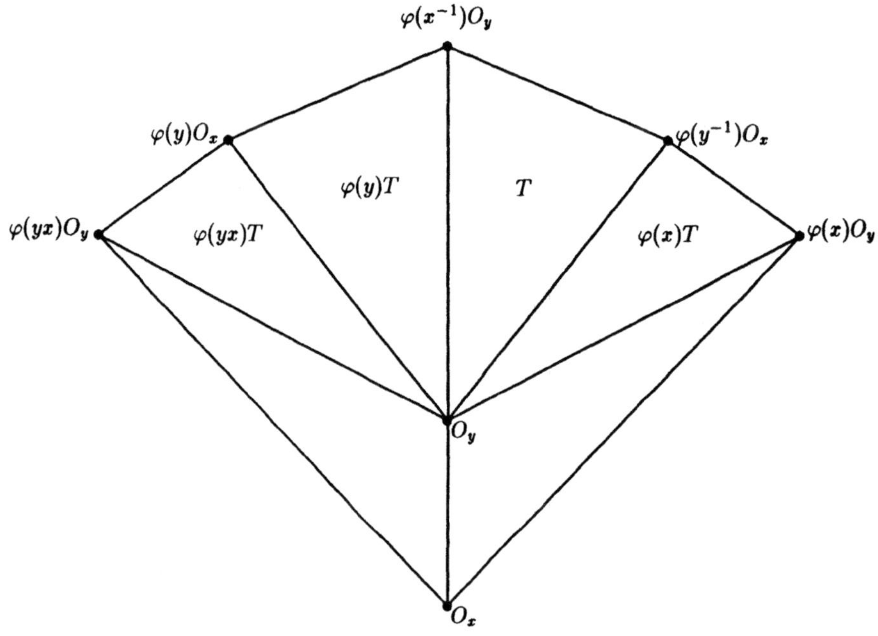

Figure 3: copies of the tetrahedron T

Now we know that our tetrahedron T is fundamental for the group $\varphi(\Gamma)$. Moreover $\varphi(\Gamma)$ has exactly the relations implied in a)-c) and hence φ is faithful.

The following pictures illustrate that performing first the identification by $\varphi(x)$ and then by $\varphi(y)$ at the tetrahedron T we really get the required Hopf link graph.

$\varphi(x)$ identifies the edges $O_x\,\varphi(x^{-1})O_y$ and $O_x\,O_y$ and gives figure 4.

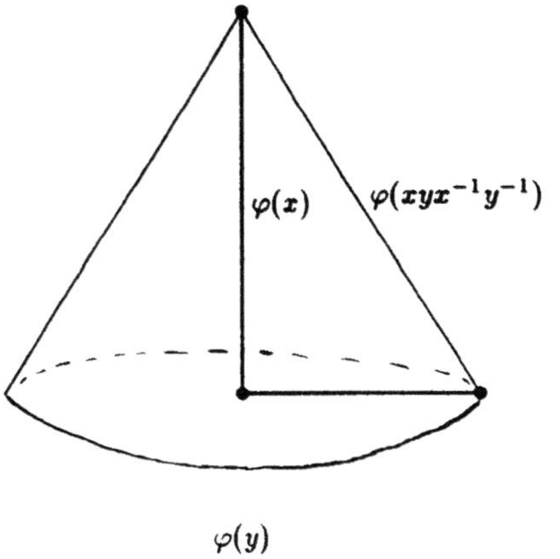

Figure 4: The identification by $\varphi(x)$

$\varphi(y)$ now identifies the edges $O_y\,\varphi(y^{-1})O_x$ and $O_y\,O_x$ yielding the Hopf link graph, see figure 5.

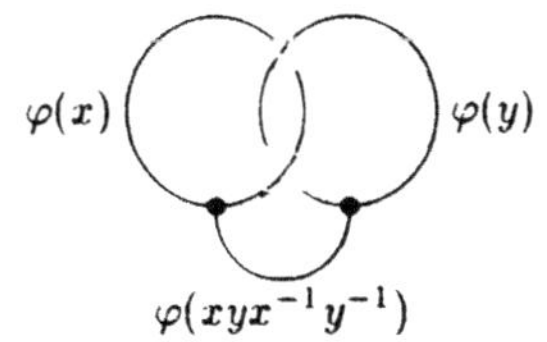

Figure 5: The identification by $\varphi(y)$

Now we have shown, that the tetrahedron T in figure 1, which was constructed by a method essentially used by Helling, Mennicke and Vinberg in [8], is really a fundamental domain for the Hopf link graph in figure 2, whose fundamental group is a generalized triangle group.

So we proved that the Hopf link with a tunnel and the branching indices k and l at the circles and m at the tunnel really is a hyperbolic orbifold if (k, l, m) is one of (3,3,3), (3,4,2) or (4,4,2). At the same time we obtained a faithful presentation of a generalized triangle group as a discrete subgroup of $PSL_2(\mathbb{C})$.
The same result holds for similar orbifolds [8],[7].

References

[1] Apanasov,B.N., *Discrete groups in space and uniformization problems,* Mathematics and Its Applications (Soviet Series), **40**, Dordrecht,Boston,London 1991.

[2] Baumslag,G., Morgan,J., Shalen,P., *Generalized triangle groups,* Math. Proc. Camb. Phil. Soc, **102** (1987), 25-31.

[3] Burde,G., Zieschang,H., *Knots,* de Gruyter Studies in Mathematics, Berlin New York 1985.

[4] Coxeter, H.S.M. *The groups determined by the relations* $S^l = T^m = (S^{-1}T^{-1}ST)^p = 1$, Duke Math. J. , **2** (1936), 61-73.

[5] Dunbar,W.D., *Geometric orbifolds,* Revista Math., **1** (1988), 67-99.

[6] Haeflieger,A., Quach,N.D., *Une présentation du groupe fondamental d'une orbifold,* Asterisque, **116** (1984), 98-107.

[7] Hagelberg,M., *Generalized triangle groups and 3-dimensional orbifolds,* SFB 343 Bielefeld, Diskrete Strukturen in der Mathematik, Preprint 92-049 (1992).

[8] Helling,H., Mennicke,J., Vinberg,E.B., *On some generalized triangle groups and 3-dimensional orbifolds,* SFB 343 Bielefeld, Diskrete Strukturen in der Mathematik, Preprint 91-057 (1991). To appear in russian.

[9] Helling,H., Kim,A.C., Mennicke,J., *A geometric study of Fibonacci groups,* Preprint Bielefeld 1990.

[10] Maskit,B., *Kleinian groups* Die Grundlehren der mathematischen Wissenschaften in Einzeldarstellungen **287** (1988).

[11] Montesinos,J.M., Hilden,H.M., Losano,M.T., *The arithmeticity of the figure eight knot,* Preprint 1989.

[12] Scott,P., *The geometries of 3-manifolds,* Bull. London Math. Soc. **15** (1983), 401-487.

[13] Thurston,W.P., *The Geometry and Topology of Three-Manifolds,* Lecture Notes, Princeton University 1980.

[14] Thurston,W.P., *Three dimensional manifolds, kleinian groups and hyperbolic geometry,* Bull. Amer. Math. Soc. (NS) **6** (1982), 357-381.

[15] Zhuk,I.K., *Regular decompositions of spaces of spaces of constant curvature in congruent tetrahedra,* Preprint/AN BSSR, Inst. Math., **12**(169) (1983)(russian) announced in *Fundamental tetrahedra in the Euclidian and Lobachevsky spaces,* Doklady Akad. Nauk SSSR, **270** (1983), NI, 35-37.

IRREGULAR DIHEDRAL BRANCHED COVERINGS OF KNOTS

Toshio HARIKAE
School of Science
Kwansei Gakuin University
Nishinomiya, Hyogo 662
Japan

Yoshiaki UCHIDA
Department of Mathematics
Kobe University
Nada, Kobe 657
Japan

ABSTRACT. We present a formula to give the number of equivalence classes of representations of a knot group onto a dihedral group and characterize the associated irregular dihedral branched coverings of knots, especially Montesinos knots.

1. Introduction

Throughout this paper we work in the piecewise linear category. A branched covering of a knot means a covering space of a 3-sphere S^3, branched along the knot. Hilden [5] and Montesinos [7] independently showed that every orientable closed 3-manifold is a 3-fold irregular branched covering of a knot. We have the problem which 3-manifold is obtained as a 3-fold irregular branched covering of a given knot. It is known that each 3-fold irregular branched covering of a 2-bridge knot, if it exists, is homeomorphic to S^3. Murasugi [8] showed that a simple 3-fold irregular branched covering of a closed 3-braid, if it exists, is homeomorphic to a lens space $L(n, 1)$ of type $(n, 1)$ for some integer $n \geq 0$, where $L(0, 1)$ is homeomorphic to $S^2 \times S^1$ and $L(1, 1)$ is homeomorphic to S^3. Further, Hosokawa and Nakanishi [6] showed that each 3-fold irregular branched covering of a pretzel knot, if it exists, is homeomorphic to $L(n, 1)$ or a connected sum of those spaces for some integer n.

In this paper we consider p-fold irregular dihedral branched coverings of knots. When $p = 3$, the coverings are 3-fold irregular branched coverings. Note that there exists a 3-manifold which is not a p-fold irregular dihedral branched covering of a knot for any prime $p > 3$. See Chumillas–Montesinos [3]. In Section 2 we present a formula that gives the number of equivalence classes of epimorphisms of a knot group onto a dihedral group. In Section 3 we investigate 5-fold irregular dihedral branched coverings of knots. The second named author gets a technique to investigate 3-fold irregular branched coverings of Montesinos knots. The first named author applies the technique to 5-fold irregular dihedral branched coverings of those knots and characterizes the coverings in Theorem 3.1. The relation between the homology group of a 3-fold irregular branched covering of a knot and that of the 2-fold branched covering of the knot has already been known (Theorem 3.5). We conjecture that the similar relation between the homology group of a 5-fold irregular dihegral branched covering of a knot and that of the 2-fold branched covering of the knot exists, but show that the conjecture is false for several Montesinos knots.

M.E. Bozhüyük (ed.), Topics in Knot Theory, 269–276.

2. D_p-representations of Knots

Let D_p be the dihedral group of symmetries of a regular, p-sided polygon for an odd integer $p \geq 3$. Note that D_p is expressed as

$$D_p = < a, b \mid a^2,\ b^p,\ abab >$$

and is a subgroup of Σ_p, the symmetric group of degree p. Let K be a knot in S^3 and $G = \pi_1(S^3 - K)$. We call an epimorphism μ of G onto D_p a *D_p-representation* of G (or of K).

Definition 2.1. Let μ and μ' be D_p-representations of K. Then, μ is called *equivalent to* μ', if there exists an inner automorphism θ of Σ_p such that $\mu' = \theta\mu$. Note that θ is also an automorphism of D_p.

Let p be a prime integer. We denote by $\widetilde{M}_2(K)$ the 2-fold branched covering of K and by ν the rank of $H_1(\widetilde{M}_2(K); Z_p)$. Using the argument of Section 10 of Fox [4], we have

Theorem 2.2. *The number of equivalence classes of D_p-representations of K is equal to $(p^\nu - 1)/(p-1)$.*

Let $\Delta_K(t)$ be the Alexander polynomial of K. Then, $\Delta_K(-1)$ is equal to the product of torsion numbers of $H_1(\widetilde{M}_2(K))$. Therefore we have the following.

Corollary 2.3. *There exists a D_p-representation of K if and only if $\Delta_K(-1) \equiv 0$ mod p.*

Remark. Corollary 2.3 holds even if p is nonprime.

3. Main Results

Suppose that K admits a D_p-representation μ. Then, we denote by $\widetilde{M}_\mu(K)$ the p-fold irregular dihedral branched covering of K associated with μ. We also call $\widetilde{M}_\mu(K)$ the *D_p-branched covering* of K associated with μ.

Remark. Let K be the 2-bridge knot $S(\alpha, \beta)$. If $\alpha \equiv 0$ mod p, then there exists a D_p-representation μ of K and $\widetilde{M}_\mu(K)$ is homeomorphic to S^3.

We need to investigate the covering of a knot whose bridge index is greater than two. From now on, we restrict knots to Montesinos knots and set $p = 5$. A Montesinos knot K has a diagram which is given as in Figure 1 and K is denoted by $M(e; (\alpha_1, \beta_1), (\alpha_2, \beta_2), \ldots, \alpha_r, \beta_r))$, where (α_i, β_i) is the Schubert's notation for a rational tangle T_i for each i and e the number of half twists. See [1] or [2] for Montesinos knots. Let ν' be the number of i's such that $\alpha_i \equiv 0$ mod 5.

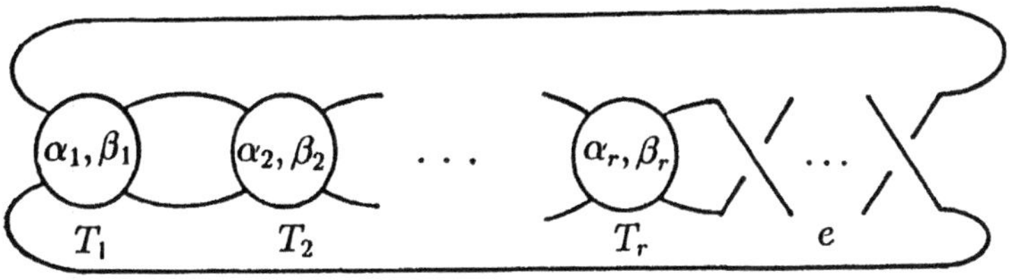

Figure 1

Theorem 3.1. *Let K be a Montesinos knot.*
(1) *If $\nu' = 0$ and there exists a D_5-representation of K, then the number of equivalence classes of D_5-representations is one and the associated D_5-branched covering is homeomorphic to S^3.*
(2) *If $\nu' = 1$, then there does not exist a D_5-representation of K.*
(3) *If $\nu' \geq 2$, then there exists a D_5-representation of K and the number of equivalence classes is equal to $(5^{\nu'-1}-1)/4$. Each of the associated D_5-branched coverings is homeomorphic to $L(p_i,q_i)\#L(p_i,q_i)$ or a connected sum of those spaces, where $(p_i,q_i) = (\alpha_j,\beta_j), 1 \leq j \leq r$, $(0,1)$ or $(1,1)$.*

Let $\nu = \nu(K)$ be the rank of $H_1(\widetilde{M}_2(K); Z_5)$. For a Montesinos knot K, we have

$$H_1(\widetilde{M}_2(K)) = < S_1, \ldots, S_r, h \mid S_i^{\alpha_i} h^{\beta_i}, 1 \leq i \leq r, S_1 S_2 \cdots S_r h^e > .$$

Therefore we obtain the following lemma.

Lemma 3.2. (1) *If $\nu' = 0$, then $\nu = 0$ or 1.*
(2) *If $\nu' \geq 1$, then $\nu = \nu' - 1$.*

To prove Theorem 3.1, we introduce some operations for a diagram of a knot, where each operation does not change the associated D_5-branched covering. Suppose that a knot K admits a D_5-representation μ and the group G of K is presented by a Wirtinger presentation of K. For a generator x_i corresponding to an overpass of a diagram of K, $\mu(x_i)$ is one of (25)(34), (12)(35), (13)(45), (14)(23) and (15)(24). For convenience, we shall introduce symbols $\overline{1}$, $\overline{2}$, $\overline{3}$, $\overline{4}$ and $\overline{5}$ for (25)(34), (12)(35), (13)(45), (14)(23) and (15)(24). A *trivial tangle* is a pair of a 3-ball and two proper arcs which are trivial and separated in the 3-ball. Since a simple D_5-branched covering of a 3-ball branched along a trivial tangle is a 3-ball, we have

Lemma 3.3. *Let K admit a D_5-representation μ. If we apply the following operations I and II to a diagram of K and obtain a link K' and a D_5-representation μ' of K', then $\widetilde{M}_{\mu'}(K')$ is homeomorphic to $\widetilde{M}_\mu(K)$, where $\overline{u}$ and $\overline{v}$ are disjoint elements of $\{\overline{1},\overline{2},\overline{3},\overline{4},\overline{5}\}$ in Figure 2 and Figure 3.*

I.

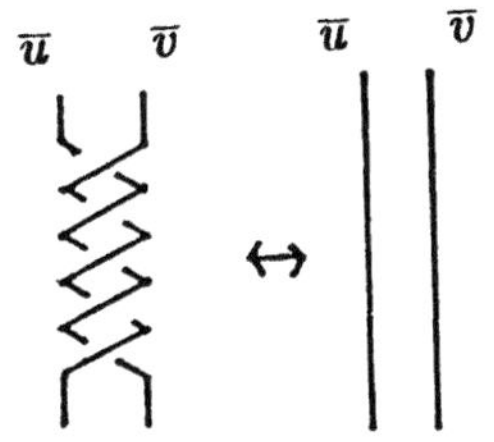

Figure 2

II.

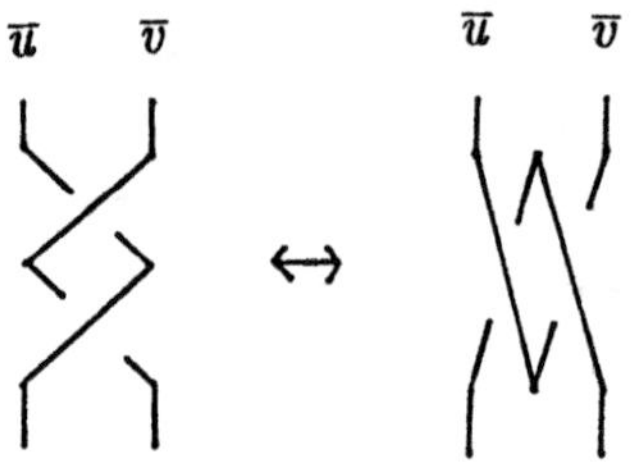

Figure 3

Remark. (1) A operation I is derivable from II.
(2) When $\overline{u} = \overline{v}$ in Lemma 3.3, $\widetilde{M}_{\mu'}(K')$ is not homeomorphic to $\widetilde{M}_{\mu}(K)$. We have $\nu(K) = \nu(K')$, even if $\overline{u} = \overline{v}$.

Proof of Theorem 3.1. By Lemma 3.2, we have the number of equivalence classes of D_5-representations of a Montesinos knot K.

Let T be a rational tangle of type (α, β) in K which admits a D_5-representation μ. By Lemma 3.3, we only need to know $\mu(x_1)$, $\mu(x_2)$, $\mu(x_3)$ and $\mu(x_4)$ where x_i corresponds to one overpass of four ends of T. When $\alpha \equiv 0 \bmod 5$, there are two cases which $\mu(x_i)$ admits, up to equivalence (see Figure 4). When $\alpha \not\equiv 0 \bmod 5$, there are also two cases which $\mu(x_i)$ admits, up to equivalence (see Figure 5). Hence, if $\nu' = 0$ and K admits a D_5-representation μ, then $\mu(x_i)$ satisfies a representation as in Figure 5 (ii) for each tangle in K. We may transform K so that any tangle in K admits a representation with $\overline{u} = \overline{1}$ in Figure 5 (ii), where such a tangle is changed to a trivial tangle of horizontal arcs by applying a sequence of operations I and II of Lemma 3.3. Therefore, the associated D_5-branched covering is homeomorphic to S^3, which is the D_5-branched covering of trivial 2-component link.

Figure 4

Suppose that $\nu' \geq 2$. If $\alpha \not\equiv 0 \bmod 5$ for a tangle T in K, then T admits a representation as in Figure 5 (i). If $\alpha \equiv 0 \bmod 5$ for T, then T admits a representation as in Figure 4 (i) or (ii). Note that a tangle in Figure 4 (ii) is changed to a trivial tangle of vertical arcs by applying a sequence of operations I and II of Lemma 3.3. Let $L = L_1 \cup L_2 \cup \cdots \cup L_n$ be the resulting link, where $2 \leq n \leq \nu'$. Suppose that $n = 2$. Then, each of L_1 and L_2 is a connected sum of 2-bridge knots $S(\alpha_j, \beta_j)$, respectively. Note that disjoint element $\overline{u}$ and $\overline{v}$ of D_5 are given to diagrams of L_1 and L_2, repectively. Hence, $\widetilde{M}_{\mu}(L)$ is homeomorphic to a connected sum of two $\widetilde{M}_2(L_1)$'s and two $\widetilde{M}_2(L_2)$'s, which is $\#(L(\alpha_j, \beta_j)\#L(\alpha_j, \beta_j))$. If $n = 3$ and $\overline{u}$, $\overline{v}$ and $\overline{w}$ are given to diagrams of L_1, L_2 and L_3,

Figure 5

then not all of $\overline{u}$, $\overline{v}$ and $\overline{w}$ are same. Assume that $\overline{u} \neq \overline{v}$. Let μ_1 (resp. μ_2) denote the resulting representation of $L_1 \cup L_2$ (resp. L_3) induced by $\overline{u}$ and $\overline{v}$ (resp. $\overline{w}$). Choose 3-balls B_1 and B_2 such that $L_1 \cup L_2 \subset B_1, L_3 \subset B_2, B_1 \cup B_2 = S^3$ and $\partial B_1 = \partial B_2$. The lift of B_2 in $\widetilde{M}_{\mu_1}(L_1 \cup L_2)$ consists of five distinct 3-balls $\widetilde{B}_{21}, \widetilde{B}_{22}, \ldots, \widetilde{B}_{25}$. Similarly the lift of B_1 in $\widetilde{M}_{\mu_2}(L_3)$ consists of five distinct 3-balls $\widetilde{B}_{11}, \widetilde{B}_{12}, \ldots, \widetilde{B}_{15}$. Then, $\widetilde{M}_\mu(L)$ is given by $(\widetilde{M}_{\mu_1}(L_1 \cup L_2) - \bigcup_{i=1}^{5} \widetilde{B}_{2i}) \cup (\widetilde{M}_{\mu_2}(L_3) - \bigcup_{i=1}^{5} \widetilde{B}_{1i})$, where $\partial \widetilde{B}_{1i}$ and $\partial \widetilde{B}_{2i}$ are identified for $i = 1, 2, \ldots, 5$. Since $\widetilde{M}_{\mu_2}(L_3)$ consists of three disjoint spaces, i.e. two $\widetilde{M}_2(L_3)$'s and one S^3, we have $\widetilde{M}_\mu(L) \cong \#(L(\alpha_j, \beta_j)\#L(\alpha_j, \beta_j))\#(S^2 \times S^1)\#(S^2 \times S^1)$. Similarly when $n \geq 4$, $\widetilde{M}_\mu(L)$ is homeomorphic to a connected sum of $\#(L(\alpha_j, \beta_j)\#L(\alpha_j, \beta_j))$ and $S^2 \times S^1$'s, where there are $2(n-2)$ $(S^2 \times S^1)$-type terms.

Let ρ_μ be the rank of $H_1(\widetilde{M}_\mu(K); Z_5)$.

Collorary 3.4. *If a Montesinos knot K admits a D_5-representation μ, then we have*

$$\rho_\mu = 2(\nu - 1).$$

Proof. If $\nu' = 0$ and K admits a D_5-representation μ, then $\nu = 1$ and $\rho_\mu = 0$ by Theorem 3.1 (1). If $\nu' \geq 2$, then $\nu = \nu' - 1$ and $\rho_\mu = 2(\nu' - 2)$ by the proof of Theorem 3.1 (3).

Suppose that K admits a D_3-representation μ. Then, the following theorem, mentioned by Sakuma, expresses the relation between $\widetilde{M}_\mu(K)$ and $\widetilde{M}_2(K)$. The theorem is easily proved by comparing a presentation matrix of $H_1(\widetilde{M}_\mu(K); Z_3)$ with that of $H_1(\widetilde{M}_2(K); Z_3)$.

Theorem 3.5. ([9]) *If a knot K admits a D_3-representation μ, then we have*

$$H_1(\widetilde{M}_\mu(K); Z_3) \oplus Z_3 \cong H_1(\widetilde{M}_2(K); Z_3).$$

On the analogy of Theorem 3.5, we expect the following conjecture.

Conjecture 3.6. If a knot K admits a D_5-representation μ, then we have

$$H_1(\widetilde{M}_\mu(K); Z_5) \oplus Z_5 \cong H_1(\widetilde{M}_2(K); Z_5).$$

Corollary 3.4 gives a counter-example for Conjecture 3.6 when $\nu \geq 2$, i.e. $\nu' \geq 3$.

Remark. Corollary 3.4 is false for several knots, for example $9_{40}, 9_{49}, 10_{103}$ and 10_{155}, each of which is not a Montesinos knot.

For any odd prime p, let $\nu_p(K)$ be the rank of $H_1(\widetilde{M_2}(K); Z_p)$ and $\rho_{\mu,p}(K)$ the rank of $H_1(\widetilde{M_\mu}(K); Z_p)$ for a D_p-representation μ of K. Sakuma also mentions the following conjecture.

Conjecture 3.7. If K admits a D_p-representation μ, then we have

$$\nu_p(K) - 1 \leq \rho_{\mu,p}(K) \leq \frac{p-1}{2}(\nu_p(K) - 1).$$

Appendix. The following table shows the knots of less than eleven crossings and more than 2 bridges that admit D_5-representations. The homology group of 2-fold branched covering, and the associated D_5-branched covering are also given. The following table also points out several errors in the table compiled in [3]. For example, the D_5-branched covering of 10_{122} is a prism manifold $M_{1,-4}$ and that of 10_{129} is S^3. Furthermore, there exist six equivalence classes of D_5-representations of 9_{40} and the associated D_5-branched coverings are mutually homeomorphic. Similar results happen for $9_{49}, 10_{103}$ and 10_{155}.

K	$H_1(\widetilde{M_2}(K))$	$\widetilde{M_\mu}(K)$
8_{16}	Z_{35}	S^3
8_{18}	$Z_{15} \oplus Z_3$	$RP^3 \# RP^3$
8_{21}	Z_{15}	S^3
9_{24}	Z_{45}	S^3
9_{37}	$Z_{15} \oplus Z_3$	S^3
9_{39}	Z_{55}	S^3
9_{40}	$Z_{15} \oplus Z_5$	$L(5,4)$
9_{49}	$Z_5 \oplus Z_5$	$L(5,2)$
10_{56}	Z_{65}	S^3
10_{58}	Z_{65}	$RP^3 \# RP^3$
10_{59}	Z_{75}	$RP^3 \# RP^3$
10_{60}	Z_{85}	$RP^3 \# RP^3$
10_{62}	Z_{45}	S^3
10_{66}	Z_{75}	S^3
10_{81}	Z_{85}	S^3
10_{83}	Z_{85}	S^3
10_{100}	Z_{65}	$L(11,7)$
10_{101}	Z_{85}	$L(11,3)$
10_{103}	$Z_{15} \oplus Z_5$	$L(5,2)$
10_{106}	Z_{75}	S^3
10_{109}	Z_{85}	$RP^3 \# RP^3$
10_{116}	Z_{95}	$L(11,3)$
10_{120}	Z_{105}	$L(11,7)$
10_{121}	Z_{115}	$L(19,8)$
10_{122}	Z_{105}	$M_{1,-4}$
10_{129}	Z_{25}	S^3
10_{132}	Z_5	S^3
10_{136}	Z_{15}	$RP^3 \# RP^3$
10_{137}	Z_{25}	$RP^3 \# RP^3$
10_{138}	Z_{35}	$RP^3 \# RP^3$
10_{142}	Z_{15}	S^3
10_{155}	$Z_5 \oplus Z_5$	$L(5,2)$
10_{156}	Z_{35}	S^3
10_{158}	Z_{45}	$L(11,7)$
10_{161}	Z_5	$L(11,7)$
10_{162}	Z_5	$L(11,3)$
10_{163}	Z_{35}	$L(11,7)$
10_{165}	Z_{45}	$L(11,3)$

Table

References

[1] Boileau, M. and Zimmermann, B. (1987) "Symmetries of nonelliptic Montesinos links," Math. Ann. 277, 563–584.

[2] Burde, G. and Zieschang, H. (1985) Knots, Walter de Gruyter & Co., Berlin–New York.

[3] Chumillas, V. and Montesinos, J. M. (1988) "The homology of cyclic and irregular dihedral coverings branched over homology spheres," Math. Ann. 280, 483–500.

[4] Fox, R. H. (1962) "A quick trip through knot theory," in M. K. Fort Jr.(ed.), Topology of 3-manifolds and related topics, Prentice–Hall, Englewood Cliffs, New Jersey, pp. 120–167.

[5] Hilden, H. M. (1974) "Every closed orientable 3-manifold is a 3-fold branched covering space of S^3," Bull. Amer. Math. Soc. 80, 1243–1244.

[6] Hosokawa, F and Nakanishi, Y. (1986) "On 3-fold irregular branched covering spaces of pretzel knots," Osaka J. Math. 23, 249–254.

[7] Montesinos, J. M. (1974) "A representation of closed, orientable 3-manifolds as 3-fold branched coverings of S^3," Bull. Amer. Math. Soc. 80, 845–846.

[8] Murasugi, K. (1980) "On dihedral coverings of S^3," C. R. Math. Rep. Acad. Sci. Canada 2, 99–102.

[9] Sakuma, M. (1989) Personal communication.

2-DIMENSIONAL BRAIDS AND CHART DESCRIPTIONS

Dedicated to Professor Yoko Tao on her sixtieth birthday

SEIICHI KAMADA*
Department of Mathematics
Osaka City University
Osaka, 558, Japan

ABSTRACT. A simple 2-dimensional m-braid is a compact oriented surface F embedded in a bidisk $B^2 \times D^2$ satisfying a certain condition. It is described by an immersed graph on D^2, which is called a chart. Using chart descriptions, we define numerical invariants of a simple 2-dimensional m-braid and of a closed oriented surface embedded in R^4.

1. Introduction

Let B^2 and D^2 be oriented 2-disks and let X_m be a fixed set of distinct m points on the interior of B^2. An m-fold branched covering map between surfaces is said to be *simple* if the inverse image of each branch point consists of $m-1$ points.

DEFINITION: A *2-dimensional m-braid* is a compact oriented surface F properly embedded in a bidisk $B^2 \times D^2$ such that (1) the composition $F \subset B^2 \times D^2 \to D^2$ of the inclusion and the projection is an m-fold branched covering and (2) the restriction of F to $B^2 \times \partial D^2$ is the product $X_m \times \partial D^2$. If the branched covering is simple, then we call it a *simple* 2-dimensional m-braid.

Throughout this paper all 2-dimensional m-braids are simple unless otherwise stated.

Let F be a 2-dimensional m-braid in $B^2 \times D^2$ and S^2 an oriented 2-sphere obtained from D^2 by identifying ∂D^2 with a point. Since the boundary of F is $\partial F = X_m \times \partial D^2$, F naturally induces a closed oriented surface $\widehat{F}$ embedded in $B^2 \times S^2$. Identify $B^2 \times S^2$ with a tubular neighborhood of a standard 2-sphere in R^4, and we have a closed oriented surface $\widehat{F}$ embedded in R^4. We call it the *closure* of F or a *closed 2-dimensional m-braid.*

THEOREM 1.1 (VIRO/KAMADA[4]). *For a closed oriented surface S embedded in R^4, there is a 2-dimensional m-braid for some m such that the closure is ambient isotopic to S in R^4.*

A 2-dimensional m-braid is described by an immersed graph on D^2, which is called a *chart* (cf.[5]). We devote Sect.2 to explaining how to describe a 2-dimensional m-braid by a chart. Each vertex of a chart is colored white or black.

*Partially supported by Grant-in-Aid for Scientific Reserch, Ministry of Education, Science and Culture: A Fellow of the Japan Society for the Promotion of Science for Japanese Junior Scientists.

M.E. Bozhüyük (ed.), Topics in Knot Theory, 277–287.

A black vertex corresponds to a branch point of the branched covering $F \to D^2$ associated with 2-dimensional m-braid F, and hence the number of black vertices is $m - \chi$, where χ is the Euler characteristic of F.

DEFINITION:

(1) Let F be a 2-dimensional m-braid. The *w-index* of F, denoted by $w(F)$, is the minimum number of white vertices of charts which describe F.

(2) Let S be a closed oriented surface embedded in R^4. The *w-index* of S, denoted by $w(S)$, is the minimum number of w-indices of 2-dimensional braids whose closures are ambient isotopic to S.

By the above theorem, the w-index of S is defined.

Let S be a compact oriented surface properly embedded in a 4-manifold W^4. A 3-ball B^3 embedded in W^4 is said to be a *1-handle attaching to S* if the intersection of B^3 and S is a pair of 2-disks on the boundary of B^3 and the closure of $S \cup \partial B^3 - S \cap B^3$ in W^4 is an orientable surface. We assign the resultant surface in W^4 an orientation induced from $S - S \cap B^3$ and call it a *surface obtained from S by surgery along B^3* or simply the *surgery result*. A closed oriented surface embedded in R^4 is said to be *ribbon* if it is obtained from a trivial 2-link in R^4 by surgery along some 1-handles.

THEOREM 1.2 (cf. [**5**]). *Let S be a closed oriented surface embedded in R^4. The w-index is zero if and only if S is ribbon.*

For a closed oriented surface embedded in R^4, there are mutually disjoint 1-handles attaching to it such that the surgery result is ribbon. (In fact there are mutually disjoint 1-handles such that the surgery result is *unknotted* i.e. bounds mutually disjoint handlebodies in R^4, [**3**].) Denote by $r(S)$ the minimum number of such 1-handles. We denote by $t(S)$ the minimum number of triple points of projections of S (cf. Sect.3).

THEOREM 1.3. *For any non-negative integer N, there is a 2-knot such that*

$$w(S) \geq t(S) \geq r(S) > N.$$

2. Chart description of a 2-dimensional m-braid

In [**5**] it is introduced a method to describe a 2-dimensional m-braid by an immersed graph in D^2. Here we explain it through a slightly different way from that in [**5**]. In this paper we use the term "graph" for the underlying space of a finite 1-complex such that each 0-simplex has degree 1, 2 or 6. A "vertex" is a 0-simplex of degree 1 or 6. A connected component of the "graph" removed "vertices" is called an "edge" or a "loop" according as it is homeomorphic to an open interval or a 1-sphere.

DEFINITION: An *m-chart* Γ is an immersed graph (possibly, the empty) in the interior of D^2 satisfying the following:

(1) Each edge or loop is oriented and labeled an integer in $\{1, \cdots, m-1\}$.

(2) For each degree 6 vertex, the edges attaching to it are labeled alternately i and $i+1$ ($i \in \{1, \cdots, m-2\}$), consecutive three edges are oriented toward the outside of the vertex and the other consecutive three edges toward the inside.

(3) The singularity is empty or consists of double points where two edges or loops whose difference in labels is more than one intersect transversally.

We call a vertex of degree 1 a *black vertex* and a vertex of degree 6 a *white vertex*. An m-chart may be called a *chart* if we need not specify m.

For a given m-chart Γ, we define a 2-dimensional m-braid associated with Γ. Identify D^2 with $I_1 \times I_2$, where I_i ($i = 1, 2$) is the interval $[0, 1]$. Deform Γ by an isotopy of D^2 such that the map $\Gamma \subset I_1 \times I_2 \to I_2$ is a "Morse function", namely the restriction to $\Gamma - \{\text{vertices}\}$ is a Morse function in the ordinal sense and the restriction to a neighborhood of each vertex is as in Figure 1.

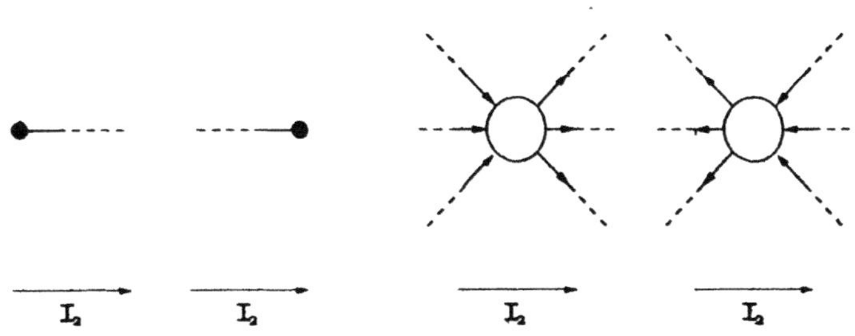

Fig. 1

A point of Γ is called an *exceptional point* if it is a critical (maximal or minimal) point of the Morse function, a double point or a vertex of Γ. A number $t \in I_2$ is called an *exceptional value* if there is an exceptional point of Γ in $I_1 \times \{t\}$. We assume that for each exceptional value t, there is just one exceptional point of Γ in $I_1 \times \{t\}$. A number $t \in I_2$ is called an *ordinary value* if it is not an exceptional value.

Let $\gamma_t : [0,1] \to I_1 \times I_2$ ($t \in I_2$)be a path defined by $\gamma_t(s) = (s, t)$ for $s \in [0, 1]$. If t is an ordinary value, then the path meets Γ transversally (or does not meet) on some edges and loops. Using orientations and labels of those edges and loops, we can assign each intersection a sign in $\{\pm 1\}$ and a label in $\{1, \cdots, m-1\}$. For an ordinary value t, let W_t be a word (possibly, the empty) in $\{\sigma_1^{\pm 1}, \cdots, \sigma_{m-1}^{\pm 1}\}$ such that the i-th letter is $\sigma_j^{\varepsilon_i}$ if the i-th intersection of γ_t with Γ has a sign ε_i and a label j.

Let $t_1, \cdots, t_s$ ($t_1 < \cdots < t_s$) be exceptional values and put $t_0 = 0$ and $t_{s+1} = 1$. For each i ($i = 0, 1, \cdots, s$), take a value t in (t_i, t_{i+1}) and define a word W^i by W_t. W^i does not depend on t. W^0 and W^s are empty words. Since we assume that there is one exceptional point of Γ in $I_1 \times \{t_i\}$ for each i ($i = 1, \cdots, s$), according to the kind of the exceptional point, the word W^i ($i = 1, \cdots, s$) is obtained from W^{i-1} by one of the following:

(a) Insertion of $\sigma_j^{\varepsilon}\sigma_j^{-\varepsilon}$.
(b) Deletion of $\sigma_j^{\varepsilon}\sigma_j^{-\varepsilon}$.
(c) Replacement of $\sigma_j^{\varepsilon_1}\sigma_k^{\varepsilon_2}$ by $\sigma_k^{\varepsilon_2}\sigma_j^{\varepsilon_1}$ ($|k-j|>1$).
(d) Replacement of $\sigma_j^{\varepsilon}\sigma_k^{\varepsilon}\sigma_j^{\varepsilon}$ by $\sigma_k^{\varepsilon}\sigma_j^{\varepsilon}\sigma_k^{\varepsilon}$ ($|k-j|=1$).
(e) Insertion of σ_j^{ε}.
(f) Delation of σ_j^{ε}.

Here $j,k \in \{1,\cdots,m-1\}$ and $\varepsilon,\varepsilon_1,\varepsilon_2 \in \{\pm 1\}$. [(a) and (b) correspond to a minimal point and a maximal point. (c) or (d) occurs when the exceptional point is a double point or a white vertex respectively. (e) and (f) correspond to a black vertex.]

Let b^i ($i=0,1,\cdots,s$) be a geometric m-braid in $B^2 \times I$ corresponding to the word W_i. Since W^0 and W^s are empty words, we may assume that b^0 and b^s are the trivial m-braid $X_m \times I \subset B^2 \times I$.

Let δ be a sufficiently small positive number. For a value t in $[t_i+\delta, t_{i+1}-\delta]$ ($i=0,1,\cdots,s$), let b_t be a geometric m-braid in $B^2 \times I_1 \times \{t\} \subset B^2 \times I_1 \times I_2 = B^2 \times D^2$ that is a copy of b^i. If W^i is obtained from W^{i-1} by one of (a)-(d), then b^i and b^{i-1} are equivalent and hence there is a continuous sequence of geometric m-braid from b^{i-1} to b^i. Using this sqeuence we define b_t in $B^2 \times I_1 \times \{t\}$ for $t \in (t_i-\delta, t_i+\delta)$. In cases (e) and (f), we use an *intercommutation* (cf.[**5**]) to define b_t in $B^2 \times I_1 \times \{t\}$ for $t \in (t_i-\delta, t_i+\delta)$ which is indicated locally as one of Figure 2. For t in $[0,\delta] \cup [1-\delta, 1]$, let b_t be the trivial m-braid $X_m \times I_1 \times \{t\}$ in $B^2 \times I_1 \times \{t\} \subset B^2 \times I_1 \times I_2$.

In this way we have a 1-parameter family of geometric m-braids $\{b_t \subset B^2 \times I_1 \times \{t\} | 0 \le t \le 1\}$ with a finite number of exceptions. The trace of this family forms a 2-dimensional m-braid in $B^2 \times I_1 \times I_2$, [**5**].

DEFINITION: Let F_1 and F_2 be 2-dimensional m-braids in $B^2 \times D^2$. F_1 is *equivalent to* F_2 if there is a fiber-preserving isotopy of $B^2 \times D^2$, when we regard $B^2 \times D^2$ as a B^2-bundle over D^2, carrying F_1 to F_2 and keeping $B^2 \times \partial D^2$ fixed.

The construction stated above has ambiguity and of course the obtained 2-dimensional m-braid is not unique. But we note that the equivalence class of the 2-dimensional m-braid constructed above is unique and depends only on the ambient isotopy class of Γ in D^2. A 2-dimensional m-braid is said to be *described by* Γ if it is equivalent to a 2-dimensional m-braid obtained from Γ by the above construction.

LEMMA 2.1 ([**5**, Theorem 14]). *For any 2-dimensional m-braid F, there is an m-chart which describes F.*

DEFINITION ([**5**]): Let Γ and Γ' be m-charts. Γ' is obtained from Γ by a *C-move of type 1, type 2* or *type 3* if they satisfy the following (1),(2) or (3) respectively:

(1) There is a 2-disk E on D^2 such that Γ and Γ' intersect ∂E transversally or do not intersect ∂E, they have no black vertices on E and they coincide on $D - E$.
(2) Γ and Γ' differ locally as in Figure 3, where $|i-j|>1$.

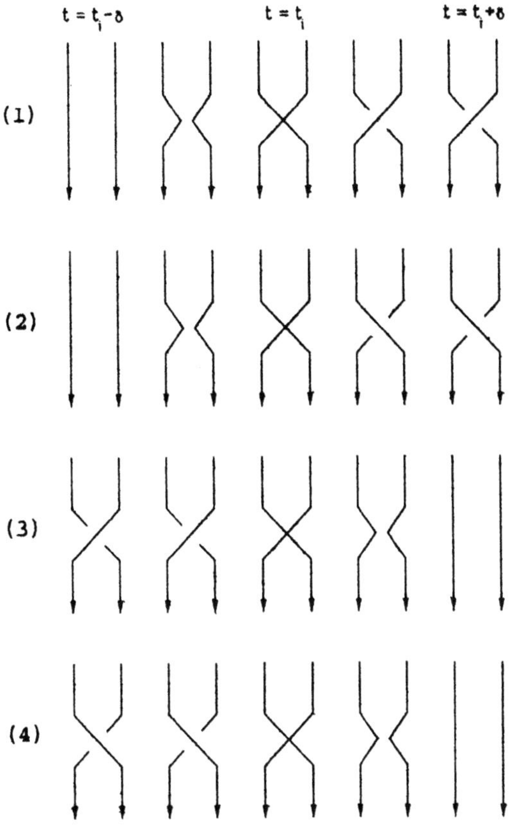

Fig. 2

(3) There are a white vertex W and a black vertex B in Γ connected by an edge, say α, such that α is not the middle of a set of consecutive three edges attaching to W which are oriented in the same direction. Γ' is obtained from Γ as follows: Delete $W \cup \alpha$, attach B to the edge opposite to α and join the other four edges in the obvious way. (cf. Figure 4, where $|i-j|=1$.)

If an m-chart Γ_2 is obtained from an m-chart Γ_1 by a finite number of applications of C-moves and their inverses (up to ambient isotopy of D^2), we say that Γ_2 is *C-move equivalent to* Γ_1.

LEMMA 2.2 ([**5**, Lemma 16]). *If an m-chart Γ_2 is C-move equivalent to an m-chart Γ_1, then the 2-dimensional m-braids described by them are equivalent.*

3. Remark on the w-index

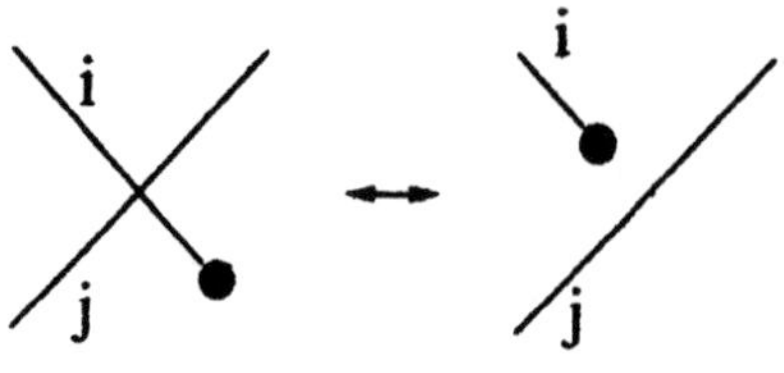

Fig. 3

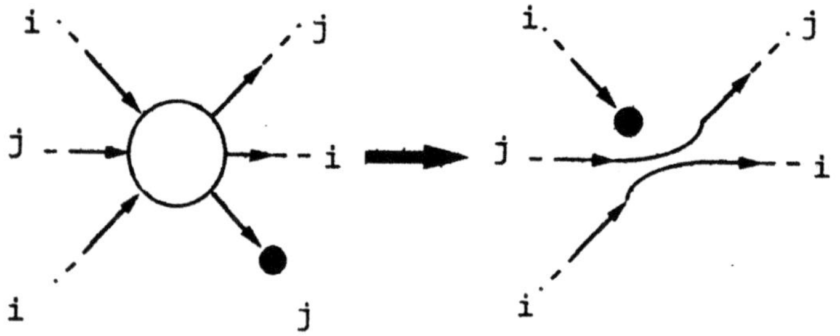

Fig. 4

DEFINITION: A 2-dimensional m-braid F is ***ribbon*** if it is equivalent to a 2-dimensional m-braid F' such that for an identification of D^2 with $I \times [0,1]$, F' is symmetric with respect to the hyperplane $B^2 \times (I \times \{1/2\})$.

Theorem 1.2 is a direct consequence of the following two lemmas.

LEMMA 3.1 ([**5**, Proposition 20]). *Let F be a 2-dimensional m-braid. F is ribbon if and only if there is an m-chart without white vertices which describes F.*

LEMMA 3.2 ([**5**, Lemma 22]). *(1) Let F be a ribbon 2-dimensional m-braid. Then the closure is a ribbon surface in R^4.*

(2) Let S be a ribbon surface in R^4. There is a ribbon 2-dimensional m-braid for some m whose closure is ambient isotopic to S in R^4.

Let $f : S \to R^3$ be a (piecewise-linear) map from a closed orientable surface S into R^3. The set $\Sigma = \{P \in R^3 | f^{-1}(P)$ consists at least two points$\}$ is called the ***singularity set*** of $f(S)$. A point in $f(S)$ is called a ***regular point***, a ***double point*** or a ***triple point*** if it has a neighborhood $N(P)$ in R^3 satisfying the following (1),(2) or (3) respectively:

(1) $N(P) \cap \Sigma$ is empty, see Fig. 5 (A).

(2) $N(P) \cap \Sigma$ is a proper arc in $N(P)$ which contains P, and $f^{-1}(N(P) \cap f(S))$ is a pair of disks on S whose images under f meet transversally each other, see Fig. 5 (B).

(3) $N(P) \cap \Sigma$ is three proper arcs in $N(P)$ which meet on P and

$f^{-1}(N(P) \cap f(S))$ consists of three disks on S whose images under f meet transversally each other, see Fig. 5 (C).

A map $f : S \to R^3$ is said to be *normal* if each point of the singularity set Σ is a double point or a triple point. In this case a point in $cl(\Sigma) - \Sigma$ is called a *branch point*, see Fig. 5 (D). A *projection* of a closed oriented surface S embedded in R^4 is the image under the projection $p : R^4 \to R^3$ of a surface S' in R^4 such that S' is ambient isotopic to S and the restriction $p|_{S'} : S' \to R^3$ is normal.

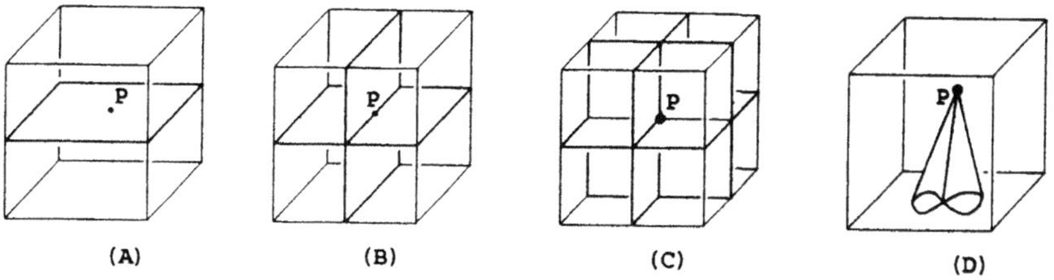

Fig. 5

LEMMA 3.3. *Let F be a 2-dimensional m-braid described by a chart Γ. The closure $\widehat{F}$ of F has a projection such that the closure of the singularity set is homeomorphic to Γ.*

PROOF: Let $\{b_t \subset B^2 \times I_1 \times \{t\} | t \in I_2 = [0,1]\}$ be a family of geometric m-braid with a finite number of exceptions $b_{t_1}, \cdots, b_{t_n}$ where intercommutations occur. Let $\widehat{b_t}$ be a closed m-braid in $B^2 \times S^1 \times \{t\}$ obtained from b_t by identifying $S^1 = I_1/\partial I_1$. Let C be the unit circle on the xy-plane and $N(C)$ a tubular neighborhood of C in the xyz-space R^3_{xyz}. By the definition of $\{b_t\}$, we assume that for each ordinary value $t \in I_2$, b_t has a projection in the xy-plane such that it is an immersed m circles with double point singularities corresponding to letters in the word W_t defined in Sect.2. By an argument of [2] we see that the compact surface which is the trace $\widehat{b_t} \times \{t\}$ $(0 \le t \le 1)$ in $N(C) \times [0,1] \subset R^3_{xyz} \times [0,1]$ has a projection in $R^2_{xy} \times [0,1] \subset R^2_{xy} \times R^1_t$, by degenerating the z-axis direction, whose singularity set has the closure homeomorphic to Γ. The closure $\widehat{F}$ of F is obtained from the above surface in $R^3_{xyz} \times [0,1] \subset R^4$ by capping the upside in $R^3_{xyz} \times [1,\infty)$ and the downside in $R^3_{xyz} \times (-\infty, 0]$ by two families of m 2-disks in the obvious way, cf. [4,§ 3][5,§ 2]. Degenerating the z-axis, we have a projection of $\widehat{F}$ in $R^2_{xy} \times R^1_t = R^3_{xyt}$ such that the closure of the singularity set is homeomorphic to Γ. This completes the proof of Lemma 3.3.

We note that in Lemma 3.3 each triple point of the singularity set corresponds to a white vertex of Γ and each branch point corresponds to a black vertex.

Therefore, for a closed oriented surface S in R^4, $w(S) \geq t(S)$. Cater and Saito **[1]** showed that if S has a projection with t triple points, then there are mutually disjoint t 1-handles attaching to S such that the surgery result is a ribbon surface in R^4. Hence we have the following:

COROLLARY. *Let S be a closed oriented surface in R^4. Then $w(S) \geq t(S) \geq r(S)$.*

The following lemma is a refinement of the above result of Cater and Saito to respect the 2-dimensional braid condition.

LEMMA 3.4. *Let F be a 2-dimensional m-braid described by a chart Γ with t white vertices. (Hence F has a projection with t triple points.) Then there are mutually disjoint 1-handles attaching to F such that the surgery result is a ribbon 2-dimensional m-braid. (Hence it has a projection without triple points).*

We need an observation before proving this lemma. A *free edge* is an edge in a chart both endpoints of which are black vertices.

LEMMA 3.5. *Let F be a 2-dimensional m-braid described by a chart Γ. Let Γ' be a chart such that Γ is obtained from Γ' by removing a free edge. Then a 2-dimensional m-braid F' described by Γ' is obtained from F by surgery along a 1-handle attaching to F.*

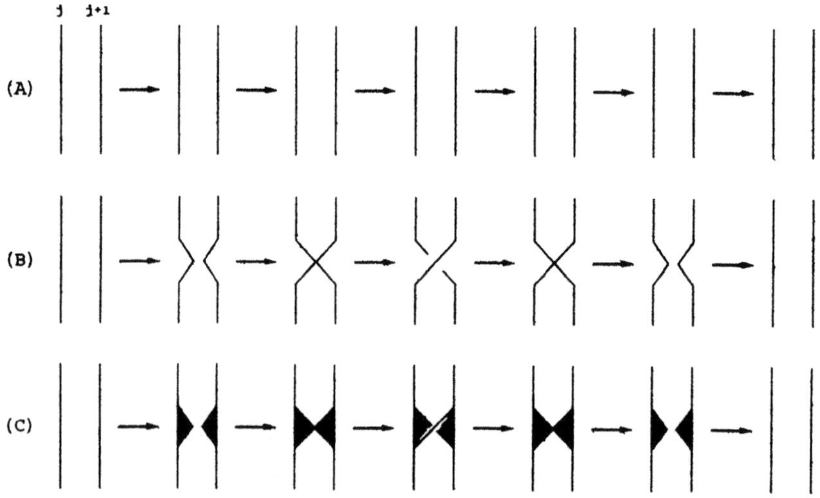

Fig. 6

PROOF: Let α be the free edge such that Γ is obtained from Γ' by removing it. If necessary applying some C-moves of type 2 we may assume that α does not intersect Γ. By an ambient isotopy of D^2, we may furthermore assume that it is parallel to I_2-axis and its orientation coincides with that of I_2. Let $N = J_1 \times J_2 \subset I_1 \times I_2$ be a neighborhood of α in D^2 which is disjoint from Γ, where J_i $(i = 1, 2)$ is a small interval in I_i. Let $\{b_t \subset B^2 \times I_1 \times \{t\} | t \in I_2\}$ and $\{b'_t \subset B^2 \times I_1 \times \{t\} | t \in I_2\}$ be families of geometric m-braids associated with F and F' as in Sect.2 respectively. By the definition, their restrictions over N are different locally as (A) and (B) in Figure 6 respectively, where j is the label of α. F' is obtained from F by surgery along a 1-handle illustrated in (C) in the figure. This completes the proof of Lemma 3.5.

PROOF OF LEMMA 3.4: For each white vertex of Γ, insert a free edge with label j as in Figure 7, where i and j ($|i-j|=1$) are labels of edges attaching to the vertex. Let Γ' be a chart obtained from Γ by insrting t free edges in this way and let F' be a 2-dimensional m-braid described by Γ'. By Lemma 3.5, F' is obtained from F by surgery along mutually disjoint t 1-handles attaching to F. We show that F' is a ribbon 2-dimensional m-braid. For each pair of a white vertex and a free edge of Γ', apply a C-move of type 1 as in Figure 8 and then a C-move of type 3 (Fig. 4). The result has no white vertices, which is C-move equivalent to Γ'. By Lemmas 2.2 and 3.1, F' is ribbon. This completes the proof of Lemma 3.4.

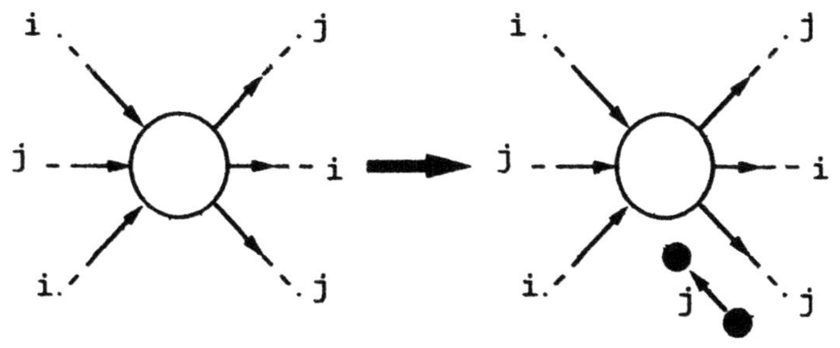

Fig. 7

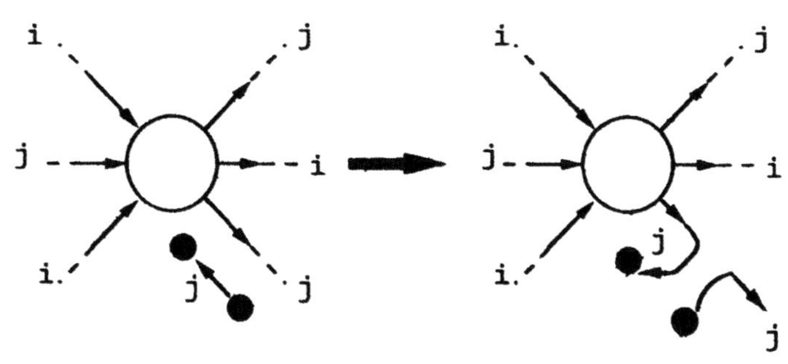

Fig. 8

PROOF OF THEOREM 1.3: Let m be an integer with $m > 2N$ and $p(> 0)$ an odd prime integer. Let K be a 2-twist spun 2-knot (cf.[12]) in R^4 of a 2-bridge knot of type (p,q), where q is an integer with $\gcd(p,q) = 1$. Consider the connected sum, say S, of m copies of K. By Corollary of Lemma 3.3, it is sufficient to show that $r(S) > N$. Let $E(K)$ and $\widetilde{E}(K)$ be the exterior of K in R^4 and its universal abelian covering space. The covering transformation group $< t >$, infinite cyclic group generated by a generator t, of $\widetilde{E}(K)$ induces a Λ-module structure on the first integral homology $\mathrm{H}_1(\widetilde{E}(K))$ of $\widetilde{E}(K)$, where Λ is the group ring $Z < t >$. Since the fundamental group of $E(K)$ has a presentation

$$< x, a | xax^{-1} = a^{-1}, a^p = 1 >,$$

the Λ-module $\mathrm{H}_1(\widetilde{E}(K))$ has a Λ-module presentation [**10**]

$$< a | ta = -a, pa = 0 >_\Lambda \cong \Lambda / I_{(t+1,p)},$$

where $I_{(t+1,p)}$ is the ideal generated by $t+1$ and p. Since S is the connected sum of m copies of K, the Λ-module $\mathrm{H}_1(\widetilde{E}(S))$ is Λ-isomorphic to the direct sum of m copies of $\Lambda / I_{(t+1,p)}$. (It is isomorphic to the direct sum of m copies of Z_p as an abelian group.) Therefore we have $e(\mathrm{H}_1(E(S))) = m$. (For a Λ-module H, we denote by $e(H)$ the minimum number of generators (as a Λ-module) of H.)

Suppose that there are mutually disjoint n 1-handles attaching to S such that the surgery result R is a ribbon surface in R^4, where n is an integer. Since R is a surface obtained from S by surgery along n 1-handles, by [**9**], we have

$$e(\mathrm{H}_1(\widetilde{E}(R))) \geq e(\mathrm{H}_1(\widetilde{E}(S))) - n = m - n.$$

On the other hand, R is a connected ribbon surface in R^4 of genus n. By Kawauchi's second duality theorem [**6**], Kawauchi [**7**] and Sekine [**11**] showed that

$$e(\mathrm{Ext}^2(\mathrm{H}_1(\widetilde{E}(R)), \Lambda)) \leq n,$$

where $\mathrm{Ext}^2(\,,\Lambda)$ stands for the second extension.

Since $\mathrm{H}_1(\widetilde{E}(R))$ is a quotient of $\mathrm{H}_1(\widetilde{E}(S))$, it is isomorphic to the direct sum of s copies of Z_p as an abelian group for some integer s with $s \leq m$. By [**8**, Cor. 4.3], $\mathrm{H}_1(\widetilde{E}(R))$ is Λ-isomorphic to $\mathrm{Ext}^2(\mathrm{H}_1(\widetilde{E}(R)), \Lambda)$. Therefore

$$m - n \leq e(\mathrm{H}_1(\widetilde{E}(R))) = e(\mathrm{Ext}^2(\mathrm{H}_1(\widetilde{E}(R)), \Lambda)) \leq n,$$

and hence $2n \geq m > 2N$. Hence we have $r(S) > N$. This completes the proof.

References

[1] J. S. Carter and M. Saito (1991), *Canceling branch points on projections of surfaces in 4-space*, preprint.

[2] J. S. Carter and M. Saito (1991), *Reidemeister moves for surface isotopies and their interpretation as moves to movies*, preprint.

[3] F. Hosokawa and A. Kawauchi (1979), *Proposals for unknotted surfaces in four-spaces*, Osaka J. Math. **16**, 233–248.

[4] S. Kamada (1990), *A characterization of groups of closed orientable surfaces in 4-space*, preprint.

[5] S. Kamada (1992), *Surfaces in R^4 of braid index three are ribbon*, Journal of Knot Theory and its Ramifications **1**, 137–160.

[6] A. Kawauchi (1986), *Three dualities on the integral homology of infinite cyclic coverings of manifolds*, Osaka J. Math. **23**, 633–651.

[7] A. Kawauchi (1990), *The Alexander modules of surfaces in 4-sphere*, Proc. KAIST Math. Workshop 5, Korea Adv. Inst. Sci. Tech., Taejon, 81–89.

[8] J. Levine (1977), *Knot modules. I*, Trans. Amer. Math. Soc. **229**, 1–51.

[9] K. Miyazaki (1986), *On the relationship among unknotting numbers, knotting genus and Alexander invariant for 2-knots*, Kobe J. Math. **3**, 77–85.

[10] D. Rolfsen (1976), *Knots and Links*, Publish and Perish, Inc..

[11] M. Sekine (1989), *Kawauchi's second duality and knotted surfaces in 4-sphere*, Hiroshima Math. J. **19**, 641–651.

[12] E. C. Zeeman (1965), *Twisting spun knots*, Trans. Amer. Math. Soc. **115**, 471–495.

ON LINKS EMBEDDED INTO SURFACES OF HEEGAARD SPLITTINGS OF S^3

O.KARALASHVILI
Departement of Mechanics and Mathematics
Tbilisi State University
University st.2
380086 Tbilisi
Georgia

ABSTRACT. This paper is concerned with coils, i.e. with the isotopy classes of embeddings $(F_g, \tilde{k}) \to S^3$, where F_g is a closed oriented surface of genus g, $\tilde{k}$ is a system of non-intersecting simple closed curves on F_g, and the embedding k maps F_g onto the surface of some Heegaard splitting of S^3. Each coil represents a link in S^3, namely, the link onto which the system $\tilde{k}$ is mapped by the embedding k. The notion of equivalence of coils is introduced and it is proved that two coils represent the same link if and only if they are equivalent.

0.Introduction

In this paper we will concern ourselves with objects which, as far as we know, haven't been systematically studied up to this day, though, in fact, they often have been regarded in connection with various problems of knot theory. We are talking about embeddings of pairs $(F_g, \tilde{k})$ into S^3, where F_g is a closed oriented surface of genus g, $\tilde{k}$ is a system of non-intersecting simple closed curves on F_g, and F_g is embedded into S^3 as the surface of some Heegaard splitting of S^3. Such an embedding, or rather an isotopy class of such embeddings, will be called a coil of genus g. Of course, a coil can also be thought of as a link embedded into the surface of some Heegaard splitting of S^3, thus making the link a primary object and the coil a secondary. Our motivation of doing the opposite is based on ideas which are yet uncertain; still we will mention some of them.

First of all, however hopeless this may seem, we were tempted by the idea of replacing links by objects which look more convenient for explicit description. Links in S^3 are defined up to ambient isotopy, so that a link can be moved freely in S^3. Since Heegaard splittings of S^3 of given genus are defined uniquely up to isotopy, we can, while studying coils, almost ignore the isotopies of the surfaces and regard coils as links pinned to fixed surfaces in S^3. This strongly reduces the degree of freedom of the movements of links and, maybe, makes them easier to describe.

M.E. Bozhüyük (ed.), Topics in Knot Theory, 289–303.

Of course, this alone wouldn't be enough to motivate our research, since there are obviously too many coils corresponding to each link. Therefore, at first glance, the study of coils doesn't provide us with any real support in link theory; but it seems promising to pick a special kind of coils to represent links in S^3, namely, coilings of the minimal genus possible for a given link. Such minimal coilings cannot vary very widely. Moreover, we are almost ready to run the risk of conjecturing that, given a link in S^3, its minimal coiling is uniquely determined. For torus knots, at least, this is known to be true (example 1.7.b.), and it leads to a full classification of torus knots. (Frankly speaking, we rather think that our conjecture is only "almost" true, and that the minimal coilings of a given link still do vary, but in a range narrow enough to make them interesting to study.)

What we can do at any rate, is to introduce a new numeric link invariant, which we will call the coiling genus of a link. The latter is defined as the genus of the minimal coiling(s) of a given link. We will see (example 1.7.c.) that the coiling genus is upper-bounded by the bridge-number.

The next step to take would be to find a simple geometric criterion for a coil to be minimal; this, as it seems, is a solvable problem, and would help us to restrict the area of our research and (why not?) try to approach the problem of classification of the minimal coils of given genus.

One more advantage, which we hope to gain from studying coils, is connected with the Neuwirth conjecture on the decomposition of knot groups to free products with amalgamation. As proved by Culler and Shalen [2], every knot group can be presented as a product of free groups amalgamated along some subgroup. Nothing is known, though, about the rank of these groups. We suggest that, if k is a minimal coiling of a knot $\mathfrak{K}$ of coiling genus g on the Heegaard surface F_g in S^3, the surface F_g will automatically prove to be incompressible in the complement of $\mathfrak{K}$, and thus, by the theorem of Seifert - Van Kampen, (see, e.g.,[1]) will give a presentation of the group of the knot $\mathfrak{K}$ as a product with amalgamation of free groups of rank g. (This would also be a simple proof of the Neuwirth conjecture itself.)

The first problem, however, which must be considered when studying coils, is to find a criterion which would allow to decide whether two given coils represent the same link. In this paper we stick to this question, and give to it a full answer in theorem 3.1.

1.Basic Definitions

We will start by introducing the main notions we will deal with in this paper.

1.1.*Definition:* A handlebody H of genus g is obtained from a 3–ball D^3 by attaching g handles $H_i \cong D_i^2 \times I, I = [0,1], i = 1, \ldots, g$, such that the boundary $F = \partial H$ is a closed orientable surface of genus g (see fig.1): $H = D^3 \cup H_1 \cup \ldots \cup H_g; H_i \cap H_j = \emptyset, i \neq j; H_i \cap D^3 = D_{i0} \cup D_{i1}, D_{i0} \cap D_{i1} = \emptyset, D_{ij} \cong D_i^2 \times \{j\} \cong D^2$.

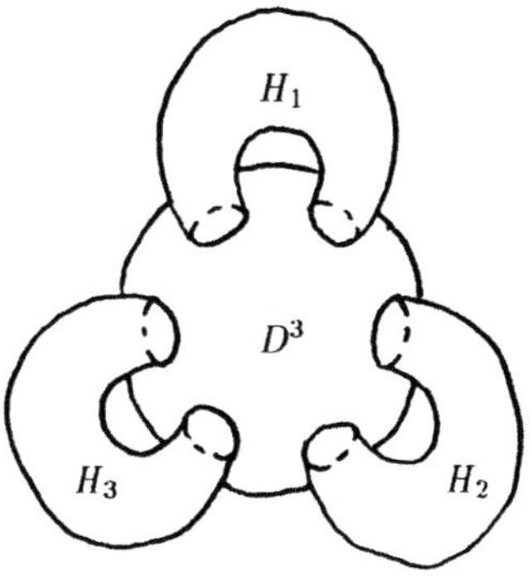

fig.1

1.2.*Definition:* The decomposition (H, H') of a closed orientable 3–manifold M into two handlebodies H, H' of genus g, $M = H \cup H', H \cap H' = \partial H = \partial H'$, is called a Heegaard splitting of M of genus g. The surface $F = \partial H = \partial H'$ is called a Heegaard surface in M.

We will be interested in embeddings of links into Heegaard surfaces in S^3. First of all, to avoid possible abuse of terminology, let's recall the following facts:

1.3.*Proposition:* For any $g \geq 1$ there exists, up to isotopy, a unique Heegaard splitting of S^3 of genus g. In other words, any two Heegaard surfaces of the same genus are ambient isotopic in S^3.

1.4.*Proposition:* Let k and k' be two embeddings of the union of m disjoint copies of the 1–sphere S^1 into a Heegaard surface $F \subset S^3$. If k and k' are isotopic in F, then they are ambient isotopic in S^3 (i.e. they are representations of the same link $\mathfrak{K}$ of multiplicity m in S^3).

Keeping 1.3-1.4. in mind, we will now introduce the notion of a coil. Let k be an embedding of the union of m disjoint copies of the 1–sphere S^1 into a Heegaard surface $F \subset S^3$ of genus g. The embedding k represents a link $\mathfrak{K}$ of multiplicity m in S^3. We will say that k is a coiling of $\mathfrak{K}$ on the surface F. If k, k' are two coils on the surface F and k is isotopic to k' in F, then k and k' are coilings of the same link $\mathfrak{K}$ on F. We will often ignore the difference between k and k' and use the term "coiling of $\mathfrak{K}$ on F" meaning "isotopy class of coilings of $\mathfrak{K}$ on F".

Also, if k and k' are coils on the Heegaard surfaces F and F' of the same genus g, and k' results from k by an isotopy connecting F and F', we will not distinguish between k and k'. Since all Heegaard surfaces of genus g in S^3 are isotopic, we will often use the term "coil of genus g" or "g–coil", meaning a coil on "some" Heegaard surface of genus g in S^3. The link represented by such a "g–coil" will be well defined.

1.5.*Example:* Two embeddings of the union of m disjoint copies of the 1–sphere S^1 into a Heegaard surface F of genus g may be not isotopic in F, but still represent the same g–coil. For example, the two realisations $t(2,3)$ and $t(3,2)$ of the trefoil knot on the surface of a standardly embedded torus $T^2 \subset S^3$ (see fig.2) give the same coiling of the trefoil knot, though they aren't isotopic in T^2. Indeed, if H and H' are the "inner" and the "outer" solid tori into which T^2 splits S^3, then $t(2,3)$ results from $t(3,2)$ by an isotopy connecting the Heegaard splittings (H, H') and (H', H).

Let's now give the precise definition:

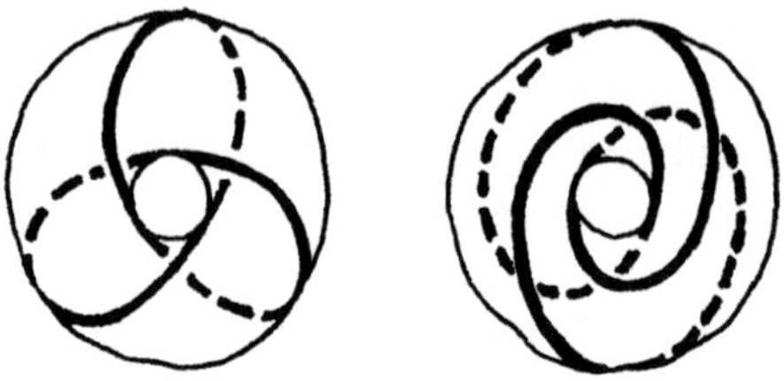

$t(2,3)$ $t(3,2)$

fig.2

1.6.*Definition:* A coil k of genus g and multiplicity m is an isotopy class of pairs $((H, H'), \tilde{k})$, where (H, H') is a Heegaard splitting of S^3 of genus g and $\tilde{k}$ is a system of m non-intersecting simple closed curves on the surface $F = \partial H = \partial H'$. If $\mathfrak{K}$ is the link represented by a g-coil k, we will say that k is a g-coiling of $\mathfrak{K}$.

1.7.*Examples:*

a. The only links which have 0-coilings are unlinks.

b. Any torus knot has a uniquely determined 1-coiling; this directly follows from the classification theorem for torus knots (see, e.g., [1], p.45).

c. Any n-bridge knot $\mathfrak{K}$ has a coiling k of genus n; k can be easily constructed from the n-bridge presentation of $\mathfrak{K}$ by placing the n lower bridges of the presentation on the surface of a 3-ball in S^3 and adding a handle to carry each of the n upper bridges (see fig.3).

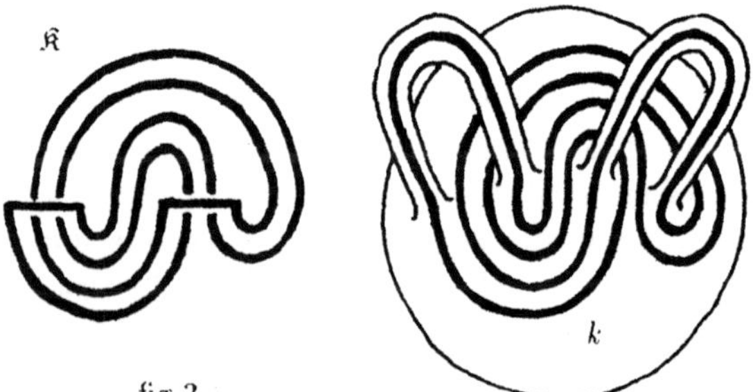

fig.3

Example 1.7.c. shows us how to construct a coiling for a given knot. Of course, the same can be done for links as well. So, to complete this section, we can state the following

1.8.*Proposition:* For every link $\mathfrak{K}$ in S^3 there exists a coil k representing $\mathfrak{K}$.

2. Elementary Operations on Coils

The first problem that naturally arises when we start to study coils as tools of link theory, is to describe the relations between different coilings of a link. In this section we will introduce some elementary operations, which, given a coil, will allow us to obtain from it new coils of the same link type. Later on we will see that, in fact, all coilings of a fixed link are related under these elementary operations.

2.1.*Definition:* Let (H, H') be a Heegaard splitting of S^3 of genus g and $D \subset H, D \cap \partial H = \partial D$ a non-separating, non-self-intersecting disc in H. Denote by $N(D)$ the regular neighborhood of D in $H, N(D) \cong D \times I, N(D) \cap \partial H \cong \partial D \times I$. The splitting $(r_D H, r_D H') = (H \backslash int N(D), H' \cup N(D))$ will be said to be obtained from (H, H') by cutting along D (see fig.4).

While $r_D H$ is a handlebody of genus $g-1$, $r_D H'$ isn't nessecarily a handlebody at all, so that $(r_D H, r_D H')$ needn't be a Heegaard splitting of S^3 (see fig.5).

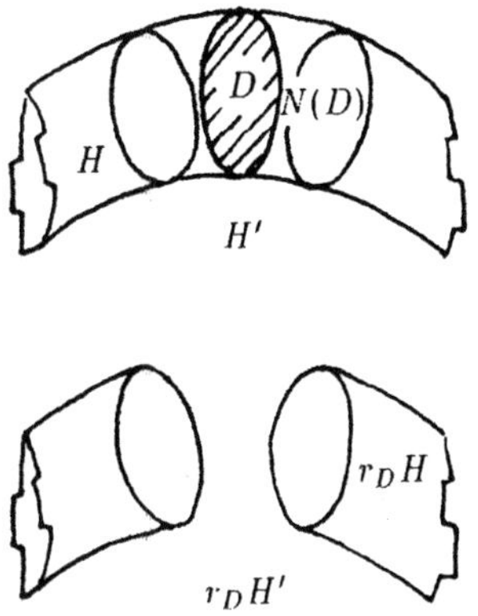

fig.4

2.2.*Definition:* Let (H, H') be a Heegaard splitting of S^3. A non-separating, non-self-intersecting disc $D \subset H, D \cap \partial H = \partial D$, will be called elementary, if the splitting $(r_D H, r_D H')$, obtained from (H, H') by cutting along D, is again a Heegaard splitting of S^3.

Now let k be a coil on the surface ∂H of a Heegaard splitting (H, H') of S^3, and let $D \subset H$ be an elementary disc such that $D \cap k = \emptyset$.

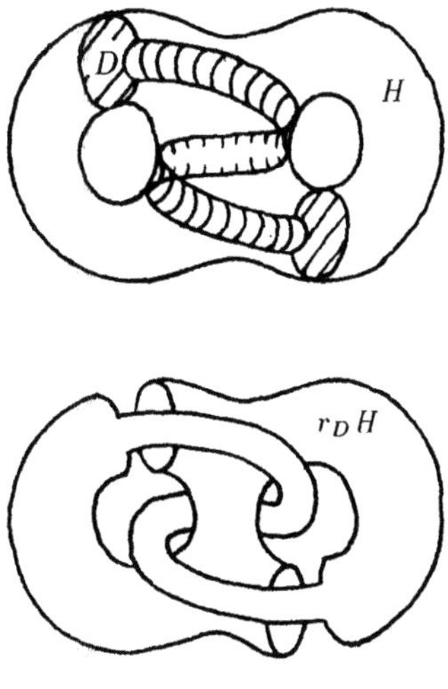

fig.5

2.3.*Definition:* The coil $r_D k$ on the Heegaard surface $\partial(r_D H)$ obtained from k by cutting the splitting (H, H') along the disc D will be called the reduction of k along D.

Obviously k and $r_D k$ represent the same link.

2.4.*Definition:* Let (H, H') be a Heegaard splitting of S^3, and let α be a simple arc in $H', \alpha \cap \partial H = \partial \alpha$. Denote by $N(\alpha) \cong \alpha \times D^2$ the regular neighborhood of α in H', $N(\alpha) \cap \partial H \cong \partial \alpha \times D^2$. The splitting $(s_\alpha H, s_\alpha H') = (H \cup N(\alpha), H' \backslash int N(\alpha))$ will be said to result from (H, H') by attaching a handle along α (see fig.6).

$(s_\alpha H, s_\alpha H')$ isn't always a Heegaard splitting of S^3; it will be one if and only if the arc α is boundary-parallel in H', i.e. if there exists an ambient isotopy $h_t : (H', \partial H) \times I \to (H', \partial H) \times I$ such that $h_0 = id(H', \partial H)$ and $h_1(\alpha) = const.$ Let k be a coil on the surface ∂H of the Heegaard splitting (H, H') of S^3, and let $\alpha \subset H'$ be a boundary-parallel arc such that $\alpha \cap k = \emptyset$.

2.5.*Definition:* The coil $s_\alpha k$ obtained from k by attaching to (H, H') a handle along the arc α, will be called the stabilisation of k along α.

Clearly, $s_\alpha k$ represents the same link as k.

Obviously, each of the operations of reduction or stabilisation has an inverse, which, accordingly, will be the stabilisation along some arc or the reduction along some disc. Indeed, let k be a coil on the surface ∂H of the Heegaard splitting (H, H') of S^3, and let $D \subset H, \alpha \subset H'$ be an elementary disc in H and a boundary-parallel arc in H' such that $D \cap k = \alpha \cap k = \emptyset$.

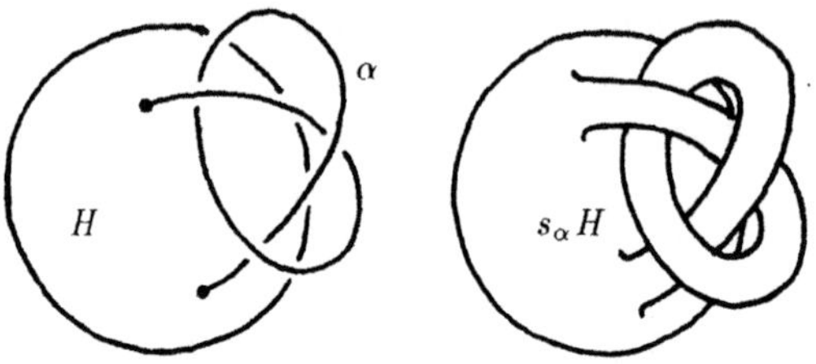

fig.6

Let x be an inner point of D; denote by α_x the arc $\alpha_x \cong x \times I \subset D \times I \cong N(D)$ in $r_D H'$. Then $(H, H') = (s_{\alpha_x} r_D H, s_{\alpha_x} r_D H')$ and $s_{\alpha_x} r_D k = k$ (see fig.7). Let y be an inner point of α; denote by D_y the disc $D_y \cong y \times D^2 \subset \alpha \times D^2 \cong N(\alpha)$ in $s_\alpha H$. Then $(H, H') = (r_{D_y} s_\alpha H, r_{D_y} s_\alpha H')$ and $r_{D_y} s_\alpha k = k$ (see fig.8).

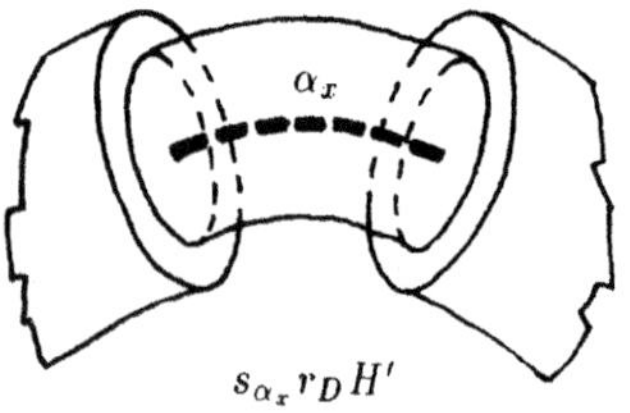

fig.7

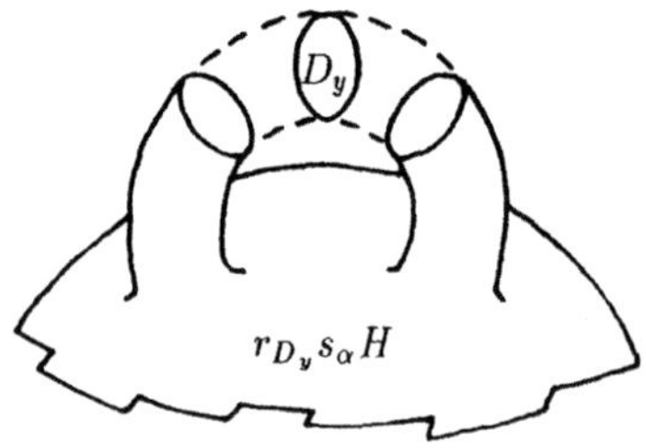

fig.8

2.6.*Definition:* Let k be a coil on the surface ∂H of the Heegaard splitting (H, H') of S^3, and let $D \subset H$ be an elementary disc intersecting k in exactly one point x_0.Denote by κ the component of k which intersects D, and fix an orientation on κ. Choose an orientation on D and denote by δ the closed path with initial point x_0, running once around ∂D in the positive direction. If the curve $\kappa\delta$ on ∂H is simple, then the coil $t_D k$ obtained from k by changing the component κ to $\kappa\delta$ will be called the translation of k along D (see fig.9). Again, $t_D k$ will represent the same link as k.

The coil obtained from k by changing the component κ to $\kappa\delta^{-1}$ will be denoted by $t_{D^{-1}} k$. At least one of the translations $t_D k, t_{D^{-1}} k$ is defined for any elementary disc $D \subset H, D \cap k = \{x_0\}$. Both of them will be defined if the intersection of k and δ is transversal.

If a translation $t_{D^\varepsilon} k, \varepsilon = 1$ or $\varepsilon = -1$, is defined for the disc D, then, as it is easy to check, the translation $t_{D^{-\varepsilon}} t_{D^\varepsilon} k$ is also defined and $t_{D^{-\varepsilon}} t_{D^\varepsilon} k = k$.

2.7.*Definition:* The operations of reduction, stabilisation and translation will be called the elementary operations on coils.

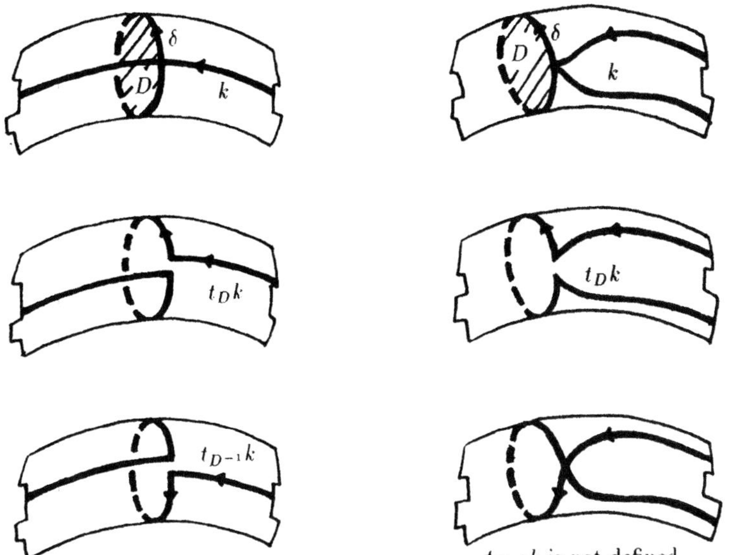

fig.9

As we have seen, each elementary operation has an inverse which is again an elementary operation. This justifies the following

2.8.*Definition:* Two coils k, k' will be called equivalent, if there exists a finite sequence of elementary operations $e_1, \ldots, e_n$ such that $k' = e_n \ldots e_1 k$.

3. Coil Diagrams

Our aim is now to prove the following main theorem:

3.1.*Theorem:* Two coils are coilings of the same link if and only if they are equivalent.

We have already seen that equivalent coils do represent the same link; now we're going to show that any two coilings of a link are equivalent.

The proof will consist of two parts. First, we shall introduce coil diagrams, which are black-and white colourings, due to certain conditions, of link diagrams. We shall prove that any two coils, whose diagrams are colourings of the same link diagram, are equivalent. After that, we will show that the Reidemeister moves by which different diagrams of a link can be transformed into one another, can be "translated" to elementary operations on the corresponding coils.

Consider the 2-sphere S^2 standardly embedded into S^3 as the equator. Let D_g be a disc with g holes embedded in S^3, $D_g = S^2 \backslash (U_0 \cup U_1 \cup \ldots \cup U_g)$, where $U_i \subset S^2, i = 0, 1, \ldots, g$ are open discs, $U_i \cap U_j = \emptyset$ for $i \neq j$. Denote by $N(S^2) \cong S^2 \times I, I = [-1, 1]$, the regular neighborhood of S^2 in S^3, and consider the handlebody $H_g \cong D_g \times I \subset S^2 \times I \cong N(S^2)$ (see fig.10). The surface $F_g = \partial H_g$ is a Heegaard surface of genus g in S^3, and the standard projection $p : N(S^2) \to S^2$ maps F_g onto D_g.

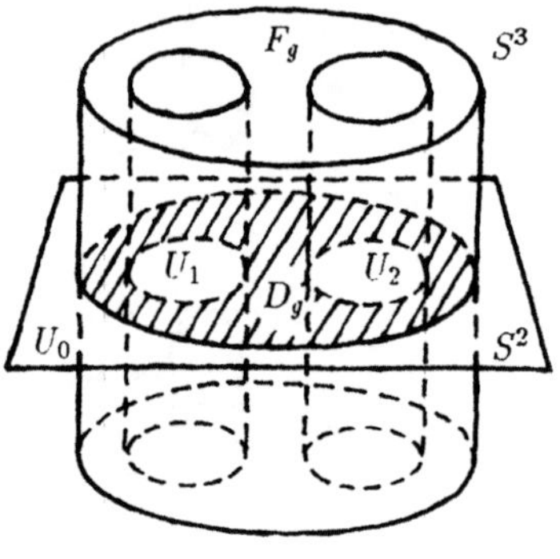

fig.10

Let k be a coiling of the link $\mathfrak{K}$ on the surface F_g, and let $p(k)$ be the projection of k on S^2.

3.2. *Definition:* The diagram of the coil k is a black-and-white colouring of the areas of the link diagram $p(k)$ obtained by painting black all areas σ^2 of $p(k)$ such that $\sigma^2 \subset D_g$, and painting white all the remaining areas (see fig.11).

3.3. *Definition:* Let K be a link diagram. A shading $\bar{K}$ of the diagram K is a black-and-white colouring of the areas of K such that, for any alternating edge σ^1 of K, at least one of the two areas adjacent to σ^1 is white-coloured.

It's easy to see that any coil diagram is a shaded link diagram in the sense of definition 3.3.

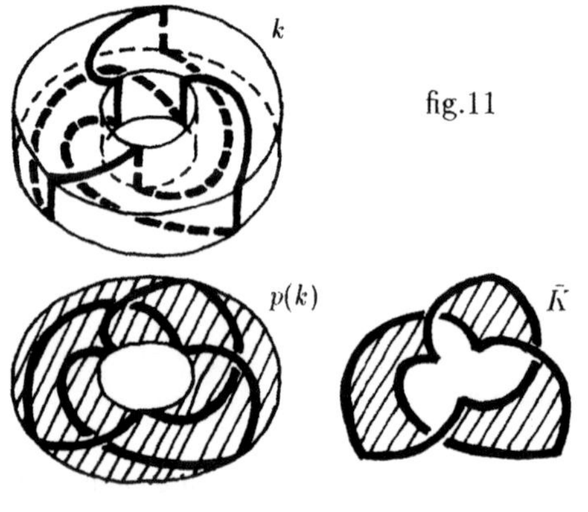

fig.11

3.4. *Examples:*

a. For any link diagram K the chess-board colouring of its areas is a shading of K.

b. For any link diagram K its trivial shading can be introduced by painting all areas of K white.

3.5. *Lemma:* Any two coils which have the same diagram are equivalent.

Proof: Let $\bar{K}$ be a shaded link diagram and suppose it is the diagram of a coil k on the Heegaard surface F_g. Let $\sigma_1^2, \ldots, \sigma_n^2$ be the areas of the diagram $\bar{K}$. By V_i we will denote the intersection of the area σ_i^2 with the holed disc $D_g = S^2 \backslash (U_0 \cup \ldots \cup U_g)$.

Consider a white-coloured area σ_i^2 of $\bar{K}$. For $j = 0, 1, \ldots, g$ we have either $U_j \cap \sigma_i^2 = \emptyset$ or $U_j \subset \sigma_i^2$. Suppose $U_j \subset \sigma_i^2$ for $j = 0, 1, \ldots, j(i)$ and $U_j \cap \sigma_i^2 = \emptyset$ for $j \geq j(i) + 1$. Obviously, $V_i = \sigma_i^2 \backslash (U_0 \cup \ldots \cup U_{j(i)})$.

Suppose $j(i) \geq 1$. Let $\gamma \subset intV_i$ be a simple arc connecting the boundaries of U_0 and U_1. The disc $p^{-1}(\gamma) \cong \gamma \times I$ will be an elementary disc in H_g, $p^{-1}(\gamma) \cap k = \emptyset$. By reducing the coil k along $p^{-1}(\gamma)$ we obtain a coil on the Heegaard surface F_{g-1} which has the same diagram $\bar{K}$. The number of the holes $U_j \subset \sigma_i^2$ will be reduced by one (see fig.12).

Repeating this operation $j(i) - 1$ times, we will obtain a coil on the Heegaard surface $F_{g-(j(i)-1)}$, whose diagram again will be $\bar{K}$ and for which the number of holes $U_j \subset \sigma_i^2$ will be equal to one.

The same procedure can be repeated for all white-coloured areas of the diagram $\bar{K}$ without changing the equivalence class and the diagram of the coil k. Therefore, we can, without loss of generality, assume that k is a coil on the Heegaard surface F_w, where $w+1$ is the number of the white-coloured areas of the diagram $\bar{K}$, and that for all white-coloured areas σ_i^2 of $\bar{K}$ we have $V_i = \sigma_i^2 \backslash U_i$.

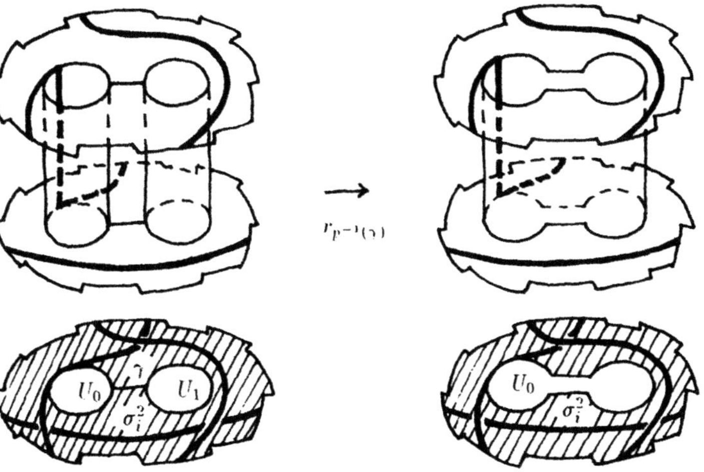

fig.12

Now, let σ be an edge of the diagram $\bar{K}$, and let $\sigma(0), \sigma(1)$ be the endpoints of σ. For $i = 0, 1$ denote by $\tilde{\sigma}(i)$ the point $(\sigma(i), 1) \in F_w \cong \partial(D_w \times I)$, if $\sigma(i)$ is an overcrossing, and the point $(\sigma(i), -1) \in F_w \cong \partial(D_w \times I)$, if $\sigma(i)$ is an undercrossing. Obviously, the part of the coil k projecting onto σ must be a simple arc $\tilde{\sigma}$ on F_w connecting $\tilde{\sigma}(0)$ and $\tilde{\sigma}(1)$.

Let σ_1^2, σ_2^2 be the areas of the diagram $\bar{K}$ adjacent to σ. By our assumption, for $i = 1, 2$ we have $V_i = \sigma_i^2 \backslash U_i$. (Of course, the area $\sigma_i^2, i = 1$ or $i = 2$, may be black- coloured, but we will still use the notation $V_i = \sigma_i^2 \backslash U_i$, putting $U_i = \emptyset$ if this is so). Consider the intersection $\sigma \cap \partial U_i$. It is a (possibly empty) collection of subsegments of the edge σ. Denote these subsegments by $\alpha_1, \ldots, \alpha_n$ for U_1 and by $\beta_1, \ldots, \beta_m$ for U_2. The intersection $\sigma \cap intD_w = \sigma \backslash (\alpha_1 \cup \ldots \cup \alpha_n \cup \beta_1 \cup \ldots \cup \beta_m)$ will be a collection of open intervals $\gamma_0, \ldots, \gamma_{n+m}$ in σ. The arc $\tilde{\sigma}$ must be a composition of arcs $\tilde{\alpha}_k, \tilde{\beta}_l, \tilde{\gamma}_j$ on F_w, each of them projecting onto the corresponding part of the edge σ. Since $\gamma_j \subset intD_w$ for all j, there are only two ways of lifting γ_j to F_w, namely, $\gamma_j^\varepsilon \cong \gamma_j \times \{\varepsilon\} \subset \partial(D_w \times I) \cong F_w, \varepsilon = +,-$. Hence, the arc $\tilde{\sigma}$ will be a composition of arcs $\tilde{\alpha}_k \subset p^{-1}(\alpha_k), \tilde{\beta}_l \subset p^{-1}(\beta_l), \gamma_j^\varepsilon, \varepsilon = +,-$ (see fig.13).

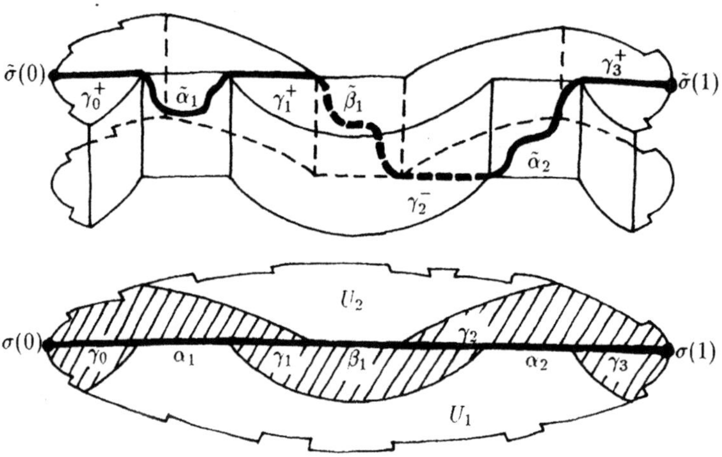

fig.13

We will now proceed to deform the Heegaard surface F_w and the arc $\tilde{\sigma}$ in order to obtain a standard form of lifting the edge σ to the surface F_w.

Step 1. (see fig.14): Contract the segments α_k, β_l towards their middle-points a_k, b_l. The rectangles $p^{-1}(\alpha_k), p^{-1}(\beta_l)$ will contract to the vertical segments $\tilde{a}_k = p^{-1}(a_k), \tilde{b}_l = p^{-1}(b_l)$, respectively. Deform the arcs, obtained, in this way, from $\tilde{\alpha}_k, \tilde{\beta}_l$, to straight-line segments along $\tilde{a}_k, \tilde{b}_l$, leaving their endpoints fixed. The arcs $\tilde{\alpha}_k, \tilde{\beta}_l$ will either contract to points, or deform to the vertical segments $\tilde{a}_k, \tilde{b}_l$, respectively. Hence, the arc $\tilde{\sigma}$ will now become a composition of the arcs $\tilde{a}_k, \tilde{b}_l, \gamma_j^\varepsilon, \varepsilon = +,-$.

Step 2. (see fig.15): If some part of the arc $\tilde{\sigma}$ has the form $\tilde{a}_k\gamma_j^\varepsilon \cdots \gamma_{j+s}^\varepsilon \tilde{a}_{k+1}$, deform it along F_w to the path $\gamma_j^{-\varepsilon} \cdots \gamma_{j+s}^{-\varepsilon}$; as a result, the arc $\tilde{\sigma}$ will now have the form $\ldots c_j^\varepsilon\, \tilde{a}_k\, c_{j+1}^{-\varepsilon}\, \tilde{b}_l\, c_{j+2}^\varepsilon\, \tilde{a}_{k+1}\, c_{j+3}^{-\varepsilon}\, \tilde{b}_{l+1}\, c_{j+4}^\varepsilon \cdots$ where $c_j^\varepsilon = \gamma_r^\varepsilon \cdots \gamma_{r+s}^\varepsilon$

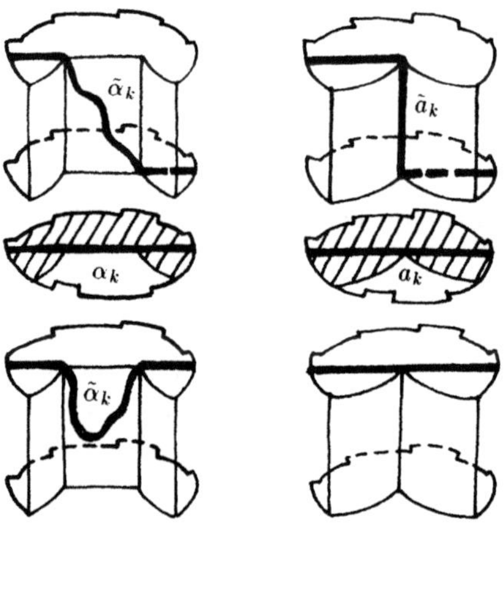

fig.14

Step 3. (see fig.16): If some part of the arc $\tilde{\sigma}$ has the form $\gamma_j^\varepsilon \gamma_{j+1}^\varepsilon$, push the boundary of the hole U_i slightly away from σ, so that it doesn't touch σ in the point deviding γ_j and γ_{j+1}. Finally, the arc $\tilde{\sigma}$ will have the standard form $\ldots \gamma_j^\varepsilon \tilde{a}_k \gamma_{j+1}^{-\varepsilon} \tilde{b}_l \gamma_{j+2}^\varepsilon \tilde{a}_{k+1} \gamma_{j+3}^{-\varepsilon} \tilde{b}_{l+1} \gamma_{j+4}^\varepsilon \cdots$ (see fig.17).

If at least one of the areas σ_1^2, σ_2^2 adjacent to σ is black-coloured, the lift $\tilde{\sigma}$ of σ to F_w will be uniquely determined. Indeed, let, for instance, σ_2^2 be black-coloured. Then $U_2 = \emptyset$, $\{b_l\}_{l=1,\ldots,m} = \emptyset$, and $\tilde{\sigma} = \gamma_0^\varepsilon$ or $\tilde{\sigma} = \gamma_0^\varepsilon \tilde{a}_1 \gamma_1^{-\varepsilon}$ is uniquely determined by its endpoints $\tilde{\sigma}(0)$, $\tilde{\sigma}(1)$.

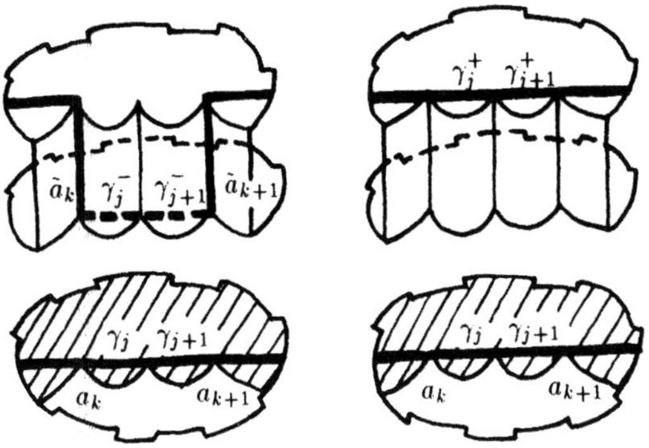

fig.15

If both σ_1^2, σ_2^2 are white-coloured, then σ has many lifts of the standard form, but, as it is easy to see, they differ only by translations along the elementary discs $p_{-1}(\gamma_j) \subset H_w$, and by applying step 3. of the deformation described above (see fig.18).

The proof of lemma 3.5. is completed.

3.6.*Proposition:* Every shaded link diagram $\bar{K}$ is the diagram of a coil k on the surface $F_w = \partial(D_w \times I)$, where $w+1$ is the number of the white-coloured areas of $\bar{K}$, such that $V_i = \sigma_i^2 \cap D_w = \sigma_i^2 \backslash U_i$ for all white-coloured areas $\sigma_0^2, \ldots, \sigma_w^2$ of $\bar{K}$.

Indeed, if the edge σ of $\bar{K}$ is nonalternating, it can be lifted to an arc $\tilde{\sigma} = \gamma_0^\varepsilon$ connecting the points $\tilde{\sigma}(0), \tilde{\sigma}(1)$ on F_w. If the edge σ is alternating, then at least one of the areas σ_i^2, σ_j^2 adjacent to σ is white-coloured. Let it be the area σ_i^2. Then we have $V_i = \sigma_i^2 \backslash U_i$, where U_i is a non-empty open disc in σ_i^2. Deform U_i so that its boundary ∂U_i touches σ in exactly one point a_1. Then σ can be lifted to the arc $\tilde{\sigma} = \gamma_0^\varepsilon \tilde{a}_1 \gamma_1^{-\varepsilon}$ connecting the points $\tilde{\sigma}(0), \tilde{\sigma}(1)$ on F_w.

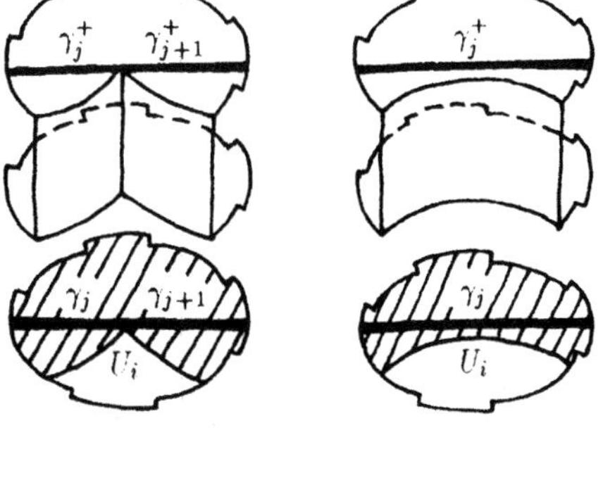

fig.16

3.7.*Lemma:* Any two coils k, k', whose diagrams $\bar{K}, \bar{K}'$ are shadings of the same link diagram K, are equivalent.

Proof: Let σ^2 be a black-coloured area of the diagram $\bar{K}$, and let $x \in int\sigma^2$ be an inner

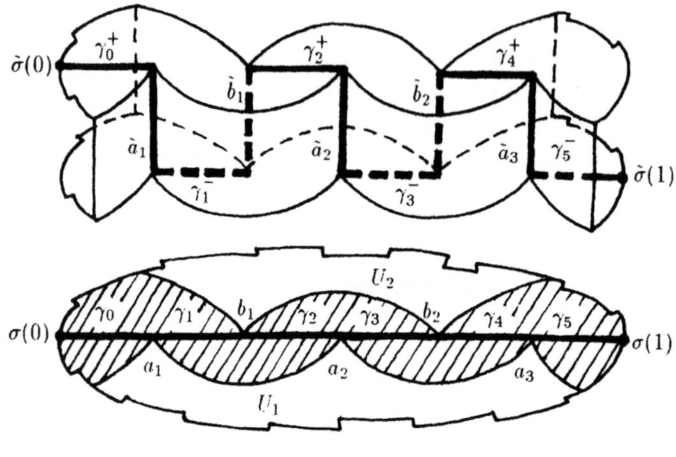
$\tilde\sigma(0)$
γ_0^+
γ_2^+
γ_4^+
$\tilde b_1$
$\tilde b_2$
$\tilde a_1$
$\tilde a_2$
$\tilde a_3$
γ_5^-
γ_1^-
γ_3^-
$\tilde\sigma(1)$
U_2
$\sigma(0)$
γ_0
γ_1
b_1
γ_2
γ_3
b_2
γ_4
γ_5
$\sigma(1)$
a_1
a_2
a_3
U_1

fig.17

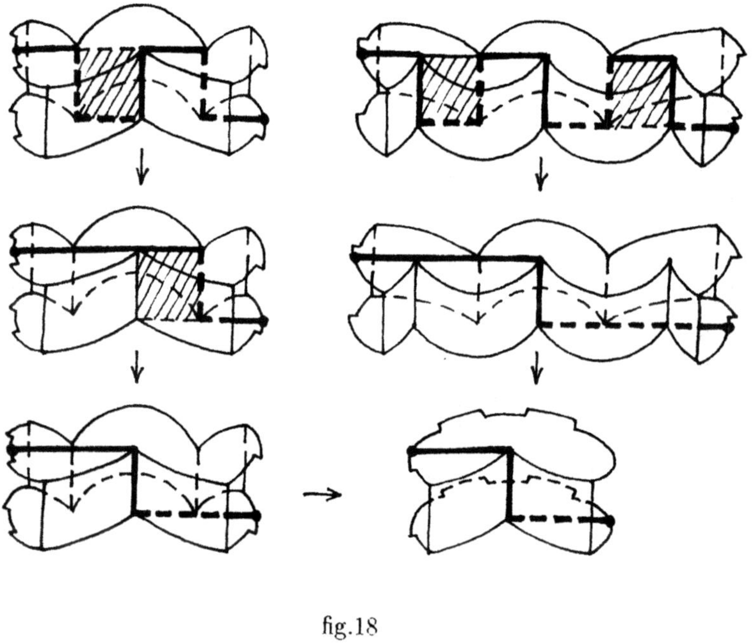

fig.18

point of σ^2. Denote by $\tilde{x}$ the arc $p^{-1}(x)$, and cosider the coil $s_{\tilde{x}}k$ obtained from k by stabilisation along $\tilde{x}$. If k is a coil on the surface $F_g = \partial(D_g \times I)$, then $s_{\tilde{x}}k$ will be a coil on the surface $F_{g+1} = \partial((D_g \backslash U(x)) \times I)$, where $U(x)$ is a small disc in σ^2 containing x. the diagram of the coil $s_{\tilde{x}}k$ will be a shading of the link diagram K which differs from $\bar{K}$ only by the colour of the area σ^2 - the latter will be white instead of black (see fig.19). By repeating the same operation for all black-coloured areas of $\bar{K}$, we will obtain a coil which is equivalent to k and whose diagram is the trivial shading of the link diagram K. The same can be done for the coil k'. Thus, we obtain two coils which are equivalent to k and to k', respectively; on the other hand, these two coils have the same diagram - the trivial shading of the link diagram K. Therefore, by lemma 3.5., they are equivalent, and so will be k and k'.

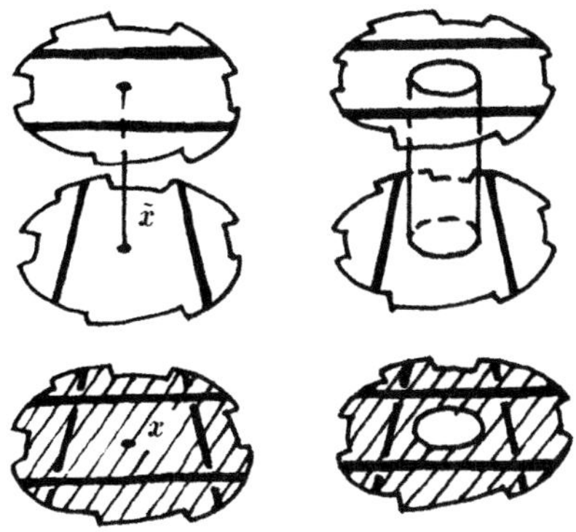

fig.19

The first part of the proof of theorem 3.1. is now completed. We have shown that:

a) Every coiling k of a link $\mathfrak{K}$ defines a coil diagram. The latter is a shading of some diagram K of the link $\mathfrak{K}$. *b*) Every shaded diagram $\bar{K}$ of a link $\mathfrak{K}$ is the diagram of some coiling k of $\mathfrak{K}$. *c*) For any diagram K of a link $\mathfrak{K}$, all coils whose diagrams are shadings of K are equivalent.

Now we must see what happens if we consider coils whose diagrams are shadings of different diagrams of a given link $\mathfrak{K}$.

The following classical theorem gives us the relation between different diagrams of a link:

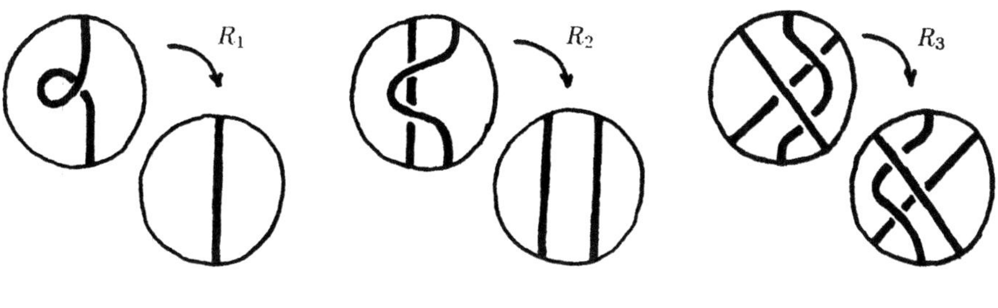

fig.20

3.8. *Theorem:* Two link diagrams represent the same link if and only if one can pass from one to the other by a finite sequence of Reidemeister moves R_1, R_2, R_3, described in fig.20, or their inverses.

According to theorem 3.8. and to what we have already shown, it suffices for our purpose to prove the following: if K_0, K_1 are two link diagrams which differ only by one of the Reidemeister moves $R_i, i = 1,2,3$, then there exist shadings $\bar{K}_0, \bar{K}_1$ of K_0, K_1, and coils k_0, k_1 such that the coils k_0 and k_1 are equivalent and their diagrams are $\bar{K}_0, \bar{K}_1$, respectively.

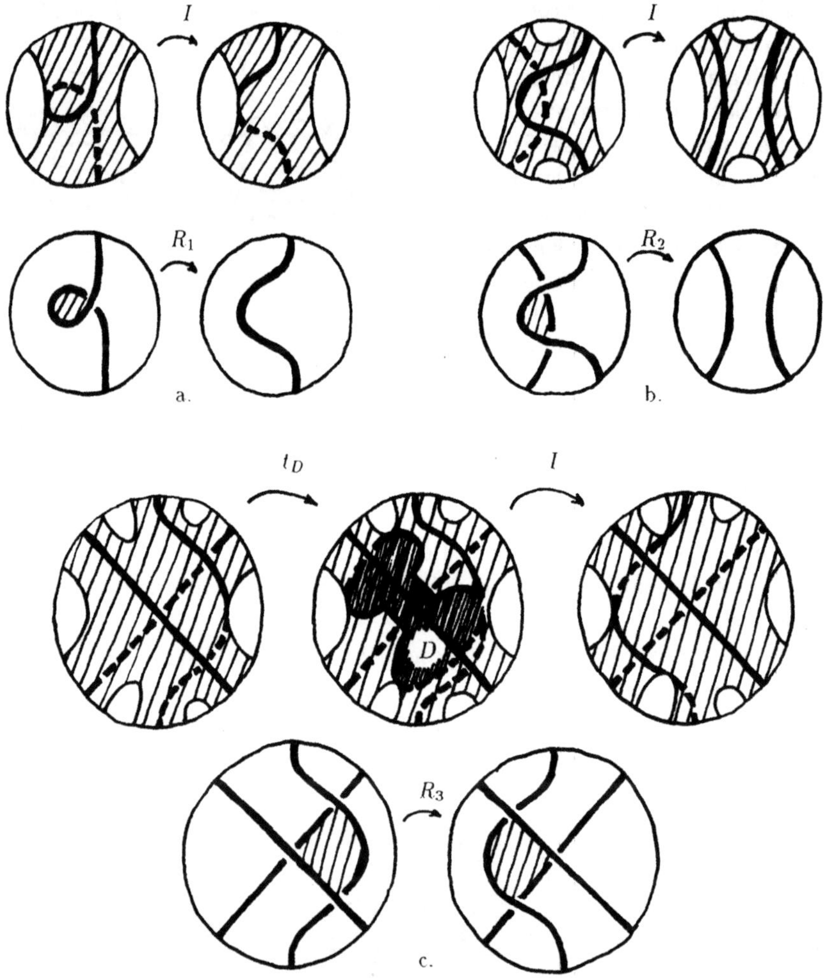

fig.21

Fig 21, a, b, c, presents pieces of the projections of the coils k_0, k_1 and the corresponding pieces of the coil diagrams $\bar{K}_0, \bar{K}_1$, having the required properties for the Reidemeister moves R_1, R_2, R_3, respectively. (We imply that all the areas of the coil diagrams not shown on the pictures are white-coloured).

The proof of theorem 3.1. is completed.

References

1. Burde, G., Zieschang, H. (1985) Knots, Walter de Gruyter, Berlin · New York

2. Culler, M., Shalen, P.B. (1984) Bounded, separating, incompressible surfaces in knot manifolds, Inventiones Mathematicae 75, 537-545

3. Rolfsen, D. (1976) Knots and links, Publish or Perish, inc. Berkeley, CA

A Search for Kernels of Burau Representations

DJUN M. KIM
Mathematics Department,
The University of British Columbia,
Vancouver, B.C., V6T 1Z2
Canada

ABSTRACT. John Moody [8], proved in 1991 that the Burau representation β_n of the n-strand braid group B_n is not faithful if $n \geq 9$. This result was improved to $n \geq 6$ by Darren Long and Michael Paton [5]. A description of this work is given. Using these results, a (so far unsuccessful!) computer search for non-trivial elements in the kernel of β_4 and β_5 was made.

1 The Braid Group B_n

Braids are defined geometrically as disjoint, possibly tangled strands. To make this precise, we need a few definitions.

Let $E = \mathsf{E}^2 \subset \mathsf{E}^3$ be the standard cartesian plane in euclidean 3-space; let E' be E translated down one unit in the z direction. Let $Q_n \subset E$ be marked points on the x-axis at $0, 1, \ldots, n-1$ and let Q'_n be the corresponding set in E'. Consider n disjoint embedded arcs $\alpha_i: [0,1] \to \mathsf{E}^3$ $(i = 1, \ldots, n)$ with α_i meeting some point of Q_n (respectively Q'_n) transversely at $t = 0$ (respectively $t = 1$) and so that every horizontal plane between E and E' meets the α_i transversely in exactly n points. An n-**braid** is an isotopy class (rel E and E') of such embedded arcs.

The set of n-strand braids forms a group which will be denoted B_n; the operation in the group is composition, defined by taking representatitives, performing the following operation, and taking the isotopy class of the result:

$$\sigma \circ \tau := \phi(\sigma) \cup \left(\phi(\tau) - (0, 0, \frac{1}{2})\right)$$

where ϕ is the homeomorphism of E^3 given by $(x, y, z) \mapsto (x, y, z/2)$.

There are several steps required to verify that this is well defined and gives a group structure on B_n; these are left as an exercise for the reader.

1.1 Artin's presentation

Every braid $\sigma \in B_n$ can be written as the composite of so-called **elementary braids** $\sigma_1, \ldots, \sigma_{n-1}$, where σ_i is the braid in which the i-th and $i+1$-st strands are interchanged

M.E. Bozhüyük (ed.), Topics in Knot Theory, 305–317.

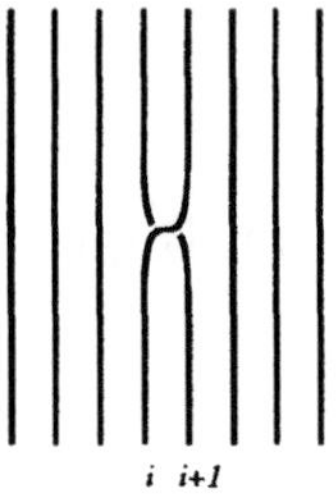

Figure 1: The i-th elementary braid

by a right-hand twist (other authors use the other convention!) and all other strands pass straight through. Hence in a planar projection, the $i+1$-st strand passes over the i-th strand. See figure 1.

Hence the set of elementary braids is a set of generators for B_n. Contemplating braids diagrams, we see that several relations hold in the braid group. First, "distant braids commute". That is,

$$\sigma_i\sigma_j = \sigma_j\sigma_i \qquad (|i-j| \geq 2) \tag{1}$$

Adjacent braids satisfy the following commutation relation:

$$\sigma_i\sigma_{i+1}\sigma_i = \sigma_{i+1}\sigma_i\sigma_{i+1}. \tag{2}$$

Figure 2 makes this perfectly clear. The proof that the two relations (1) and (2) generate all relations among elements of B_n may be found, for example, in Birman's book [2].

Note that adding the $n-1$ relations $\sigma_i^2 = 1$ gives a presentation for the symmetric group Sym(n) as a quotient group of B_n.

1.2 Automorphisms of free groups

Let $\mathcal{F}_n$ be a free group of rank n, with generators $x_1, \ldots, x_n$. An object of great interest in combinatorial group theory is Aut($\mathcal{F}_n$), the automorphism group of the free group. Many interesting groups appear naturally as subgroups of Aut($\mathcal{F}_n$), and among these is B_n, the n-strand braid group.

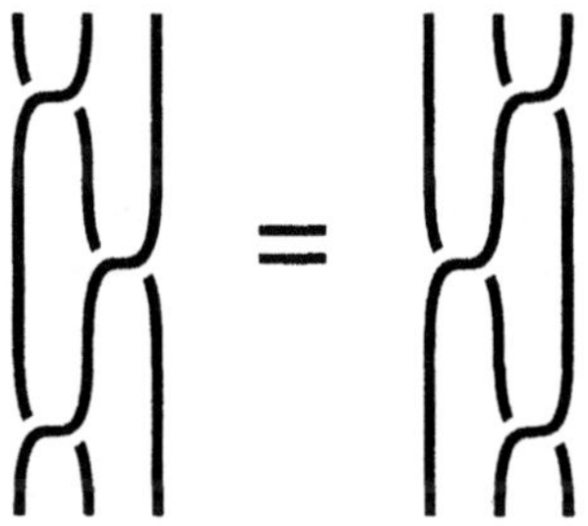

Figure 2: Commutation relations for braids

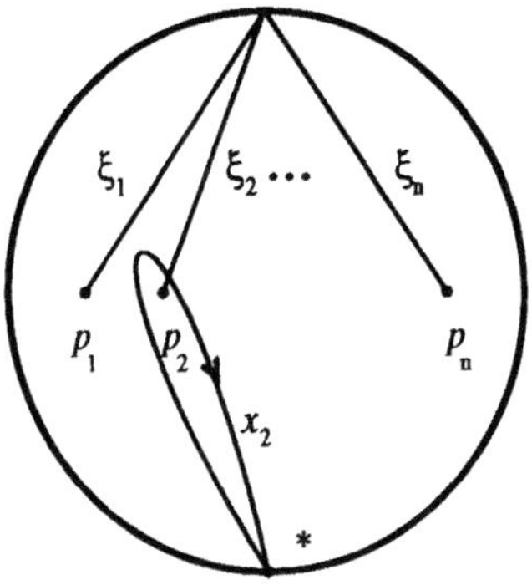

Figure 3: The punctured disk $\mathbf{D}_n$ and generators for $\mathcal{F}_n$

Let $\mathbf{D}_n := D^2 - Q_n$ denote the disk with n punctures $Q_n = p_1, \ldots, p_n$. $\mathcal{F}_n$ can be identified with $\pi_1(\mathbf{D}_n)$. To be precise, let $x_1, \ldots, x_n$ be loops in $\mathbf{D}_n$ representing generators of $\pi_1(\mathbf{D}_n)$, oriented in the clockwise sense, as shown in figure 3. For future reference, we also assume that $\mathbf{D}_n$ has been marked by n line segments ξ_i joining x_i to $\partial\mathbf{D}_n$.

Interchanging p_i and p_{i+1} by a right-hand twist diffeomorphism σ_i on a disk containing these two punctures but no others and fixed outside a neighbourhood of this disk defines an automorphism of $\mathbf{D}_n$. These $n-1$ automorphisms of the disk induce maps $\pi_1(\mathbf{D}_n) \to \pi_1(\mathbf{D}_n)$ defined on generators x_i by

$$\begin{aligned} x_i &\mapsto x_i x_{i+1} \overline{x_i}, \\ x_{i+1} &\mapsto x_i, \\ x_j &\mapsto x_j \qquad (j \neq i, i+1.) \end{aligned}$$

This action is illustrated in figure 4, with the arcs to be interpreted as in section 2.1. Any diffeomorphism which differs from σ_i by an isotopy fixed on $\partial\mathbf{D}_n$ induces the same map on $\pi_1(\mathbf{D}_n)$. In fact, E. Artin showed [1] that there is an isomorphism between the group of orientation preserving diffeomorphisms of $\mathbf{D}_n$ modulo diffeomorphisms isotopic to the identity and the braid group B_n.

2 The Burau representation

The representation of B_n which we will be discussing in this section was first defined in 1936 by W. Burau [3]; it and the related **Gassner representation** are examples of so-called **Magnus representations**, which can be defined by the use of Fox's "Free differential Calculus" — a kind of formal partial differentiation defined in group rings of free groups. The geometric approach to this representation which we will be describing avoids this construction.

To each element $g \in B(n)$ the Burau representation β_n associates an invertible matrix whose entries are Laurent polynomials (with integer coefficients) in a single indeterminate:

$$\overline{\beta_n}: B_n \to GL(n-1, \mathbf{Z}[t^{\pm 1}])$$

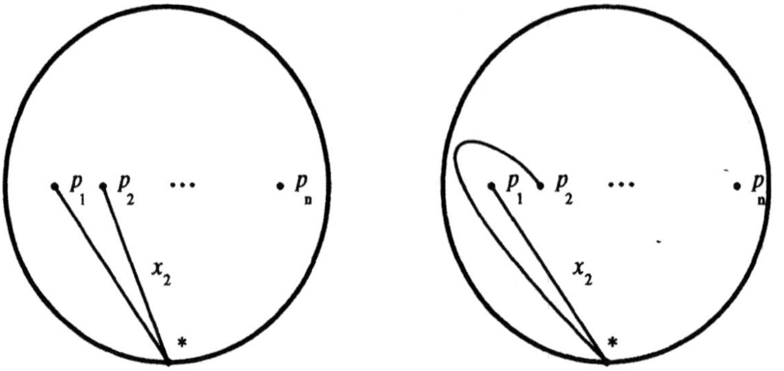

Figure 4: The action of σ_i on $\pi_1(\mathbf{D}_n)$

As we saw in section 1.2, B_n acts on $\pi_1(\mathbf{D}_n)$. This action can be extended to an action on the infinite cyclic homology of $\mathbf{D}_n$ as follows. Let $\alpha: \pi_1(\mathbf{D}_n) \to \mathbf{Z}$ be the **augmentation** homomorphism, which takes a word in $\pi_1(\mathbf{D}_n)$ to its exponent sum. The kernel of this map is a normal subgroup of $\pi_1(\mathbf{D}_n)$; corresponding to this normal subgroup is a covering space $\widetilde{\mathbf{D}}_n$ of $\mathbf{D}_n$ with deck group infinite cyclic, generated by a covering translation t. This covering space can be thought of as countably many copies of $\mathbf{D}_n$ glued together along the slits ξ_i, to give an "infinite parking garage", with n ramps between levels, each corresponding to a ξ_i. Note that there is a deformation retraction of $\mathbf{D}_n$ onto its one-skeleton, which is a wedge of n circles. Then the covering spaces of $\mathbf{D}_n$ are determined by coverings of the 1-skelton. Thus we secretly think of $\widetilde{\mathbf{D}}_n$ as the complex on the right in figure 5.

It is clear that this space has homology which is infinitely generated over $\mathbf{Z}$. But any homology generator may be written as a translate of a generator from a suitably chosen set of $n-1$ generators at "level 0" in $\widetilde{\mathbf{D}}_n$, so $H_1(\widetilde{\mathbf{D}}_n)$ may be expressed as a freely generated rank $n-1$ module over the ring of Laurent polynomials in one variable, $\Lambda = \mathbf{Z}[t^{\pm 1}]$. This construction is similar to the "geometrical" definition of the Alexander module of a link. (See [4], for example.)

Now, each element $\sigma \in B_n$ acts on $\pi_1(\mathbf{D}_n)$, and this action extends to a module automorphism

$$\overline{\beta_n}(\sigma): H_1(\widetilde{\mathbf{D}}_n) \to H_1(\widetilde{\mathbf{D}}_n),$$

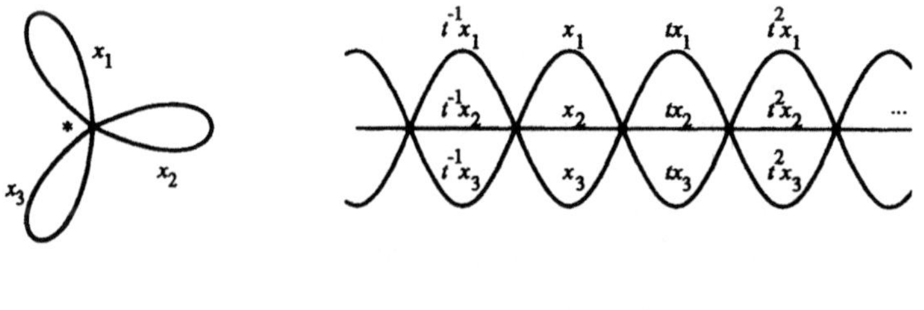

Figure 5: The infinite cyclic covering of $\mathbf{D}_n$

which defines the **reduced Burau representation** of B_n.

Suppose that we attempt to lift the given basis for $H_1(\mathbf{D}_n)$ to $\tilde{\mathbf{D}}_n$. The lifts of these loops are now non-closed *arcs* in $\tilde{\mathbf{D}}_n$, connecting the points in $\{\tilde{*}\}$, the pre-image of the basepoint; hence they do not form cycles in $H_1(\tilde{\mathbf{D}}_n)$. They are, however, cycles in the *relative* homology module $H_1(\tilde{\mathbf{D}}_n, \{\tilde{*}\})$, and in fact give a basis for $H_1(\tilde{\mathbf{D}}_n, \{\tilde{*}\})$ as an n-dimensional Λ-module. As before, B_n acts on this module, and this action defines a representation called the **unreduced Burau representation** of B_n:

$$\beta_n(\sigma)\colon H_1(\tilde{\mathbf{D}}_n, \{\tilde{*}\}) \to H_1(\tilde{\mathbf{D}}_n, \{\tilde{*}\}).$$

As the name suggests, this representation is reducible; consider the long exact sequence for relative homology:

$$\begin{array}{ccccccccc}
\cdots \longrightarrow & H_1(\{\tilde{*}\}) & \longrightarrow & H_1(\tilde{\mathbf{D}}_n) & \longrightarrow & H_1(\tilde{\mathbf{D}}_n; \{\tilde{*}\}) & \longrightarrow & H_0(\{\tilde{*}\}) & \longrightarrow 0 \\
& \downarrow\cong & & \downarrow\cong & & \downarrow\cong & & \downarrow\cong & \\
\cdots \longrightarrow & 0 & \longrightarrow & H_1(\tilde{\mathbf{D}}_n) & \longrightarrow & H_1(\tilde{\mathbf{D}}_n; \{\tilde{*}\}) & \longrightarrow & \Lambda & \longrightarrow 0
\end{array}$$

Since Λ is a free module over a ring with identity, it is projective, hence the short exact sequence above splits. That is,

$$H_1(\tilde{\mathbf{D}}_n; \{\tilde{*}\}) \cong H_1(\tilde{\mathbf{D}}_n) \oplus \Lambda$$

Thus β_n is reducible.

The representation β will now be made explicit. Let $\mathbf{x}_1, \ldots, \mathbf{x}_n$ be the generators for $H_1(\tilde{\mathbf{D}}_n, \{\tilde{*}\})$ obtained by lifting the generators $x_1, \ldots, x_n$ for $\pi_1(\mathbf{D}_n)$ defined in section 1.2 to some level, hereafter fixed as level 0.

A given arc α in $\mathbf{D}_n$ lifts to a path in $\tilde{\mathbf{D}}_n$ representing a homology class $\vec{\alpha}$. For each $i = 1, \ldots, n$, we can define the **weight** of the arc α with respect to the puncture p_i by

$$w(\alpha, i) := \sum_{\ell \in \mathbf{Z}} t^\ell (\vec{\alpha}, t^\ell \xi_i)$$

where $(\vec{\alpha}, t^\ell \xi_i)$ indicates the algebraic (signed) intersection number of the class $\vec{\alpha}$ with the translate to the ℓ-th level of ξ_i.

With this notation, the action of $\beta(\sigma)$ on $\mathbf{x}_i$ is given by

$$\beta(\sigma)\mathbf{x}_i = w(\sigma(x_i), 1)\mathbf{x}_1 + w(\sigma(x_i), 2)\mathbf{x}_2 + \cdots + w(\sigma(x_i), n)\mathbf{x}_n \tag{3}$$

2.1 The Burau representation is not faithful if $n \geq 6$

The faithfulness of the Burau representation has long been investigated. It is clear that β_2 is a faithful representation of $B_2 \cong \mathbf{Z}$. Magnus and Peluso [7] showed in 1969, while investigating work of V. I. Arnol'd, that the Burau representation of B_3 is faithful. The faithfulness of β_n for $n \geq 4$ stood open for 21 years, despite the efforts of a number of people, until John Moody [8] showed in 1990 that β_n is not faithful for $n \geq 9$. His techniques were subsequently refined by Darren Long and Michael Paton [5], who showed that β_n is not faithful for $n \geq 6$. The question of whether the Burau representation for the four and five

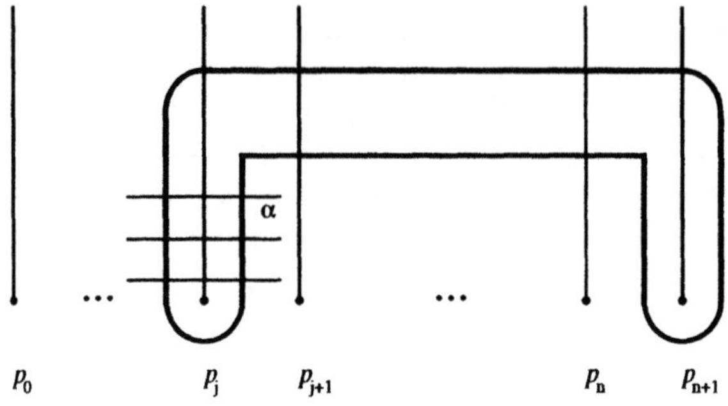

Figure 6: The twist curve C_j

strand braid groups is faithful remains open. This section discusses the results of Moody, Long and Paton.

Intrinsic to the discussion are various operations on **simple arcs** in the disk $\mathbf{D}_n$, by which we mean simple loops constructed as regular neighbourhoods of simple arcs beginning at the basepoint, avoiding punctures but finally ending at one of the punctures. Hence as loops, these arcs contain exactly one puncture.

The first result is an observation of Moody's:

Theorem 1 *(J. Moody [8])*

(a) Let n be such that $w(\alpha, j) = 0$ for a simple arc α in $\mathbf{D}_n$ implies that α can be isotoped off ξ_j. Then the Burau representation of B_n is faithful.

(b) If for some n there exists in $\mathbf{D}_n$ a simple arc α such that for some j, $w(\alpha, j) = 0$ but α cannot be isotoped off ξ_j, then the Burau representation of B_{n+2} is not faithful.

Proof.

(a) Suppose that $\beta_n(\sigma) = I_n$, then (3) tells us that

$$w(\sigma(x_i), k) = \delta_{i,k}.$$

By hypothesis, $\sigma(x_i)$ can be isotoped off ξ_k for all $k \neq i$. Hence σ is isotopic to the identity. So β_n is faithful.

(b) Consider the curve α as lying in $\mathbf{D}_{n+2}$, which is thought of as being obtained from $\mathbf{D}_n$ by removing two further points p_0 and p_{n+1}. Let C_j be a curve which encloses p_j and p_{n+1} but which encloses no other punctures, and which has minimal intersection number with α. We may assume that C_j and α are as indicated in figure 6.

Let τ_j be the Dehn twist about C_j. We may write

$$\tau_j = \sigma_j \sigma_{j+1} \sigma_{j+1} \cdots \sigma_{n-1} (\sigma_n)^2 \bar{\sigma}_{n-1} \bar{\sigma}_{n-2} \cdots \bar{\sigma}_j$$

This is an element of B_{n+2} which lies in the subgroup generated by $\sigma_1, \ldots, \sigma_n$. We now claim: for all k,

$$w(\alpha, k) = w(\tau_j(\alpha), k).$$

This implies that $\beta_n(\tau_j)\vec{\alpha} = \vec{\alpha}$, so the arc α gives rise to an eigenvector $\vec{\alpha}$ with eigenvalue equal to 1 for the matrix $\beta_n(\tau_j)$.

Observe that the action of τ_j on the word α is to replace each occurence of x_j:

$$x_j \mapsto \Big(\prod_{r=j}^{n+1} x_r\Big)\Big(\prod_{s=n}^{j+1} x_s^{-1}\Big)x_j\Big(\prod_{s=j+1}^{n} x_s\Big)\Big(\prod_{r=n+1}^{j} x_r^{-1}\Big) \tag{4}$$

which has exponent sum 1, and each occurence of x_j^{-1} by the inverse of this word, which has exponent sum -1.

To prove the claim, consider the intersection of α with a small neighbourhood of ξ_j. Since $w(\alpha, j) = 0$, for every segment of the arc crossing ξ_j, there is another segment crossing ξ_j in the opposite direction, at the same "level". More precisely, for each occurence of x_j in the word α, there is an occurence of x_j^{-1}, and the exponent sum of the prefixes to the two occurences are equal. This observation implies that any contribution to $w(\tau_j(\alpha), k)$ could arise purely from the terms substituted in for some x_j by the action of τ_j. But each such term is cancelled by the corresponding term coming from a substitution for the corresponding x_j^{-1}. This verifies the claim.

Observe that α and $\tau_j(\alpha)$ are different in $\pi_1(\mathbf{D}_n)$ since, by hypothesis, α cannot be isotoped off ξ_j, and we can assume that α is in reduced form, so there is no cancellation at the ends of the word in (4).

The proof is completed by producing a non-trivial braid which lies in the kernel of β_n. Let $\psi \in B_{n+2}$ be a diffeomorphism class (lying in the subgroup generated by $\sigma_1, \ldots, \sigma_n$) which takes $x_1 \mapsto \alpha$. (The braid ψ exists since α is a simple loop enclosing just one puncture, say x_k. Hence it is a conjugate $\rho x_k \rho^{-1}$, and if ρ is not the identity, then the word α can be shortend by applying $\sigma_k^{\pm 1}$. Then supposing $\alpha = x_k$, a sequence of elementary braids takes x_k to x_1.)

Then the image under β_{n+2} of $\psi^{-1}\tau_j\psi$ acts as the identity on the submodule generated by $\mathbf{x}_1, \mathbf{x}_2$ in $H_1(\widetilde{\mathbf{D}}_{n+2}, \{\tilde{*}\})$, because σ_0 is not involved and by construction, $\psi^{-1}\tau_j\psi$ stabilizes $\mathbf{x}_1$. So

$$\beta_{n+2}(\psi^{-1}\tau_j\psi) = \begin{bmatrix} 1 & 0 & * & * & * & * \\ 0 & 1 & * & * & * & * \\ 0 & 0 & * & * & * & * \\ \cdots & \cdots & \cdots & \cdots & \cdots & \cdots \\ 0 & 0 & * & * & * & * \\ 0 & 0 & * & * & * & * \end{bmatrix} \tag{5}$$

Since $\beta_{n+2}(\sigma_0)$ only acts on $\mathbf{x}_0, \mathbf{x}_1$, the commutator braid $[\sigma_0, \psi^{-1}\tau_j\psi]$ is represented as an upper triangular matrix:

$$\beta_{n+2}([\sigma_0, \psi^{-1}\tau_j\psi]) = \begin{bmatrix} 1 & 0 & * & * & * & * \\ 0 & 1 & * & * & * & * \\ 0 & 0 & 1 & 0 & 0 & 0 \\ \cdots & \cdots & \cdots & \cdots & \cdots & \cdots \\ 0 & 0 & 0 & 0 & 1 & 0 \\ 0 & 0 & 0 & 0 & 0 & 1 \end{bmatrix} \tag{6}$$

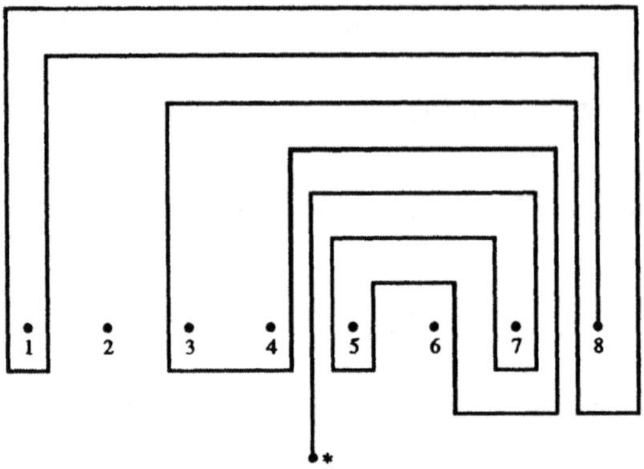

Figure 7: An arc giving a counter-example to the fidelity of β_{10}

We may check the action of the braid $[\sigma_0, \psi^{-1}\tau_j\psi]$ on the loop x_1 to verify that it is not the identity:

$$\begin{aligned}(\psi^{-1}\tau_j^{-1}\psi)\sigma_0^{-1}(\psi^{-1}\tau_j\psi)\sigma_0(x_1) &= (\psi^{-1}\tau_j^{-1}\psi)\sigma_0^{-1}(\psi^{-1}\tau_j\psi)(x_0)\\ &= (\psi^{-1}\tau_j^{-1}\psi)\sigma_0^{-1}(x_0)\\ &= (\psi^{-1}\tau_j^{-1}\psi)(x_1)\\ &\neq x_1\end{aligned}$$

Finally, the subgroup of upper triangular matrices in the image $\beta_{n+2}(B_{n+2})$ of the Burau representation is a unipotent group. (Recall that a linear group G is **unipotent** if every for every element A there is an $r > 0$ such that $(A - I_n)^r = 0$. This is equivalent to saying that the **lower central series** for G is finite: defining $G^{(0)} := G$, and $G^{(n)} := [G, G^{(n-1)}]$, there is some $r > 0$ so that $G^{(r)} = \{Id\}$.) So setting $\eta_1 := [\sigma_0, \psi^{-1}\tau_j\psi], \eta_k := [\sigma_0, \eta_{k-1}]$, we eventually obtain a non-trivial braid whose image is the identity. ∎

Moody gives an example of an arc which satisfies the hypotheses of part (b) of his theorem, for $n = 7$, showing that the Burau representation β_n is not faithful for $n \geq 9$. His example is a braid of 88 crossings. In fact, a somewhat shorter example can be found as a subarc α of Moody's example; one may check that $w(\alpha, 6) = 0$. See figure 7.

The main improvement of Long and Paton is to eliminate the need for extra points in $\mathbf{D}_n$ in the construction of their criterion: hence they have a geometric criterion on arcs in $\mathbf{D}_n$ which has consequences for the Burau representation of B_n, rather than B_{n+2} as in Moody's argument.

The first step in the proof of Long and Paton is the following observation, which simplifies the argument of the preceding theorem and is essential in what follows.

Theorem 2 *If M is a matrix in the image of the Burau representation which has ones on the diagonal and zeros below the diagonal, then M is the identity matrix.*

The proof of this theorem, which relies on linear algebra and an earlier result of D. Long [6], is omitted. For details see [5].

Corollary 2.1 *A matrix M in the image of the Burau representation which has first column identical with that of the identity matrix is a direct sum $(1) \oplus M'$.*

Proof. Suppose that M has first row $(1, a_1, a_2, \ldots, a_{n-1})$. Let $\mu \in B_n$ be the braid which M represents. Embed B_n into B_{n+1} by adding "σ_0", which is not involved with any of the braids in the subgroup $B_n < B_{n+1}$. Then the image of μ under this embedding is represented by β_{n+1} as M', an $(n+1) \times (n+1)$ matrix of shape

$$\begin{bmatrix} 1 & 0 & 0 & 0 & \cdots & 0 \\ 0 & 1 & a_1 & a_2 & \cdots & a_{n-1} \\ 0 & 0 & * & * & \cdots & * \\ \hdotsfor{6} \\ 0 & 0 & * & * & \cdots & * \\ 0 & 0 & * & * & \cdots & * \end{bmatrix}$$

The commutator $[\sigma_0, M']$ has the form of (6), hence by the theorem, it is the identity. A computation shows $[\sigma_0, M'] - I_n$ has columns $a_k(1 - \sigma_0^{-1} e_2)$ for $k \geq 1$, which implies that $a_k = 0$ for $k \geq 1$. ∎

Theorem 3 (D. Long and M. Paton [5]) *The Burau representation is not faithful on B_n if and only if n is such that there exists a simple arc α in $\mathbf{D}_n$, which for some i passes geometrically between p_i and p_{i+1} and for which $w(\alpha, i) = w(\alpha, i+1)$.*

Several lemmas are necessary for the proof.

Lemma 3.1 *For any $\gamma \in B_n$,*

$$(1, t, t^2, \ldots, t^{n-1})\beta_n(\gamma) = (1, t, t^2, \ldots, t^{n-1})$$

Proof. It suffices to check on the generators. In general, $\beta_n(\sigma_k)$ has the form

$$\left[\begin{array}{c|cc|c} I_{k-1} & 0 & 0 & 0 \\ \hline 0 & 1-t & t & 0 \\ 0 & 1 & 0 & 0 \\ \hline 0 & 0 & 0 & I_{n-k-1} \end{array}\right]$$

Then $1, t, \ldots, t^{k-2}$ are fixed, as are $t^{k+1}, \ldots t^{n-1}$. In the $k-1$ and k coordinates, we get $t^{k-1}(1-t) + t^k = t^{k-1}$ and $t^{k-1} \cdot t = t^k$ respectively. ∎

Lemma 3.2 *For any $\gamma \in B_n$, the column vector $(1, 1, \ldots, 1)^T$ is an eigenvector for $\beta_n(\gamma)$.*

Proof. Again, it is enough to check on generators. Then

$$\left[\begin{array}{c|cc|c} I_{k-1} & 0 & 0 & 0 \\ \hline 0 & 1-t & t & 0 \\ 0 & 1 & 0 & 0 \\ \hline 0 & 0 & 0 & I_{n-k-1} \end{array}\right] \cdot \begin{pmatrix} 1 \\ \vdots \\ \\ 1 \end{pmatrix} = (\underbrace{1, \ldots, 1}_{k-1}, 1 - t + t, 1, \underbrace{1, \ldots, 1}_{n-k-1})$$

The same proof shows that if a vector contains the same entry in position k as in $k+1$, then the vector will be an eigenvector for $\beta_n(\sigma_k)$. ∎

Lemma 3.3 *Suppose M is any matrix in $GL(n,\Lambda)$ which has the properties enjoyed by $\beta_n(\gamma)$ in lemmas 3.1, 3.2. Then M commutes with $\beta_n(\Delta_n^2)$, where $\Delta_n^2 = (\sigma_1\sigma_2\ldots\sigma_n)^n$.*

Proof. Check that $\beta_n(\Delta_n^2) = t^n I_n + (1-t)R_n$ where I_n is the identity and

$$R_n := \begin{bmatrix} 1 & t & t^2 & \cdots & t^{n-1} \\ 1 & t & t^2 & \cdots & t^{n-1} \\ \cdots & \cdots & \cdots & \cdots & \cdots \\ 1 & t & t^2 & \cdots & t^{n-1} \\ 1 & t & t^2 & \cdots & t^{n-1} \end{bmatrix}$$

Then $M\beta_n(\Delta_n^2) = M(t^n I_n + (1-t)R_n) = t^n M + (1-t)MR_n = t^n M + (1-t)R_n$, and $\beta_n(\Delta_n^2)M = (t^n I_n + (1-t)R_n)M = t^n M + (1-t)R_n M = t^n M + (1-t)R_n$. ∎

Proof. (Main Theorem)

($\Longrightarrow$) Suppose that β_n is not faithful. Any element σ in the kernel of the Burau representation satisfies $w(\sigma(x_i), k) = \delta_{i,k}$. But a nontrivial element of the kernel will take some arc x_i into a more complicated arc, passing geometrically between some adjacent pair of points $k, k+1$.

($\Longleftarrow$) Let α be an arc as in the hypotheses of the theorem. By the remark following lemma 3.2, the weights coming from the arc α form an eigenvector for $\beta_n(\sigma_i)$, although σ_i acts non-trivially on α. Let ψ be an element of B_n which carries x_1 into the isotopy class of α. The braid $\bar{\psi}\sigma_i\psi$ is non-trivial, but its image under β_n has the first column the same as that of the identity matrix; by corollary 2.1, the matrix $\beta_n(\bar{\psi}\sigma_i\psi)$ is a block sum $(1) \oplus M$. By lemmas 3.1 and 3.2, $\beta_n(\bar{\psi}\sigma_i\psi)$, and hence M satisfy the properties of the lemmas. Applying lemma 3.3, $[M, \beta_n(\Delta_n^2)] = I_n$, so the braid $[\bar{\psi}\sigma_i\psi, \Delta_n^2]$ is in the kernel of β_n. ∎

Long and Paton have used this criterion to show that β_6 is not faithful. Their example has

$$\psi = \bar{\sigma}_5\bar{\sigma}_4\sigma_5\bar{\sigma}_3\sigma_4\bar{\sigma}_2\bar{\sigma}_3\bar{\sigma}_3\bar{\sigma}_3\sigma_1\sigma_1\sigma_1\sigma_5\sigma_4\bar{\sigma}_3\bar{\sigma}_2\sigma_1$$

and hence $[\bar{\psi}\sigma_5\psi, \Delta_6^2]$, where $\Delta_6^2 = (\sigma_2\sigma_3\sigma_4\sigma_5)^5$ is a non-trivial element of B_6 in the kernel of β_6.

3 The search for elements of the kernels of β_4 and β_5.

It is tempting to believe that the criteria established by Long and Paton may be applied to produce counterexamples to the fidelity of the Burau representation of B_n in the remaining open cases, B_4 and B_5. A computer search for such counterexamples has been attempted by the author, and is described in this portion of the paper.

3.1 THE ALGORITHMS

In order to produce a counterexample to the faithfulness of the Burau representation from the criteria described in the previous section, one must find a simple arc α in the n-times punctured disk $\mathbf{D}_n$, running from a basepoint on the boundary of $\mathbf{D}_n$ to one of the punctures, avoiding all other punctures and satisfying the geometric criteria of theorem 3.

Such an arc α is described, up to homotopy, by the sequence of generators x_i (and inverses) corresponding to the intersections of α with the line segments ξ_i $(i = 1, \ldots, n)$ as, for example, in figure 8.

The first step in the search for a non-trivial element in the kernel of β_n $(n = 4, 5)$, therefore, is to generate reduced words in $\mathcal{F}_n$ representing simple arcs. Each such arc is then tested to see whether it has equal weights with respect to two adjacent punctures; if so, it is checked to see if it passes geometrically between these adjacent punctures.

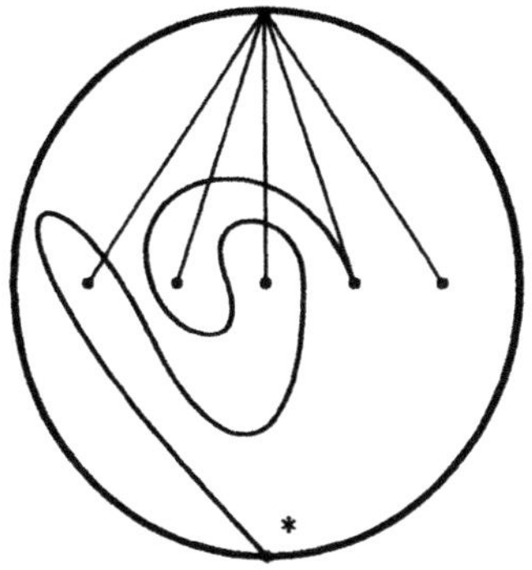

Figure 8: The punctured disk $\mathbf{D}_5$ and the realization of $x_1x_3^{-1}x_2x_3x_4$ $(aCbcd)$

3.2 Implementation details

For the sake of easy string manipulation and formatting, we represent the generators $x_1, x_2, x_3, \ldots$ by lower case letters $a, b, c, \ldots$, and inverses by the corresponding upper case letters. There are no spaces between letters. For example, the word $x_1x_3^{-1}x_2x_3x_4$ is represented by the string *aCbcd*.

The first step is carried out by a routine which reads a line representing a reduced word from $\mathcal{F}_n$ and checks whether it can be realized as an arc in the marked disk $\mathbf{D}_n$. The test is by a recursive depth-first search as follows. A simple arc of the type described separates the disk into a number of components. These components we call **chambers**. Each chamber is identified by the edges which bound it, and can hence be described by a (cyclic) list of generators and their inverses. Viewing the construction of the arc as a dynamic process, we see that there is always a **current chamber**, one that contains the "free end" of the arc. Extending an arc by piercing an edge splits the current chamber into two chambers, one of which becomes the current chamber.

For example, the initial state for $\mathbf{D}_5$ is denoted by [**aAbBcCdDeE*]. This notation indicates that there is one chamber, with walls labelled (from left to right, in circular order) $a, A, b, \ldots, E$. (These may be read off from the disk in Fig. 8 by scanning around the disk, with the edges deleted and edges of the slits labelled by a, A, b, etc.) This chamber is the current chamber, as indicated by the '*'. The list of chambers corresponding to the realization of the word *aCbcd* is $[[a]aA[Ab[bB[Bc]c]cC]C[Cd]dD * DeE]$. See figure 9 for a diagrammatic representation of these chambers.

Thus, for our search space, we take the collection of all chambers which may be obtained from the initial state by piercing edges.

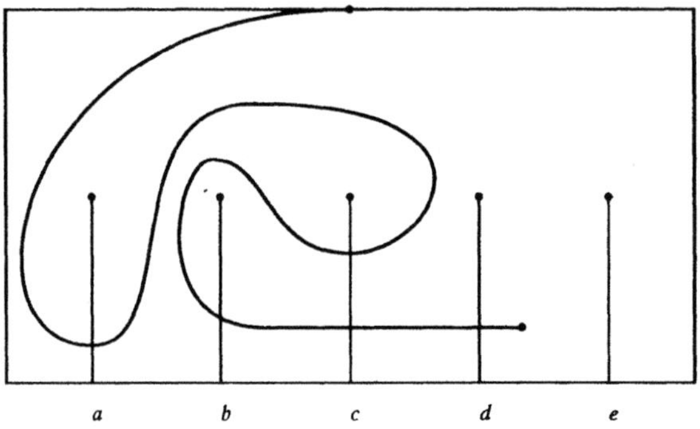

Figure 9: Six chambers arising from *aCbcd*

Computation of the weights can easily be done directly from the definition. Finally, testing whether an arc passes between two punctures is an easy exercise in string searching.

One can compute the weights associated to an arbitrary word, whether or not it comes from a simple arc. Since it is much easier to test the weight criterion than to test whether a word is realizable as a simple arc, this test is carried out first. For the examples which we have computed, this speeds up the computation by a factor of approximately 4.

3.3 Results

To date, no arcs satisfying the criteria of Long and Paton have been found, in either D_4 or D_5. It is unlikely that an exhaustive search by the method described above will succeed in producing a counter-example, since the number of words to be examined grows exponentially in the length of the word. At the same time, the number of words coming from simple arcs and satisfying the weight criterion seems to grow very slowly (see for example table 1). So a randomly chosen word is quite unlikely to yield a counterexample either!

length:	3	4	5	6	7
# words:	1,000	10,000	100,000	1,000,000	10,000,000
# reduced words:	810	7,290	65,610	590,490	5,314,410
# simple arcs:	190	570	1,512	3,558	7,654
# adj. weights equal:	154	371	832	1,595	2,932

Table 1: Growth of Words in $\mathcal{F}_5$

4 Acknowledgements

I would like to thank Dale Rolfsen and Roger Fenn for helpful conversations and comments, and Bill Casselman and Michael Cherkasoff for suggesting improvements to my computer programs.

References

[1] Emil Artin. Theorie der Zöpfe. *Abh. Math Sem., Univ. Hamburg*, 4:47–72, 1925.

[2] Joan S. Birman. *Braids, Links and Mapping Class Groups.* Number 82 in Annals of Mathematics Studies. Princeton University Press, Princton, 1975.

[3] W. Burau. Über Zopfgruppen und gleichsinnig verdrillte Verkettungen. *Abh. Math Sem., Hansischen Univ.*, 11:171–178, 1936.

[4] Gerhard Burde and Heiner Zieschang. *Knots.* de Grutyer, Inc., New York, 1989.

[5] Darren Long and Mike Paton. The Burau representation is not faithful for $n = 6$. Preprint, 1991.

[6] Darren D. Long. On the linear representation of braid groups. *Trans. A.M.S.*, 311:535–561, 1989.

[7] Wilhelm Magnus and Ada Peluso. On a theorem of V. I. Arnol'd. *Comm. Pure and App. Math.*, 22:683–692, 1969.

[8] John Moody. The Burau representation is unfaithful for large n. *Bulletin A.M.S.*, 25:683–692, 1991.

The Witten–Reshetikhin–Turaev invariant for three–manifolds

Richárd Rimányi

Department of Geometry
Eötvös Loránd University, Budapest
Rákóczi út 5. 1088, HUNGARY

ABSTRACT. An array of invariants (one for all integer $r > 1$) for closed oriented 3-manifolds has been revealed by E. Witten using the inspiration of quantum field theory. In this paper two approaches of this topic is studied: one is via quantum groups revealed by Reshetikhin and Turaev, the other — due to Lickorish — uses the Jones polynomial and the Temperley–Lieb algebra.

This paper does not state any new results. It rather gives a synopsis of the two approaches to the Witten–Reshetikhin–Turaev invariant mentioned above outlining some of the basic techniques and ideas.

We will not proof all the lemmas and theorems in this paper partly because they would lead too far from the topic partly because that whould make the paper very extensive. In these cases usually we just refer to the proof.

This work has been completed while I visited The University of Sussex with the support of TEMPUS grant. My thanks are due to this university for their kind hospitality. My sincere gratitude goes to my Hungarian and English supervisors, Dr Gábor Moussong and Dr Roger A. Fenn.

1. BASIC NOTIONS IN KNOT THEORY

1.1. Knots, links and tangles

Definition 1.1.1. *A link is an equivalence class of smooth embeddings of a finite collection of circles to $\mathbb{R}^3$. Two such embeddings L_1, L_2 are equivalent if there is an orientation preserving homeomorphism h of $\mathbb{R}^3$, which makes the following diagram commute:*

$$\begin{array}{ccc} & \overset{L_1}{\nearrow} & \mathbb{R}^3 \\ S^1 \cup \ldots \cup S^1 & & \downarrow \\ & \underset{L_2}{\searrow} & \mathbb{R}^3 \end{array}$$

The equivalence class is called link type or simply **link**.

One could define the equivalence by isotopy of the embeddings ("movement of the strings in 3-space") but this would give the same concept in our case.

An oriented link is a link with a chosen orientation on the circles.

M.E. Bozhüyük (ed.), Topics in Knot Theory, 319–347.

By abuse of language sometimes we call link a representative from the link type, or simply its image in $\mathbb{R}^3$. Sometimes it is more useful to regard the link in $\mathbb{R}^3 \cup \infty = S^3$, which does not essentially alter things.

A generalisation of the concept allows any 3-manifold instead of $\mathbb{R}^3$ but in this paper we are only concerned by the above definition.

If a link contains only one component then we call it a **knot**.

In our case (we are only concerned with smooth embeddings) a link $L : S^1 \cup \ldots \cup S^1 \to \mathbb{R}^3$ can always be extended to an embedding of solid tori $\bar{L} : S^1 \times D^2 \cup \ldots \cup S^1 \times D^2 \to \mathbb{R}^3$ so that $\bar{L}|_{S^1 \times 0 \cup \ldots \cup S^1 \times 0} = L$ and the image of $\bar{L}$ is a regular neighbourhood of the image of the link. Note that this extension is far from unique, even if we allow isotopies on them.

If K is a knot, $K \times D^2$ a solid torus (regular neighbourhood) of K, then a simple curve on the boundary of the torus is called a **longitude** if it is isotopic with K and has linking number 0 with it. We call a a simple closed curve on the boundary of the solid torus a **meridian** if it bounds a disk in the torus.

The linking number of two disjoint knots $k_1, k_2 : S^1 \to \mathbb{R}^3$ can be defined as follows e.g. by the Brouwer degree of the map : $S^1 \times S^1 \to S^2$ given by

$$(x, y) \mapsto \frac{k_1(x) - k_2(y)}{||k_1(x) - k_2(y)||}$$

A framed link L is a link with each component L_i being provided with an integer $fr(i)$, the framing. Let L_i be a component of the link, i.e. $L_i : S^1 \to \mathbb{R}^3$ is a knot. Fix an orientation on L_i. We call a circle on the boundary of the regular neighbourhood of L_i a **framing curve** if homotopically it is the sum of one longitude — which is oriented as L_i is — and $fr(i)$ meridians. By the orientation of the meridian we mean the one which has linking number +1 with L_i.

In this special case a figure is enough to show which orientation of the meridian we mean:

Observe that the definition of the framing curve formally depends on the orientation. But one can check that this dependence is only formal: the reverse orientation would give the same framing curve (actually a homotopic one, because the definition is only up to homotopy).

Definition 1.1.2. *A* **tangle** *T is a 1-manifold smoothly, properly embedded in the unit cube I^3 in $\mathbb{R}^3 = S^3 - \infty$ with $\partial T \subset \frac{1}{2} \times I \times \partial I$. Define $\partial_- T = T \cap (I^2 \times 0)$ and $\partial_+ T = T \cap (I^2 \times 1)$ and call T an (m, n)-tangle if $m = |\partial_- T|$ and $n = |\partial_+ T|$.*

If we fix an orientation on the 1-manifold then we call the tangle oriented. Note that a $(0,0)$-tangle is a link. Framing and framing curves can be defined for tangles as they were defined for links, we leave the explicit definition to the reader.

An (m,n)-tangle is called special if $m = n$. Special tangles will play an important role in chapter 5.

1.2. Planar diagrams

We will often consider planar diagrams of knots, links and tangles, that are images of orthogonal projections to a plane. This plane should be the (y,z)-plane in the case of tangles.

Having assumed earlier that all the maps concerned were *smooth*, we are now ready to state the following theorem (for a proof we refer to any introductory knot theory book e.g. [R]):

Theorem 1.2.1. *Every link- (knot-, tangle-) type has a representative whose projection on an appropriate plane (in case of tangles let it be the (y,z)-plane) has only finitely many multiple points (whose preimage contains more than one points), and all multiple points being double points.*

In a neighbourhood of the double points the diagrams are drawn as:

which shows the under- and overcrossing components.

Let L be a framed link. Consider a diagram of the ith component L_i. Given an arbitrary orientation to L_i we can count the *positive* and *negative crossings* in the diagram of L_i as follows

positive crossing negative crossing

Let the **writhe** be the sum of the signs $w(i)$ of the diagram of L_i. It is independent of the given orientation. If $w(i) = fr(i)$ for all i then we call the diagram a *good* diagram of the framed link. From now on if we are talking about diagrams of the framed link, then we will always mean a good one. They exist, because we can always add positive and negative kinks to a "bad" diagram to change its writhe.

positive kink negative kink

By the writhe $w(D)$ of a diagram D we mean the sum of the signs of crossings in D, or the linking number of the link (which is represented by D) and the link of the framing curves of its components. These two can be proved to be the same.

1.3. Reidemeister–type theorems

The following theorems are typical examples of a translation of a topological problem — the classification of some embeddings (knots, links, tangles) — to a combinatorial one.

We call these theorems Reidemeister–type theorems in honour of K. Reidemeister whose work is the basis for all of them.

Theorem 1.3.1. *Any two planar diagrams which represent the same link (i.e. are projections of equivalent links) are related by a finite sequence of the following moves:*

Ω_1 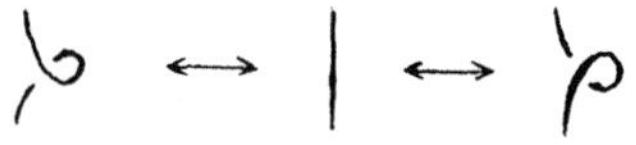

Ω_2 

Ω_3 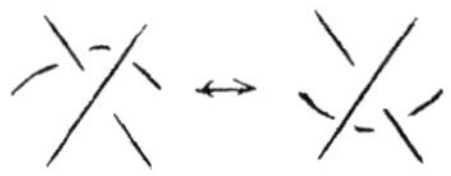

The moves take place within a disc and do not alter the rest of the plane. (Of course between the moves we allow appropriate plane homeomorphisms.) We call the moves Reidemeister moves.

Unfortunately this is only an existence theorem. It does not give an algorithm to decide whether two diagrams represent the same link or not. This also refers to all the theorems of this section.

Theorem 1.3.2. *Any two (good !) planar diagrams of the same framed link are related by a finite sequence of the following moves :*

Ω_1^* 

Ω_2 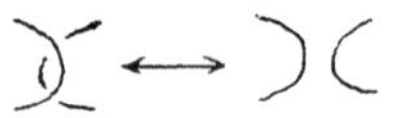

Ω_3 

We note that we may have defined Ω_1^* to be

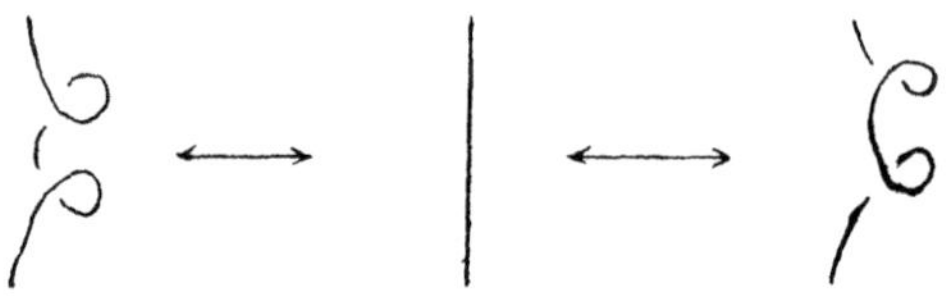

This would give the same equivalence.

We can define $w(\ )$ for any component of a framed tangle the same way as we did for framed links, and we can define the tangle diagram *good* if $w = fr$. The existence of good diagrams is again just a question of adding kinks to the diagram.

Theorem 1.3.3. *Every tangle diagram can be factored into the elementary diagrams I, r, l, $\cup$, $\cap$ as shown:*

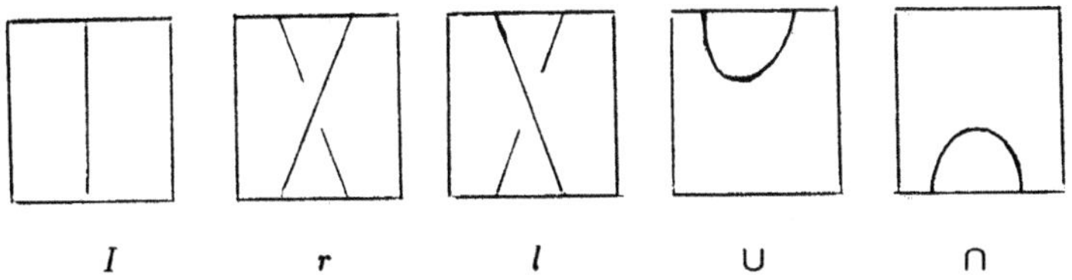

(with all possible orientations in case of oriented tangles) using the composition (when defined) and the tensor product of diagrams :

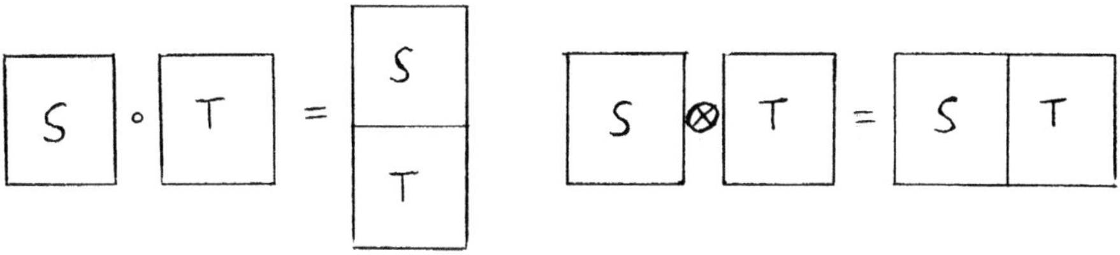

The Reidemeister–type theorem for tangles is the following.

Theorem 1.3.4. *Any two factored good diagrams of a framed tangle are related by a finite sequence of the following moves:*

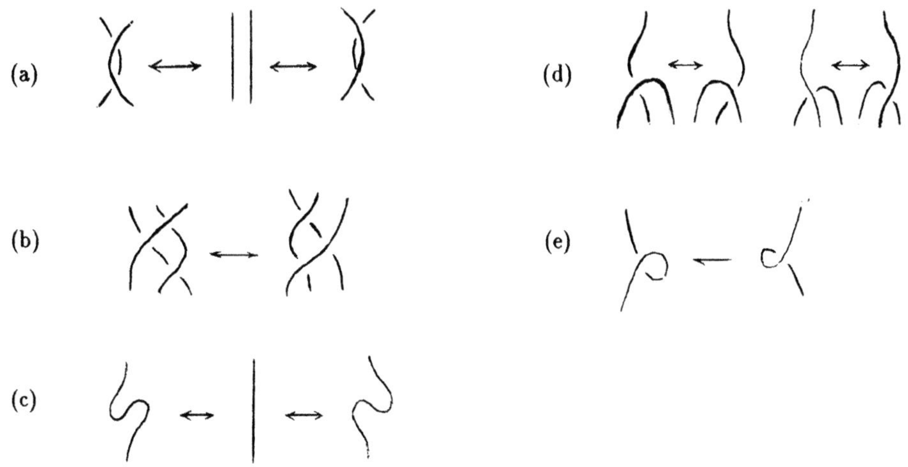

together with the implicit associativity and identity relations and $(S \circ T) \otimes (S' \circ T') = (S \otimes S') \circ (T \otimes T')$.

Of course the theorems above concerning links, framed links, framed tangles could be stated for *oriented* links, framed links, framed tangles with the only change that the moves must be meant with all possible orientations.

2. THE JONES POLYNOMIAL

Let us consider unoriented planar diagrams. The Kauffman bracket polynomial which we will now define is an invariant of them:

$$< >: \{\textit{unoriented planar diagrams}\} \rightarrow \mathbb{Z}[A^{\pm 1}]$$

The main idea in defining it is to consider the following *skein relation* :

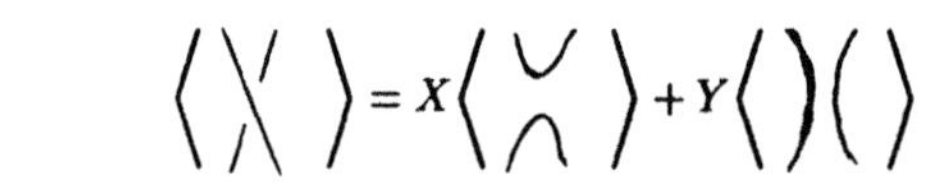

where $X, Y \in \mathbb{Z}[A^{\pm 1}]$ and the parts of diagrams represent any kind of, but identical diagrams except where it is shown. Observe that the idea is to *unknot* the diagram at the crossings by eliminating the double points in the two different ways.

The following easy calculation shows that if X and Y fulfill some simple requirements, then the bracket becomes invariant under the second Reidemeister move.

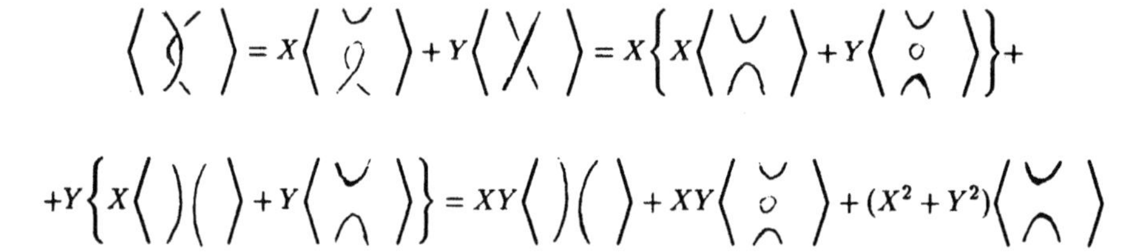

Therefore the conditions for < > being invariant under Ω_2-move are:

(1)

$$XY = 1$$

(2)

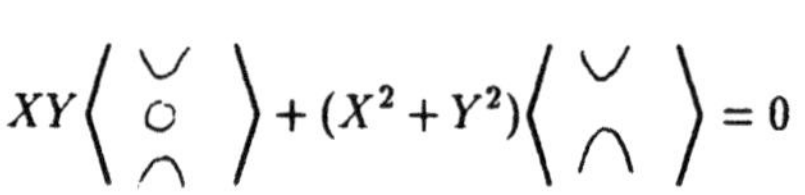

According to (1) let us identify $X = A$, $Y = A^{-1}$. Using (1) we see that (2) is equivalent to

(2*)

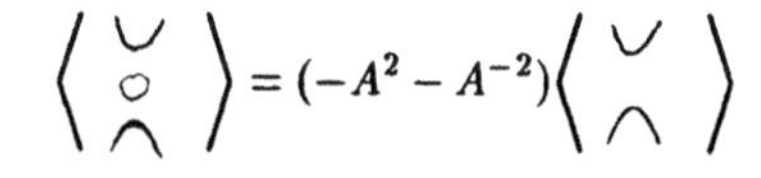

that is adding a disjoint unknot to any diagram multiplies the bracket by $\delta = (-A^2 - A^{-2})$.

Surprisingly enough the bracket is invariant under the Ω_3-move using only the invariance under Ω_2:

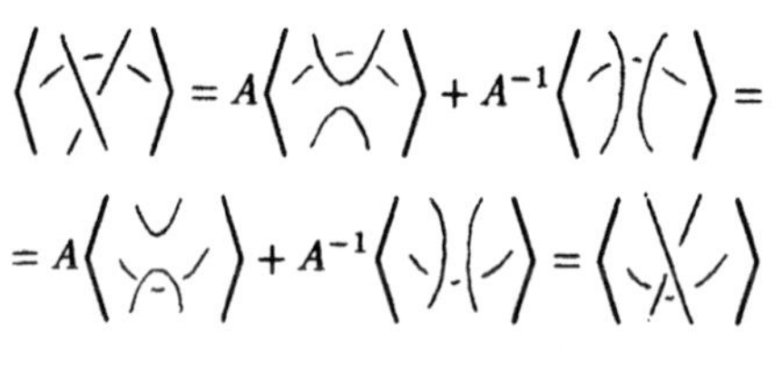

Under Ω_1 the bracket is not invariant but behaves nicely:

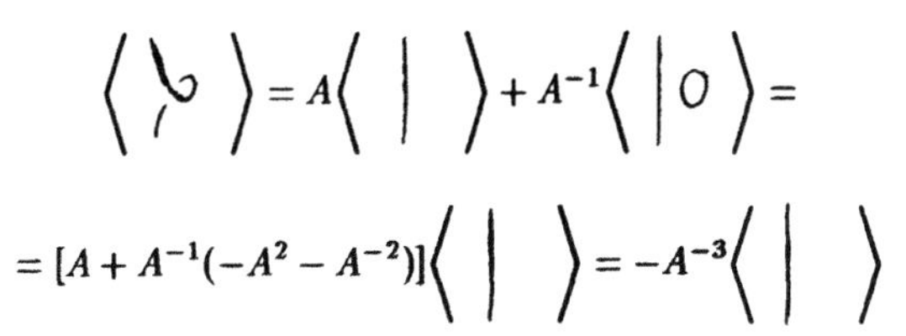

and similarly

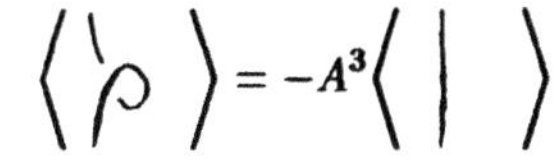

In order to obtain an invariant under all three Reidemeister moves the only thing to do is to balance this change:

Theorem 2.0.1. *There is an invariant — the bracket polynomial — for unoriented planar diagrams satisfying*

(i)

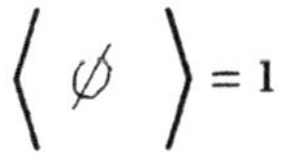

(ii)

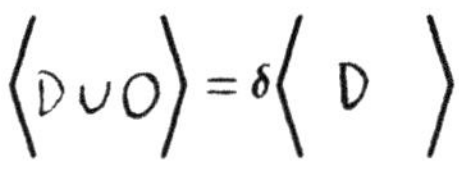

where O *is a component with no crossing at all and* $\delta = -A^2 - A^{-2}$

(iii)

where this refers to the three diagrams identical except where shown.
Then

$$\check{V}_D(A) := (-A)^{-3w(D)} < D >$$

is invariant under all three types of Reidemeister moves, and so is an invariant of the oriented link, whose projection is D.

We obtained an **oriented** link invariant because the definition of the writhe of the diagram (see section 1.1) depends on the orientation. (In case of a knot it does not depend on it.)

Definition 2.0.2. *The Jones polynomial of a link,* $V_L(t)$, *is the invariant* $\check{V}_L$ *above expressed with variable* $t = A^{-4}$.

An easy exercise is to prove that $< D_1 \cup D_2 >=< D_1 >< D_2 >$ if D_1 and D_2 are disjoint diagrams (i.e. they can be separated by a circle).

Definition 2.0.3. *Let us call three oriented links* (L_+, L_-, L_0) *a skein triple if they have diagrams identical except locally they show :*

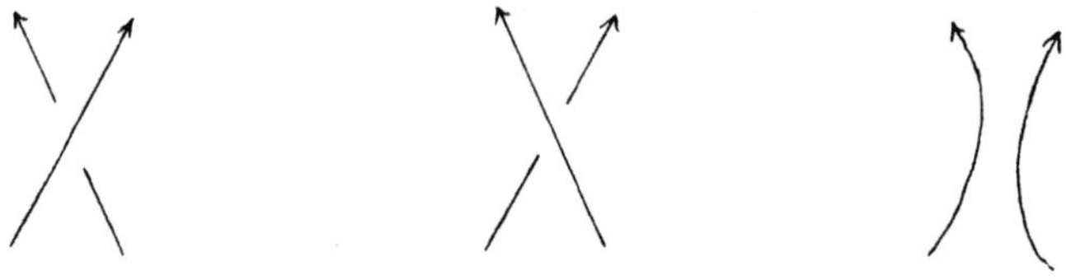

L_+ L_- L_0

We call a linear connection between some fixed invariant of (L_+, L_-, L_0) *a skein relation.*

Due to Conway we know that the most classical invariant, the Alexander polynomial, can also be defined by a skein relation this way :

$$\nabla_O(z) = 1$$

$$\nabla_+(z) - \nabla_-(z) = z\nabla_0(z)$$

and then the Alexander polynomial $\Delta(t) = \nabla(z = t^{\frac{1}{2}} - t^{-\frac{1}{2}})$,of course, up to a multiple of a power of $t^{\frac{1}{2}}$.

Observe that the skein relation — together with a normalisation is enough to define the polynomial (see e.g. the Conway polynomial above). So the following theorem, together with a normalisation of V could be an alternative definition of the Jones polynomial.

Theorem 2.0.4. *The Jones polynomial satisfies the following skein relation :*

$$t^{-1}V_{L_+}(t) - tV_{L_-}(t) = (t^{\frac{1}{2}} - t^{-\frac{1}{2}})V_{L_0}(t)$$

for any skein triple (L_+, L_-, L_0).

Proof By definition

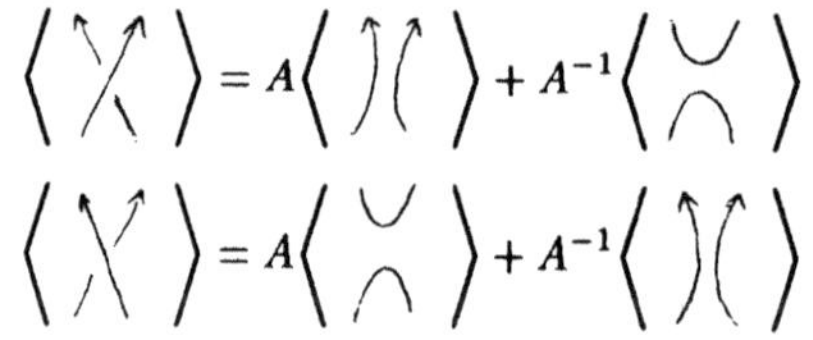

Where we did not show orientation, there is no meaningful one. Let us eliminate those terms. Then we obtain:

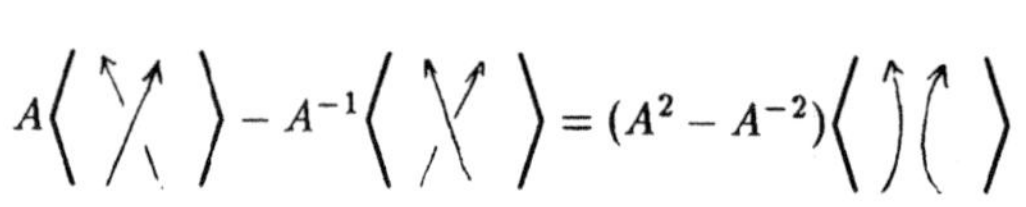

Now the total writhe for the three links are $w+1$, $w-1$, w for some integer w. Multiplying by $(-A)^{-3w}$ gives the result:

$$A^{-4}V_{L_+}(A) - A^4V_{L_-}(A) = (A^2 - A^{-2})V_{L_0}(A)$$

from which the theorem follows.

The proofs of the following properties of the Jones polynomial are left to the reader:

1. $V_{L_1 \cup L_2} = V_{L_1} V_{L_2}$ where $L_1 \cup L_2$ means that the two links can be separated by a sphere.
2. $\delta V_{L_1 \# L_2} = V_{L_1} V_{L_2}$ where $L_1 \# L_2$ means the connected oriented sum of the two links.
3. $V_{-K} = V_K$ where K is a knot. By $-K$ we denote the same knot with the opposite orientation.
4. $V_{K^*}(t) = V_K(t^{-1})$ where K is a knot. By K^* we denote the "mirror image" of the knot in the 3-space.
5. (Jones' reversing theorem) $V_{L-K\cup(-K)} = t^{3\lambda} V_L$ where K is a component of L, 2λ is the sum of the crossings of K with the other components of $L - K$.

3. SURGERY

Surgery is an operation that assigns an oriented closed 3-manifold to a framed link in S^3. Lickorish [L1] and Wallace [Wa] showed that every closed oriented connected 3-manifold can be obtained by surgery. So through this operation we can pass from the study of 3-manifolds to the study of framed links in S^3.

We hereby give two equivalent definitions of surgery on the framed link L.

Definition 3.0.1. *Remove a small open solid torus neighbourhood of each L_i. On each resulting toral boundary component consider a framing curve; attach new solid tori so that each of these curves now bounds a disk.*

Definition 3.0.2. *The framed link gives rise to a 4-manifold W_L the following way. Attach 2-handles ($B^2 \times B^2$) to the 4-ball B^4 along $L \subset S^3 = \partial B^4$, that is, identify $\partial B^2 \times B^2 \subset B^2 \times B^2$ (the two handle) to a small tubular neighbourhood of each link component so that a longitude of $\partial B^2 \times B^2$ is attached to a relevant framing curve (which is unique up to homotopy). Then take M_L to be the boundary of W_L.*

The process given in the second definition is a special case of *handlebody decomposition* studied thoroughly by Kirby [K2].

The two definitions are easy to show to be equivalent, observing that if k is a longitude of $\partial B^2 \times B^2$ then k is a meridian — so bounds a disk — in $B^2 \times \partial B^2$.

If some 2-sphere separates the components of a link then surgery on the link yields a 3-manifold which is a connected sum of the manifolds resulting from surgery on the two sublinks.

Without proofs we give some examples of surgery:

Example 3.0.3. Surgery on the unknot with framing n gives the lens space $L(n, 1)$ (with the convention that $L(0, 1)$ is $S^2 \times S^1$ and $L(1, 1)$ is S^3).

Example 3.0.4. Surgery on a simple link

where p and q are the framings gives $L(pq - 1, p)$.

Example 3.0.5. Surgery on

gives the 3-torus.

Example 3.0.6. The five links pictured here all give the Poincaré dodecahedral space

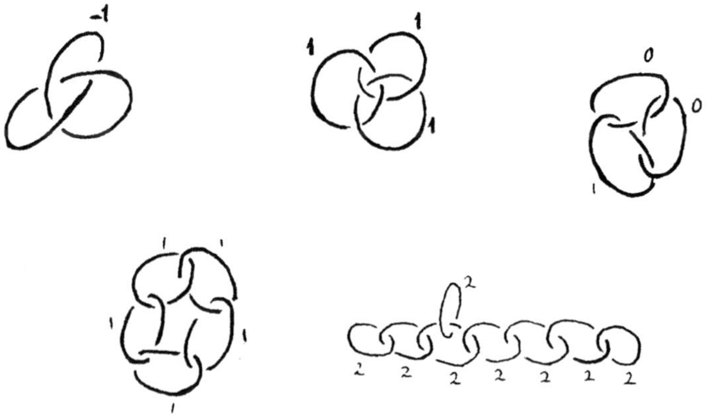

As we can see from the examples the "link representation" of a 3-manifold is far from unique.

Now, let us describe certain moves on framed links that do not change M_L.

The band move or β-move takes the band connected sum of L_i with a framing curve of L_j. Of course the band is disjoint from the link except that one end of it is an arc of L_i and the other is the framing curve of L_j. Replace $fr(i)$ by $fr(i) + fr(j) + 2lk(L_i, L_j)$. Regarding Definition 3.0.2 this move corresponds to sliding the ith 2-handle over the jth one. (For more details see Kirby [K2].)

Note that if a move above is done on a diagram, then after the move one must add or delete kinks to adjust the diagram to be "good" again.

Lemma 3.0.7. *An unknotted circle L_i with framing ± 1 can always be moved away from the rest of the link with effect of giving all arcs going through L_i a full ∓ 1 twist and changing the framings by adding $\mp lk(L_i, L_j)^2$ to each other component L_j.*

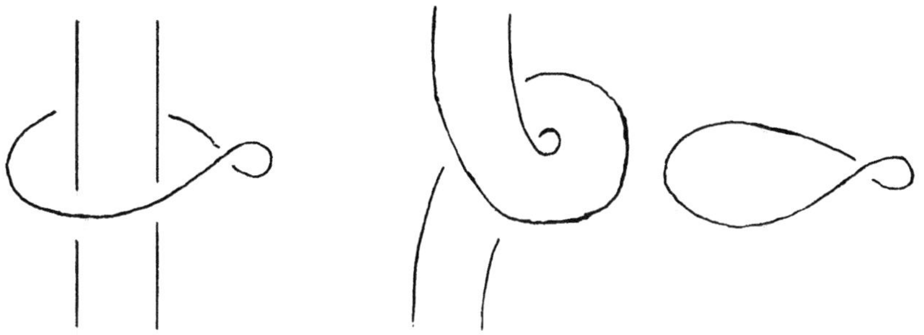

Proof First we do the case for one arc, by sliding the arc once over the circle (β-move) and one can check that now L_i can be moved away from the rest of the link.

In general slide all arcs over the circle once. It is not difficult to calculate the relevant change of the framings (See Rolfsen [R]).

Using this kind of moves it can be shown that e.g. 

and 

give the same manifold, and 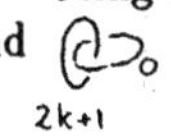

and 

also.

Now we present an other move that does not alter M_L.

Special κ-move: Consider an unknotted component L_i with framing ± 1. If this is disjoint from the rest of the link (i.e. can be separated by a sphere) then L_i can be deleted from the link without altering M_L.

An explicit homeomorphism can be found in the paper of Fenn and Rourke [FR].

From the lemma above one can easily see that the following move does not alter M_L either.

κ-move If L_i is an unknotted component with framing ± 1, we can delete it and add a full ∓ 1 twist (see fig.) to the arcs crossing L_i; and add $\mp lk(L_i, L_j)^2$ to all framings.

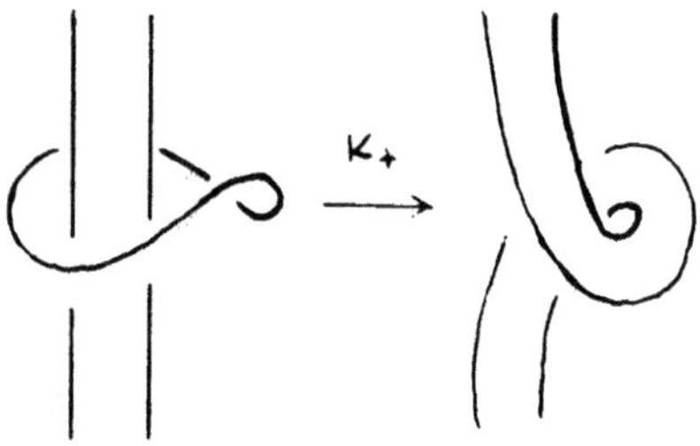

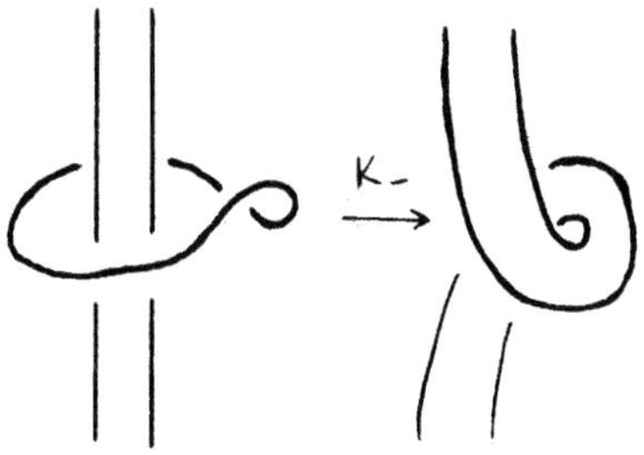

Clearly the κ-move is equivalent to the following.

Locally the following are interchangeable:

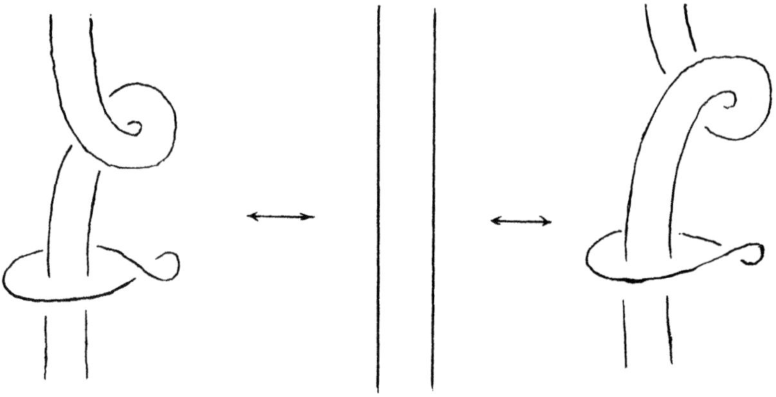

The changes of the relevant framings are signed by the "goodness" of the diagrams.
From now on if we refer to a κ-move we can mean either of the above.

We can now formulate the theorem that is the major tool constructing 3-manifold invariants.

Theorem 3.0.8. *M_L is homeomorphic to $M_{L'}$ if and only if L' can be (up to isotopy) obtained from L by a finite number of the moves given in one of (i), (ii) or (iii):*
(i) (Kirby [K1]) β, special κ and their inverses
(ii) (Fenn, Rourke [FR]) κ and its inverse
(iii) (Turaev [L3]) κ_+, special κ_- and their inverses

Here κ_ε is a κ-move with ε-framed unknotted component involved ($\varepsilon = \pm 1$).

Unfortunately we do not know an algorithm that decides whether two framed links are equivalent under the above equivalence. If one can, however, assign a complex number (or anything else) to a link so that it is invariant under
(1) link isotopy
(2) κ_+ and special κ_--moves
then it is in fact a 3-manifold invariant.

4. THE 3-MANIFOLD INVARIANT

In this chapter we will use the following notations (for $r > 1$ fixed)

$$e(a) = e^{2\pi a}, \quad q = e(\frac{1}{r}), \quad s = e(\frac{1}{2r}), \quad t = e(\frac{1}{4r})$$

and for any integer k

$$[k] = \frac{s^k - \bar{s}^k}{s - \bar{s}} = \frac{\sin\frac{\pi}{r}k}{\sin\frac{\pi}{r}}$$

and in Theorem 4.1.1. $\left[{n \atop k}\right] = \frac{[n]!}{[k]![n-k]!}$ where $[n]! = [n][n-1]\ldots[1]$ $(= 1$ if $n = 0)$. By $\frac{[p]!}{[q]!}$ we mean $[p][p-1]\ldots[q+1]$.

4.1. The quantum group

In this section a specific algebra A_r will be defined for all integer $r > 1$. First recall that a **Hopf algebra** over $\mathbb{C}$ is an additive Abelian group A with homomorphisms :

$$m : A \otimes A \to A$$

$$i : \mathbb{C} \hookrightarrow A$$

$$\Delta : A \to A \otimes A$$

$$\varepsilon : A \twoheadrightarrow \mathbb{C}$$

$$S : A \to A$$

satisfying the following relations:

(i) Multiplication is associative
(ii) $A \simeq \mathbb{C} \otimes A \xrightarrow{i \otimes 1} A \otimes A \xrightarrow{m} A$ is the identity
(iii) Co-multiplication is co-associative
(iv) $A \xrightarrow{\Delta} A \otimes A \xrightarrow{1 \otimes \varepsilon} A \otimes \mathbb{C} \simeq A$ is the identity
(v) Δ, ε are $\mathbb{C}$-algebra homomorphisms
(vi) $i\varepsilon = m(1 \otimes S)\Delta = m(S \otimes 1)\Delta$

where we call m the *multiplication* Δ the *co-multiplication and* S the *antipode map*. In fact (i) and (ii) say A is an algebra, (iii) and (iv) say A is a co-algebra, (i)–(v) say A is a bialgebra.

The great advantage of using a Hopf algebra instead of an ordinary algebra is that this allows one to define A-module structure on the duals $V^* = Hom_{\mathbb{C}}(V, \mathbb{C})$ and tensor products $V \otimes W = V \otimes_{\mathbb{C}} W$ of A-modules V and W. In particular $(\alpha f)(v) = f(S(\alpha)v)$ and $\alpha(v \otimes w) = \Delta\alpha \cdot (v \otimes w)$ where $\cdot$ is the diagonal action, for $\alpha \in A, f \in V^*, v \in V, w \in W$.

The algebra we will need is generated by $X, Y, K, \bar{K}$ (as algebra generators over $\mathbb{C}$) with the relations

$$\bar{K} = K^{-1}$$

$$KX = sXK$$

$$KY = \bar{s}YK$$

$$XY - YX = \frac{K^2 - \bar{K}^2}{s - \bar{s}}$$

$$X^r = Y^r = 0$$

$$K^{4r} = 1$$

The Hopf algebra structure is given by

$$\Delta(X) = X \otimes K + \bar{K} \otimes X \qquad S(X) = -sX \qquad \varepsilon(X) = 0$$

$$\Delta(Y) = Y \otimes K + \bar{K} \otimes Y \qquad S(Y) = -\bar{s}Y \qquad \varepsilon(Y) = 0$$

$$\Delta(K) = K \otimes K \qquad S(K) = \bar{K} \qquad \varepsilon(K) = 1$$

Without any background information this definition may seem to be mysterious so let us make it clear how we can obtain it from the Lie algebra $sl_2(\mathbb{C})$. Recall that

$$sl_2(\mathbb{C}) =< H,X,Y \mid [H,X] = 2X, \quad [H,Y] = -2Y, \quad [X,Y] = H >$$

The property for which we use this algebra to start with is that it has a unique (up to isomorphism) k-dimensional representation V^k for all $k > 0$. Explicitly $sl_2(\mathbb{C})$ acts on V^k (with basis $e_{-m}, e_{-m+1}, \ldots, e_m$ where $k = 2m+1$) by :

$$Xe_j = (m+j+1)e_{j+1}$$

$$Ye_j = (m-j+1)e_{j-1}$$

$$He_j = 2je_j$$

We used the convention that $e_{m+1} = e_{-m-1} = 0$. Note that if k is even then the indices are half integers.

We would like to modify this so that the resulting algebra also has k-dimensional representations V^k for all $k > 0$ in which we replace $(m \pm j + 1)$ by $[m \pm j + 1]$. This is sometimes called the *quantized* action.

This can be done in the following way:

Take the universal enveloping algebra of $sl_2(\mathbb{C})$ which has presentation

$$U[sl_2(\mathbb{C})] =< H,X,Y \mid HX - XH = 2X, \quad HY - YH = -2Y, \quad XY - YX = H >$$

as an associative algebra presentation. The representations of $sl_2(\mathbb{C})$ evidently extend to representations of $U[sl_2(\mathbb{C})]$ so there are unique irreducible representations of $U[sl_2(\mathbb{C})]$ in each dimension.

Having an associative algebra at hand the question arises as to whether one can define a Hopf algebra structure on it. The answer turns out to be positive if we let

$$\Delta(\alpha) = \alpha \otimes 1 + 1 \otimes \alpha \qquad S(\alpha) = -\alpha \qquad \varepsilon(\alpha) = 0$$

for all $\alpha \in sl_2(\mathbb{C})$. This can easily be checked to be a Hopf algebra structure.

Following Kulish and Reshetikhin we can define the algebra of formal power series over $U[sl_2(\mathbb{C})]$ modified by replacing the relation $XY - YX = H$ with

$$XY - YX = \frac{sinh(\frac{hH}{2})}{sinh\frac{h}{2}} = \frac{s^H - \bar{s}^H}{s - \bar{s}} = H + \frac{H^3 - H}{24} + \ldots$$

(using h for the indeterminate is refering to Planck's constant, showing that the theory has echoes of some quantum statistical physics theories). Introducing the element

$$K = t^H = 1 + \frac{H}{4}h + \frac{H^2}{2!4^2}h^2 + \ldots$$

we obtain the associated quantized relations :

$$KX = sXK$$

$$KY = \bar{s}YK$$

$$[X,Y] = \frac{K^2 - \bar{K}^2}{s - \bar{s}}$$

In associative algebras the Lie bracket is always interpreted as the difference of the two ordered products i.e. $[X,Y] = XY - YX$

Although we have modified the relations Sklyanin discovered a Hopf algebra structure on this modification as a module over the ring $\mathbb{C}[[h]]$, the algebra of formal power series in h over $\mathbb{C}$, given by

$$\Delta(X) = X \otimes K + \bar{K} \otimes X \qquad S(X) = -sX \qquad \varepsilon(X) = 0$$

$$\Delta(Y) = Y \otimes K + \bar{K} \otimes Y \qquad S(Y) = -\bar{s}Y \qquad \varepsilon(Y) = 0$$

$$\Delta(K) = K \otimes K \qquad S(K) = \bar{K} \qquad \varepsilon(K) = 1$$

The last step to get the algebra A_r that we proposed is to specialize the series at particular values of h, namely $h = \frac{2\pi i}{r}$. For this purpose, of course, we have to get rid of the divergent series , so we take the subalgebra of convergent power series (entire functions) first, and now we can have the factor algebra quotionted by $h = \frac{2\pi i}{r}, X^r = 0, Y^r = 0, K^{4r} = 0$. The proposed A_r algebra is therefore defined.

It is easy to check that our goal about the representations of this algebra is satisfied, namely that there is a k-dimensional A_r-module V^k in each dimension, on which A_r acts by

$$Xe_j = [m + j + 1]e_{j+1}$$

$$Ye_j = [m - j + 1]e_{j-1}$$

$$Ke_j = t^{2j}e_j$$

The only difficulty in checking the relations might be to check $[X,Y] = \frac{K^2 - \bar{K}^2}{s - \bar{s}}$, but it is implied by the identity $[a][b] - [a+1][b-1] = [a-b+1]$.

From this action it can be seen that the module V^k is reducible for $k > r$. Indeed, in this case the submodule generated by e_{r-m-1} does not contain e_{r-m} (If $k > r$ then $m - 1 \geq r - m - 1 \geq -m$ so both e_{r-m-1} and e_{r-m} are in the module), since one can easily check that $Xe_{r-m-1} = [m + r - m - 1 + 1]e_{r-m} = [r]e_{r-m} = 0$.

Interestingly enough in the opposite case V^k is irreducible.

Theorem 4.1.1. *If $k \leq r$ then V^k are irreducible and self dual. In particular, the map $D : (V^k)^* \to V^k$ given by $D(e_j) = \begin{bmatrix} 2m \\ m-j \end{bmatrix}^{-1} (-s)^j e_{-j}$ is an A_r-linear isomorphism.*

An easy proof is given in [KM]. We will need one more theorem describing the structure of the V^ks.

Theorem 4.1.2.

(i) If $k + k' \leq r + 1$ then

$$V^k \otimes V^{k'} = \oplus V^p$$

where the sum goes for $p = k + k' - 1, k + k' - 3, \ldots, |k - k'| + 1$.

(ii)

$$V^{n+1} = \sum_{j=0}^{\frac{n}{2}} (-1)^j \binom{n-j}{j} (V^2)^{\otimes n - 2j}$$

holds in the representation ring (i.e. sum means direct sum and e.g. $2V - U = W$ means $V \oplus V = U \oplus W$).

One can find the proof of (i) in [KM] and (ii) can be proved from (i) by an easy inductive argument.

4.2. The universal element

Theorem 4.2.1. *There is an invertible element R in $A_r \otimes A_r$ satisfying the following properties*

$$R\Delta(\alpha)R^{-1} = \breve{\Delta}(\alpha)$$
$$(\Delta \otimes 1)(R) = R_{13}R_{23}$$
$$(1 \otimes \Delta)(R) = R_{13}R_{12}$$

where $\breve{\Delta}(\alpha) = P(\Delta(\alpha))$, P is the permutation endomorphism of $A_r \otimes A_r$ given by $P(\alpha \otimes \beta) = \beta \otimes \alpha$, and $R_{12} = R \otimes 1$, $R_{23} = 1 \otimes R$, $R_{13} = (P \otimes 1)(R_{23})$.

An element with this property will be called a universal R-matrix of A_r. Universal R-matrices have been found in several Hopf algebras. Without proof — which is just a question of calculation — we give an R in $A_r \otimes A_r$:

$$R = \frac{1}{4r}\Sigma\frac{(s-\bar{s})}{[n]!}t^{ij+(j-i)n+n}X^n K^i \otimes Y^n K^j$$

The sum is over all $0 \leq n \leq r$ and $0 \leq i, j < 4r$.

To get familiar with the universal R-matrix we make some useful calculations.
The universal element R satisfies the Yang–Baxter equation:

$$R_{12}R_{13}R_{23} = R_{23}R_{13}R_{12}$$

Indeed it follows from the defining properties :

$$R_{12}R_{13}R_{23} = R_{12}(\Delta \otimes 1)(R) = (\breve{\Delta} \otimes 1)(R)R_{12} = R_{23}R_{13}R_{12}$$

If we define $\breve{R}$ as an operator from $V \otimes W \to W \otimes V$ where V and W are A_r-modules by $\breve{R} = P \circ R$ (P is the permutation (flip-) matrix) then $\breve{R}$ is A_r-linear and satisfies the Yang–Baxter equation :

$$\breve{R}_{23}\breve{R}_{12}\breve{R}_{23} = \breve{R}_{12}\breve{R}_{23}\breve{R}_{12}$$

Indeed A_r-linearity follows from

$$\breve{R}(\alpha X) = P(R\Delta(\alpha)\cdot x) = P(P(\Delta\alpha)R\cdot x) = \Delta\alpha P(R\cdot x) = \alpha(\breve{R})(x)$$

where $x \in V \otimes W$.

This second Yang–Baxter equation is implied by the first one if we multiply on the left by the operator $P_{23}P_{12}P_{23} = P_{12}P_{23}P_{12}$. Let us note the following important equalities

$$(\varepsilon \otimes 1)(R) = (1 \otimes \varepsilon)(R) = 1$$

$$(S \otimes 1)(R) = (1 \otimes S^{-1})(R) = R^{-1}$$

$$(S \otimes S)(R) = R$$

From the counit axiom we have $(\varepsilon \otimes 1)\Delta = 1 = (1 \otimes \varepsilon)\Delta$, therefore

$$R = (\varepsilon \otimes 1 \otimes 1)(\Delta \otimes 1)(R) = (\varepsilon \otimes 1 \otimes 1)(R_{13}R_{23}) = (\varepsilon \otimes 1)(R) \cdot R$$

$$R = (1 \otimes \varepsilon \otimes 1)(\Delta \otimes 1)(R) = (1 \otimes \varepsilon \otimes 1)(R_{13}R_{12}) = (1 \otimes \varepsilon)(R) \cdot R$$

Since R is invertible these equalities imply our first statement and we leave the others as exercises to the reader.

Those who like formal calculations can be happy checking that the action of R on the module $V^k \otimes V^{k'}$ is:

$$R(e_i \otimes e_j) = \sum \frac{(s-\bar{s})^n}{[n]!} \frac{[m+i+n]!}{[m+i]!} \frac{[m'-j+n]!}{[m'-j]!} t^{4ij-2n(i-j)-n(n+1)} e_{i+n} \otimes e_{j-n}$$

The special case when R acts on $V^2 \otimes V^2$ is important, and turns out to be

$$(t) \oplus \begin{pmatrix} \bar{t} & \bar{t}(s-\bar{s}) \\ 0 & \bar{t} \end{pmatrix} \oplus (t)$$

with respect to the basis

$$e_{\frac{1}{2}} \otimes e_{\frac{1}{2}}, \quad e_{\frac{1}{2}} \otimes e_{-\frac{1}{2}}, \quad e_{-\frac{1}{2}} \otimes e_{\frac{1}{2}}, \quad e_{-\frac{1}{2}} \otimes e_{-\frac{1}{2}}$$

and thus the corresponding $\check{R}$-matrix is :

$$(t) \oplus \begin{pmatrix} 0 & \bar{t} \\ \bar{t} & \bar{t}(s-\bar{s}) \end{pmatrix} \oplus (t)$$

Theorem 4.2.2. *There is an element μ in the Hopf algebra A_r satisfying the following conditions*
μ is invertible: $\exists \bar{\mu}$ that $\mu\bar{\mu} = 1$
$\mu\alpha\bar{\mu} = S^2(\alpha)$ for all $\alpha \in A_r$
$\sum \alpha_i \bar{\mu} \beta_i = \sum \beta_i \mu \alpha_i$ where $\sum \alpha_i \otimes \beta_i = R$
$S(\mu) = \bar{\mu}$

In [KM] one can find the proof that $\mu = K^2$ is satisfying the conditions. Actually the second and fourth properties are trivial, the rest is not easy and can be found as an appendix in [KM].

4.3. Coloured tangle invariants

We can now define the invariant for 3-manifolds. As we mentioned before we will define a function on *unoriented* framed links in S^3 and prove that it is independent on the chosen orientation. First we will define coloured tangle invariants.

Now recall the definition of an oriented tangle (chapter 1.) and our quasitriangular Hopf algebra A_r (chapter 4.1). Define a colouring of a tangle T (or one of its diagrams) to be an assignment of an A_r-module to each component of T. This induces a colouring on ∂T as follows: if S is an arc of colour V then assign V to each endpoints of S where S is oriented down, and the dual module V^* to each endpoint where it is oriented up. Tensoring from left to right this gives *boundary* A_r-modules $T_\pm$ assigned to $\partial_\pm T$. As usual the empty tensor product is $\mathbb{C}$, so for a link $T_\pm = \mathbb{C}$.

In the next result we present an oriented coloured framed tangle invariant.(Theorem (3.6) in [KM].)

Theorem 4.3.1.
There exist unique A_r-linear operators

$$J_T = J_T^{A_r,R,\mu} : T_- \to T_+$$

assigned to each coloured framed tangle T which satisfy

$$J_{T\otimes T'} = J_T \otimes J_{T'}$$

$$J_{T\circ T'} = J_T \circ J_{T'}$$

and for the elementary diagrams defined in chapter 1:

$$J_I = id$$

$$J_r = \check{R} \qquad J_l = \check{R}^{-1}$$

$$J_{\curvearrowright} = E \qquad J_{\curvearrowleft} = E_\mu$$

$$J_{\cup} = N \qquad J_{\cup} = N_{\bar{\mu}}$$

where $E(f\otimes x) = f(x)$, $E_\mu(x\otimes f) = f(\mu x)$, $N(1) = \sum e_i \otimes e^i$, $N_{\bar{\mu}}(1) = \sum e^i \otimes (\bar{\mu} e_i)$ (for any basis e_i). Note that for a link $J_L : \mathbb{C} \to \mathbb{C}$ is just a scalar.

To prove this theorem one has to check that the given operators are really A_r-linear. Then one can associate the above operators to the elementary tangle diagrams, then extend it to all tangle diagrams by the first two equations of the theorem (see theorem 1.3.3). Then one has to verify that the Reidemeister–type moves from theorem 1.3.4 do not alter the associated operation. For example the invariance of the associated operator under the move (b) from theorem 1.3.4 follows from the Yang–Baxter equation proved for $\check{R}$. The detailed proof can be found in [KM].

Let us present now some basic properties of the J_T operators as a series of lemmas.

Lemma 4.3.2. *If $V = X \oplus Y$ are A_r-modules and S is a closed component of the tangle T then*

$$J_T = J_{TX} + J_{TY}$$

where TX (TY) stands for the tangle obtained by changing the colour on S to X (Y).

Lemma 4.3.3. *If $V = X \otimes Y$ are A_r-modules and S is an arbitrary component of the tangle T then*

$$J_T = J_{TXY}$$

where TXY stands for the tangle obtained by replacing S by two parallel pushoffs of itself (using the framing) and coloured X and Y respectively.

Lemma 4.3.4. *Let S, a preferred component of the tangle T, be coloured by V.If we replace S by $-S$ with colour V^* (the dual module) to get T' then*

$$J_T = J_{T'}$$

where $T_\pm$ and $T_\pm^*$ may be identified by the isomorphism $E_\mu : V \to V^{**}$ given by $E_\mu(x)(f) = f(\mu x)$.

It is indeed A_r-linear: $E_\mu(\alpha x) = (\mu\alpha x)^{**} = (S^2(\alpha)\mu x)^{**} = \alpha(\mu x)^{**} = \alpha E_\mu(x)$.

Lemma 4.3.5. *Let T be a link with a preferred component S. Let the colouring be restricted to $V^1, V^2, \ldots, V^r$ mentioned in section 4.1. If T' is the link obtained from T by replacing S by $-S$ without changing its colour, then*

$$J_T = J_{T'}$$

This lemma strongly depends on the self–duality of the A_r-modules $V^1, V^2, \ldots, V^r$ (Theorem 4.1.1) and the previous lemma.

Lemma 4.3.6. *Let S be a preferred component of T with colour s (i.e. V^s with $s \leq r$). Let T' be T with a k-coloured unknotted meridian adjoined to S. Then*

$$J_{T'} = \frac{[sk]}{[s]} J_T$$

Lemma 4.3.7. *Let S be a preferred component of T with colour s. Let T' be T with the framing on S increased by 1. Then*

$$J_{T'} = t^{j^2-1} J_T$$

Lemma 4.3.8. *One can remove any 1-coloured component without changing J.*

Indeed, recall that K^2 acts by the identity on V^1, as does $\check{R}$ on $V^1 \otimes V^k$ and $V^k \otimes V^1$.

Let us consider coloured tangles which have all colour V^2. It turns out that our invariant is essentially the same as Kauffman's bracket polynomial or the Jones polynomial (see chapter 2).

Theorem 4.3.9. (1) *(oriented skein relations) Suppose that all colours are V^2. For a skein triple (L_+, L_-, L_0) (see Definition 2.0.3) we have:*

$$tJ_{L_+} - \bar{t}J_{L_-} = (s - \bar{s})J_{L_0}$$

(2) (unoriented skein relations) For tangles R, H, V identical except

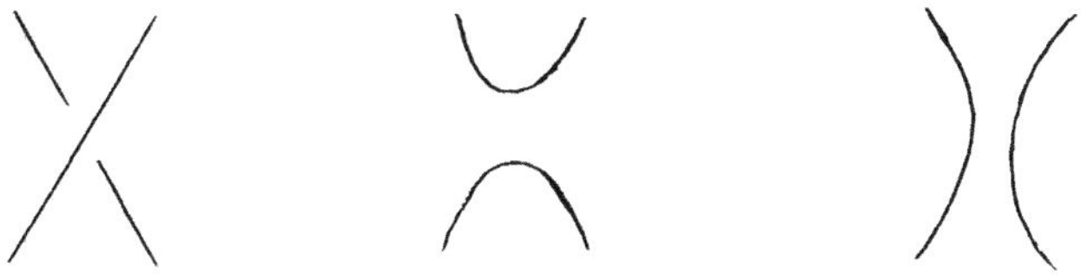

R $\qquad$ H $\qquad$ V

$$J_R = tJ_V + \bar{t}J_H$$

Proof Recall that the action of $\check{R}$ on $V^2 \otimes V^2$ is given (in the relevant basis) by

$$(t) \oplus \begin{pmatrix} 0 & \bar{t} \\ \bar{t} & \bar{t}(s - \bar{s}) \end{pmatrix} \oplus (t)$$

We find (e.g. by computing the characteristic polynomial of $\check{R}$) that

$$t\check{R} - \bar{t}\check{R}^{-1} = (s - \bar{s})I$$

and (1) follows.

For (2) orient R so that the crossing looks like L_+ and then (1) yields

$$tJ_R - \bar{t}J_L = (s - \bar{s})J_V$$

where we get L from R by locally changing the under- and over crossings. Now reverse the orientation on one strand so that the same crossing looks like L_- when rotated by a right angle, and so

$$-\bar{t}J_R + tJ_L = (s - \bar{s})J_H$$

by (1) again. Multiplying the above equalities by t and $\bar{t}$ respectively. and adding gives (2).

Theorem 4.3.10.

$$J_L = (-i)^{w(L)} < L > (A = it)$$

For the proof of this theorem one just has to check that the two sides behave the same way under the relevant skein relations — and they agree on the trivial link. The essence of this behaviour is the preceeding theorem. As the bracket polynomual is essentially the same as the Jones polynomial the above theorem states the connection between the V^2-coloured tangle invariant J and the Jones polynomial.

Now we present two more properties of J. One is that J_T can be calculated by calculating only Js of 2-coloured tangles (i.e. where all colours are V^2) — of course not only J of T. The other states an interesting *duality*.

Theorem 4.3.11. *(The cabling theorem) Let the framed link L be coloured by the modules $V^1, \ldots, V^r$. Denote this colouring* **k**, *and the constant 2-colouring* **2**. *Then*

$$J_{L,\mathbf{k}} = \sum_{\mathbf{j}=0}^{\frac{\mathbf{n}}{2}} (-1)^{\mathbf{j}} \binom{\mathbf{n}-\mathbf{j}}{\mathbf{j}} J_{L^{\mathbf{n}-2\mathbf{j}}}$$

where the multi-index notation means : $f(\mathbf{c}) = \Pi f(c_i)$ and the sum is over all $\mathbf{j}$ with $1 \leq j_i \leq \frac{n_i}{2}$. The link $L^{\mathbf{c}}$ is the one obtained by replacing each L_i with c_i parallel pushoffs of itself using the framing.

This theorem, in fact, is the consequence of the decomposition theorem (4.1.2) and lemmas 4.3.2, 4.3.3.

Theorem 4.3.12. *Let $L \cup K$ be a framed link where K has framing a, and $I = (l_1, \ldots, l_n)$ be a colouring of $L = L_1 \cup \ldots \cup L_n$ by the modules $V^1, \ldots, V^r$. If $0 < k < r$, then:*

$$J_{L \cup K, I \cup (r-k)} = i^{(r-2k)a+2\lambda} J_{L \cup K, I \cup k}$$

where $\lambda = \sum lk(K, L_i)$ for all L_i that has l_i even.

4.4. The 3-manifold invariant

The main theorem asserts that if we use $V^1, \ldots, V^{r-1}$ in the colouring of the link (as a tangle), and take an appropriate linear combination of the coloured tangle invariants, then we obtain an invariant for the 3-manifold that comes by doing surgery along the link.

Theorem 4.4.1. *(Theorem 1.6 in [KM]) If M is obtained by surgery along the framed link L, then*

$$\tau_r(M) = \alpha_L \sum_{\mathbf{k}=1}^{\mathbf{r}-1} [\mathbf{k}] J_{L,\mathbf{k}}$$

is an invariant of the 3-manifold M. Where by the constants we mean $\alpha_L = b^{n_L} c^{\sigma_L}$, $b = \sqrt{\frac{2}{r}} sin\frac{\pi}{r}$, $c = e(\frac{-3(r-2)}{8r})$, n_L is the number of components ofB L, σ_L is the signature of its linking matrix, by $[\mathbf{k}]$ we mean Πk_i, and for the sum we use the multi-index notation (see above).

Proof According to chapter 3 the only thing to check is the invariance under κ-move, i.e. checking the invariance under the following local change of the link:

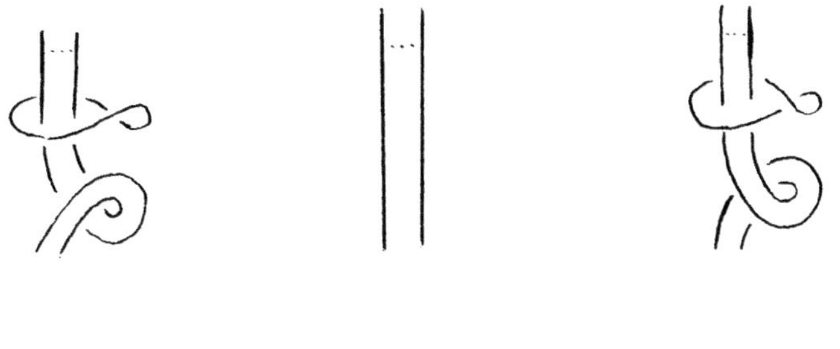

L^+ $\qquad$ L $\qquad$ L^-

Let I be a colouring of L (by that from now on we mean colouring with the modules $V^1, \ldots, V^{r-1}$). Let $I \cup k$ denote the induced colouring of L^ε with new component K coloured $k < r$. Then $[\mathbf{I} \cup \mathbf{k}] = [\mathbf{I}][\mathbf{k}]$. Since $n_{L^\varepsilon} = n_L + 1$, $\sigma_{L_\varepsilon} = \sigma_L + \varepsilon$ and so $\alpha_{L^\varepsilon} = bc^\varepsilon \alpha_L$ we have

$$\tau_L = \alpha_L \sum_{\mathbf{I}=1}^{r-1} [\mathbf{I}] J_{L,\mathbf{I}}$$

$$\tau_{L^\varepsilon} = \alpha_{L^\varepsilon} \sum_{\mathbf{I}=1}^{r-1} \Big(\sum_{k=1}^{r-1} [\mathbf{I} \cup k] J_{L^\varepsilon, \mathbf{I}\cup k} \Big) = \alpha_L \sum_{\mathbf{I}=1}^{r-1} [\mathbf{I}] bc^\varepsilon \sum_{k=1}^{r-1} [k] J_{L^\varepsilon, \mathbf{I}\cup k}$$

Thus to prove $\tau_{L^\varepsilon} = \tau_L$ it suffices to establish the identity

$$bc^\varepsilon \sum_{k=1}^{r-1} [k] J_{L^\varepsilon, \mathbf{I}\cup k} = J_{L,\mathbf{I}} \qquad (*)$$

for any fixed colouring $\mathbf{I}$ on L. We prove $(*)$ by induction on the number of strands m.

(1) $m = 0, 1$

The proof for $m = 0$ is a special case of $m = 1$ when the colour j on the strand of L passing through K is 1, by lemma 4.3.8. So assume $m = 1$. Then

$$J_{L^\varepsilon, \mathbf{I}\cup k} = \frac{[jk]}{[j]} t^{\varepsilon(j^2 - 1 + k^2 - 1)} J_{L,\mathbf{I}}$$

by lemma 4.3.6 and 4.3.7 and so $(*)$ reduces to

$$bc^\varepsilon \sum_{k=1}^{r-1} [k] \frac{[jk]}{[j]} t^{\varepsilon(j^2 - 1 + k^2 - 1)} = 1$$

This is purely analytical equation, whose proof depends on some "Gauss–sums"; the exact proof can be found in [KM].

Actually the constants b, c in τ were set (see Theorem 4.4.1) so that this identity holds.

(2) If $(*)$ holds for n-strand κ-moves for all $n < m$, then it holds for m-strand κ-moves also.

(2a) Let us consider the following special case

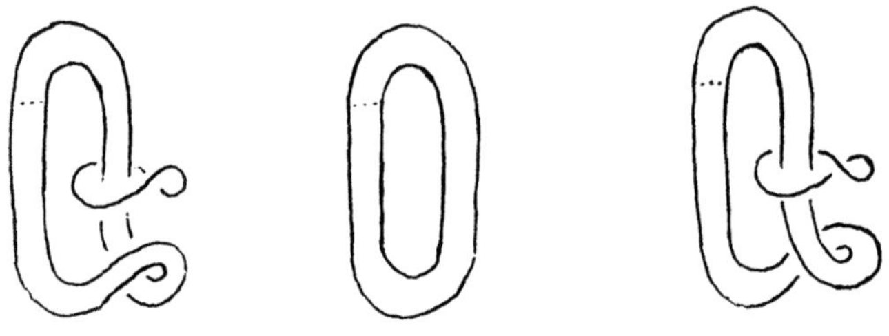

Using the Symmetry Principle (theorem 4.3.12), we may assume that all colours j of components J of L satisfy $j \leq \frac{r}{2}$. Indeed, if $j > \frac{r}{2}$, then change j to $r - j \leq \frac{r}{2}$ on J (and on the corresponding component J^ε of L^ε). This changes the left side of $(*)$ by

$$i^{\varepsilon(r-2j)+2(lk(J^\varepsilon,S)+k-1)}$$

where S is the even-coloured sublink of $L^\varepsilon - (J^\varepsilon \cup K)$, and leaves the right side unchanged. Next change k to $r-k$ on K. Then using $[k]=[r-k]$, the left side changes by

$$i^{\varepsilon(r-2k)+2(lk(K,S)+(r-j)-1)}$$

while the right side remains unchanged. Noting that $lk(J^\varepsilon, S) = lk(K,S) = |S|$, we see that the net change on the left side is

$$i^{2\varepsilon(r-j-k)+4(lk(K,S)-1)+2(r-j+k)} = 1$$

as it is on the right side.

Now by lemma 4.3.3, we may replace two components L_1 and L_2 of L, coloured by l_1 and l_2, with a single component coloured by $V^{l_1} \otimes V^{l_2}$ which is $V^{l_1+l_2-1} \oplus \ldots \oplus V^{|l_1-l_2|+1}$ by the decomposition theorem 4.1.2. (This is why we needed $l_1, l_2 \leq \frac{r}{2}$). Thus by lemma 4.3.2 and distributivity it is enough to prove $(*)$ when L_1 and L_2 are replaced by a single j-coloured component for $j < r$. This is, however, covered by the induction hypothesis.

(2b) The general case is shown

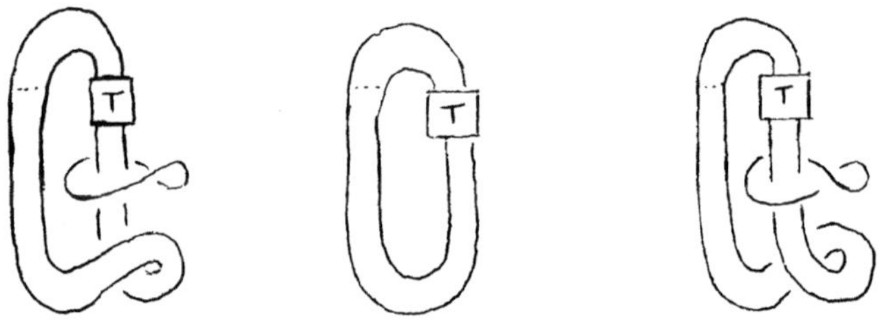

where T is an arbitrary tangle. We will reduce to the special case (2a) using cabling and skein theory.

First suppose that $\mathbf{I} = 2$, the constant 2-colouring. Then we prove $(*)$ by induction on the number of crossings in T. The induction begins with zero crossings, which is covered by (2a). In general, we may smooth a crossing of T in two ways in both L and L^{ϵ}, and $(*)$ follows by induction for each smoothing, using theorem 4.3.9.

Finally, for general I, $J_{L,\mathbf{I}}$ and $J_{L^{\epsilon},\mathbf{I}\cup k}$ can be computed using the cabling formula of theorem 4.3.11 applied to L and $L^{\epsilon} - K$, respectively. This reduces the proof of $(*)$ to the case $\mathbf{I} = 2$ proved above.

The proof of the theorem is now complete.

There are not too many results describing the properties of the invariant τ_r, but for small values of r or "simple" manifolds one can work out some formulas. Without proof we present some results here:

(i) $\tau_r(M\#N) = \tau_r(M)\tau_r(N)$
(ii) $\tau_r(-M) = \overline{\tau_r(M)}$
(iii) $\tau_r(S^3) = 1$

(iv) $\tau_2(M) = 1$ for all M

(v) $\tau_r(S^2 \times S^1) = \sqrt{\frac{r}{2}}csc(\frac{\pi}{r})$
(vi) $\tau_r(\mathbb{R}P^3) = \frac{1}{\sqrt{2}}sec(\frac{\pi}{2r})$ if r is even

(vii) τ_3 is a homotopy invariant, $t_3(M_L) = \frac{1}{\sqrt{2^n}}\frac{1-i}{\sqrt{2}}\sum i^{lk(S,S)}$ where the sum is over all the 2-coloured sublinks S of L.
(viii) τ_4 is *not* a homotopy invariant

The hope is that τ_r for greater r really is a new invariant. This can be a step towards finding a "complete" invariant for 3-manifolds.

5. AN ALTERNATIVE DEFINITION BASED ON THE JONES POLYNOMIAL

The bracket polynomial (see chapter 2) can be manipulated via linear skein theory and interpreted via the Temperley–Lieb algebra. These techniques give a short, direct proof of the existence of the 3-manifold invariants.

In the previous chapter we saw that the 3-manifold invariant is a linear combination of some coloured, framed link invariants. But a coloured framed link invariant itself is a linear combination of some terms that are essentially Kauffmann's bracket polynomial (see theorem 4.3.10). Formal calculation gives

$$\tau_r(M) = b^n c^{\sigma_L} \sum_{k=1}^{r-1} \sum_{j=0}^{\frac{k-1}{2}} [k](-1)^j \binom{k-1-j}{j} (-i)^{-3w(L)} < k-1-2j * L > (it)$$

(See the definition of the $< \; * \; >$ below.) The complication of this formula can be avoided if we consider the approach given by Lickorish [L2], [L3], [L4] for the same purpose: giving $\tau_r(M)$ as a linear combination of bracket polynomials of some appropriate modification of the link. This is given in this chapter.

5.1. The Temperley–Lieb algebra

Consider special framed tangles as they were defined in chapter 1. For convenience now regard them horizontically, so e.g. a $(3,3)$-tangle looks like this :

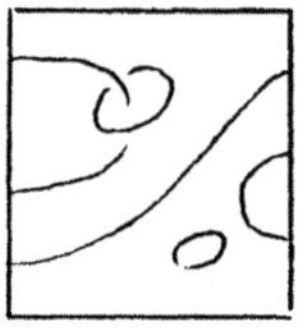

Fix an integer $r > 1$. Let A^2 be a primitive $2r$th root of unity, and consider the bracket polynomial evaluated at A. Thus the bracket polynomial takes values in $\mathbb{C}$ from now on.

Let V_m be a vector space over $\mathbb{C}$ generated by all framed (m,m)-tangle diagrams (up to isotopy) quotiened by relations of the form :

(i) $D \cup U = \delta D$ where U is a disjoint unknot with framing 0, and $\delta = -A^{-2} - A^2$.

(ii)

where these refers to diagrams identical except where it is shown.

Of course, equalities in this vector space mean partial calculations of bracket polynomials, since one can easily see the similarity of (i) and (ii) above with the definition of the bracket polynomial in chapter 2.

One can also easily check that V_m has a basis consisting of all (elements represented by) diagrams in the square with no crossing and no closed components. Let us call this basis the **standard basis** for V_m. Thus the dimension of V_m is the their number, the mth Catalan number $\frac{1}{m+1}\binom{2m}{m}$. Let 1_m be the element of the standard basis with representative consisting m straight horizontal lines.

We can define a bilinear map $V_m \times V_m \to V_m$ by placing one diagram beside another:

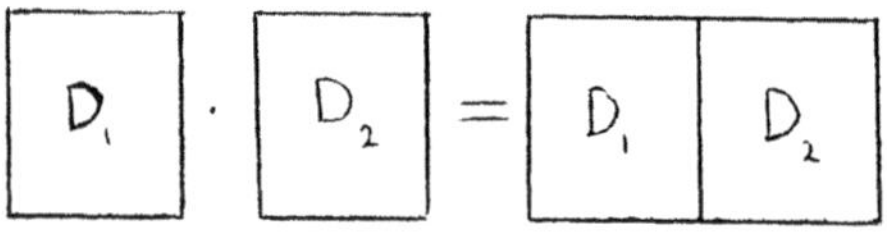

With this structure we have an algebra, that we call the mth Temperley–Lieb algebra. If D represents an element in V_m then let the Markov trace (tr) be the bracket polynomial (recall that now it takes complex numbers as values) of the link diagram formed from D by joining the points on the left edge of the square to those on the right by arcs outside the square with no crossing:

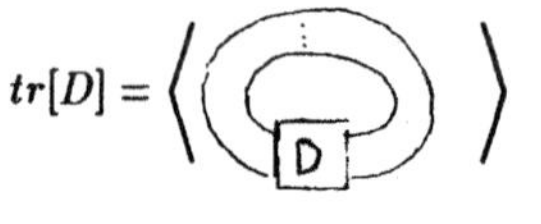

This is clearly a well defined linear map on V_m, because the relations used to define V_m are essentially the same as the defining relations of the bracket polynomial.

The following technical theorem can be proved for the linear form

$$< \, , \, >: V_m \times V_m \to \mathbb{C}$$

defined by $< x, y >= tr[xy]$.

Theorem 5.1.1. *([L3] Proposition 5) If $m \geq r-1$ then the $< \ , \ >$ bilinear form is degenerate, and there exists an element $q(m) \in V_m$ such that*

$$< \ , q(m) >=< \ , 1_m >$$

and the coefficient of $q(m)$ in the standard basis corresponding to 1_m is 0.

5.2. Definition of the invariant

To formulate the theorem stating the existence of the 3-manifold invariant it is firstly convenient to explain some notations.

Let T_{i+j} be the bracket polynomial of either of the following diagrams:

where i, j, $i+j$ besides a curve signifies the presence of i, j, $i+j$ copies of that curve *parallel in the plane.*

If D is a link diagram with $D_1, \ldots, D_n$ corresponding to the link components, and c is a function, $c : \{1, 2, \ldots, n\} \rightarrow \mathbb{Z}_+$, let $c * D$ be the diagram in which each D_s has been replaced by $c(s)$ copies all parallel *in the plane* to D_s. Note that if D and D' are related by Ω_1^*, Ω_2, Ω_3 Reidemeister moves then $c * D$ and $c * D'$ are also related by them. Usually c will be restricted to $C(n, r)$, the set of all functions $c : \{1, 2, \ldots, n\} \rightarrow \{0, 1, \ldots, r-2\}$. If the framed link L is given an orientation, the linking numbers of pairs of its components form a symmetric matrix in which $fr(s)$ is taken to be the linking number of L_s with itself. The signature and nullity of this matrix are independent of the choice of orientations. The nullity of the matrix is, in fact, the first Betti number of the 3-manifold obtained by surgery along L.

Theorem 5.2.1. *(Lickorish [L3]) Let r be a fixed integer, $r \geq 3$, then*

(i) There is a unique solution $\lambda_0, \ldots \lambda_{r-2}$ in the complex numbers to the set of linear equations

$$\sum_{i=0}^{r-2} \lambda_i T_{i+j} = \delta^j \qquad , j = 0, 1, \ldots r-2$$

(ii) Suppose that M_L is the manifold that we get by surgery from S^3 along L (see chapter 3). Let D be a good diagram of L. Let σ and μ be the signature and nullity of the linking matrix of L. Then the expression

$$\tau_r(M_L) = \kappa^{\frac{\sigma+\mu-n}{2}} \sum_{c \in C(n,r)} \lambda_{c(1)} \lambda_{c(2)} \cdots \lambda_{c(n)} < c * D >$$

where $\kappa = \sum_{i=0}^{r-2} \lambda_i \bar{T}_i$, is an invariant of the 3-manifold, a complex number independent of the choice of L or of D.

Observe that the power of κ in the expression above is equal to the number of negative eigenvalues of the linking matrix of L.

Although the proof of the first part of the theorem is not trivial and not only technical either, we will not prove it (a proof can be found in [L3]). The ideas of that proof will come up in the proof of the second — more interesting — part of the theorem.

Let us call t_j the following element of V_j:

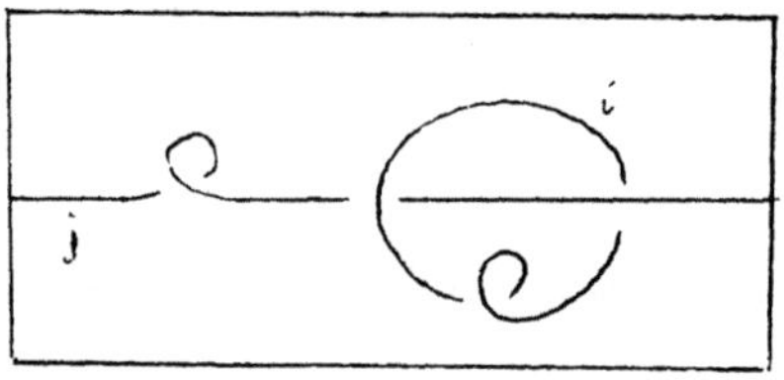

We will need the following

Lemma 5.2.2. *For every $j \geq 0$ we have the following identity of linear functions from V_j to $\mathbb{C}$.*

$$\sum_{i=0}^{r-2} \lambda_i < \ ,t_j >=< \ ,1_j >: V_j \to \mathbb{C}$$

Proof As $T_{i+j} =< 1_j, t_j >$ by definition

$$\sum_{i=0}^{r-2} \lambda_i < 1_j, t_j >= \delta^j =< 1_j, 1_j >$$

for $j = 1 \ldots r-2$.

Suppose that $j \leq r-2$ and b is a basis element of V_j and that inserting b into the square here

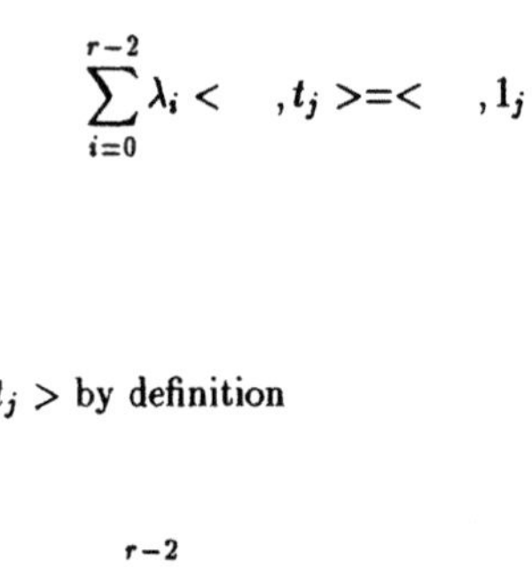

produces α curves encircling the annulus and β nullhomotopic ones, with $\alpha \leq j$. Then the isotopy in this prototype annulus induces an isotopy in an immersed annulus — a neighbourhood of the j strands and b in this figure

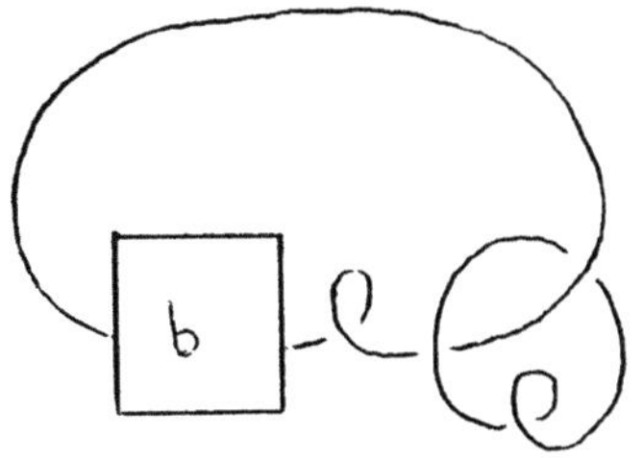

the immersed annulus

At the end of this isotopy there are α strands going around the immersed annulus and β small circles with no crossing at all. This shows that

$$\sum_{i=0}^{r-2} \lambda_i < b, t_j > = \sum_{i=0}^{r-2} \lambda_i < 1_\alpha, t_\alpha > \delta^\beta = \delta^{\alpha+\beta} = < b, 1_j >$$

Thus $\sum_{i=0}^{r-2} \lambda_i < \quad , t_j > = < \quad , 1_j >$ if $j \leq r-2$.

For $j > r-2$ we use induction. Suppose that the lemma is proved for all $j' < j$. Let b is a standard base element of V_m *other than* 1_m. Once again inserting b into the annulus above produces α curves encircling the annulus and β nullhomotopic ones. One can see that from $b \neq 1_j$ follows $\alpha < j$. Then using the trick above again and the inductive hypothesis:

$$\sum_{i=0}^{r-2} \lambda_i < b, t_j > = \sum_{i=0}^{r-2} \lambda_i < 1_\alpha, t_\alpha > \delta^\beta = < 1_\alpha, 1_\alpha > \delta^\beta = \delta^{\alpha+\beta} = < b, 1_j >$$

Now as $r-1 \leq j$ the element $q(j)$ of lemma 5.1.1 exists. It is a linear combination of elements other than 1_m. Thus by taking this linear combination $q(j)$ can be substituted for b above. Hence

$$\sum_{i=0}^{r-2} \lambda_i < 1_j, t_j > = < 1_j, 1_j >$$

and so the theorem is proved for all j.

We can now turn to the proof of the main theorem.

Proof From the definition of the Jones polynomial (chapter 2) we know that the expression in the theorem is indeed unchanged by Ω_1^*, Ω_2,Ω_3 moves (chapter 1), so to get the theorem we just have to prove the invariance under κ_+- and special κ_--moves (see chapter 3).

Let us start with κ_+-moves. Suppose now that L and L' are framed links with (good) diagrams D and D' related as shown:

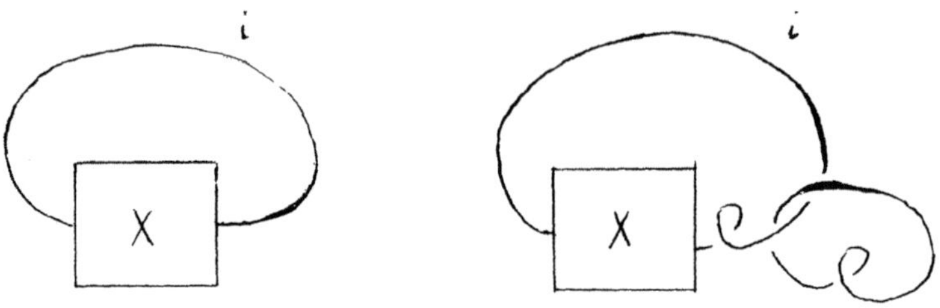

D $\qquad\qquad\qquad\qquad$ D'

Let L have n components. If $c \in C(n,r)$, let $c_i' \in C(n+1,r)$ be defined by $c_i'|_{\{1\ldots n\}} = c$, $c_i'(n+1) = i$. Then Lemma 5.2.1 part (i) really states

$$\sum_{i=0}^{r-2} \lambda_i < c_i' * D' > = < c * D >$$

Multiplying this by $\lambda_{c(1)} \ldots \lambda_{c(n)}$ and adding gives

$$\sum_{c' \in C(n+1,r)} \lambda_{c'(1)} \ldots \lambda_{c'(n+1)} < c' * D' > = \sum_{c \in C(n,r)} \lambda_{c(1)} \ldots \lambda_{c(n)} < c * D >$$

The number of negative eigenvalues of the linking matrix does not change with a κ_+-move, so from this the theorem follows for κ_+-moves.

Now if L and L' are framed links with (good) diagrams D and D' related as shown

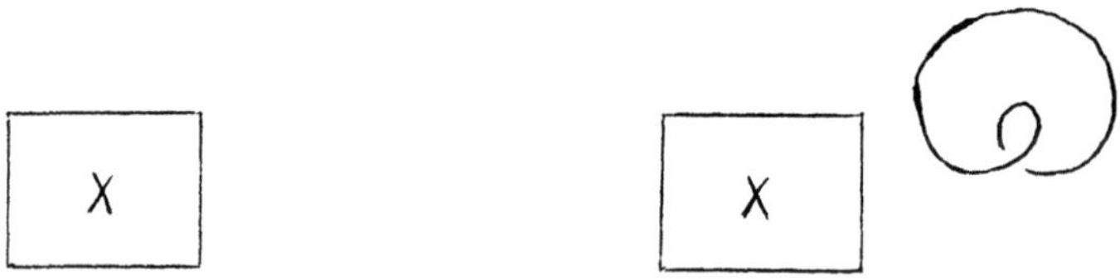

D $\qquad\qquad\qquad\qquad$ D'

Then they are related by a special κ_--move. However if D_1 and D_2 are disjoint diagrams (separated by a circle) then $< D_1 \cup D_2 > = < D_1 >< D_2 >$. Hence

$$\sum_{c' \in C(n+1,r)} \lambda_{c'(1)} \ldots \lambda_{c'(n+1)} < c' * D' > = (\sum_{i=0}^{r-2} \lambda_i \bar{T}_i) \sum_{c \in C(n,r)} \lambda_{c(1)} \ldots \lambda_{c(n)} < c * D >$$

(Because one can easily check that the bracket of i parallel copies of the unknot with framing -1 is $\bar{T}_i$.) From this the theorem easily follows noting that the number of negative eigenvalues of the linking matrix increases by one in passing from D to D'. The proof of the theorem is now complete.

BIBLIOGRAPHY

[FR] Fenn, R., Rourke, C: On Kirby's calculus of links, Topology, Vol. 18, 1-15

[F] Fenn, R.: Notes on knots (preprint)

[K1] Kirby, R.: A calculus for framed links is S^3, Invent. Math. 45 (1978), 35-56

[K2] Kirby, R.: The Topology of 4-Manifolds (Lect. Notes Math., Vol. 1374) Berlin Heidelberg New York: Springer 1989

[KM] Kirby, R., Melvin, P.: The 3-manifold invariants of Witten and Reshetikhin–Turaev for $sl(2, \mathbb{C})$, Invent. Math. 105, 473-545 (1991)

[L1] Lickorish, W.B.R.: A representation of orientable combinatorial 3-manifolds, Ann. Math. 76 (1962) 531-540

[L2] Lickorish, W.B.R.: Invariants for 3-manifolds derived from the combinatorics of the Jones polynomial, Pac. J. Math., Vol. 149 No 2, 337-348

[L3] Lickorish, W.B.R.: 3-manifolds and the Temperley–Lieb algebra (to appear)

[L4] Lickorish, W.B.R.: Calculations with the Temperley–Lieb algebra (Preprint)

[R] Rolfsen, D.: Knots and Links. Publish or Perish 1977

[RT1] Reshetikhin, N.Yu., Turaev, V.G.: Invariants of 3-manifolds via link polynomials and quantum groups, Invent. Math. 103, 547-597 (1991)

[RT2] Reshetikhin, N.Yu., Turaev, V.G.: Ribbon graphs and their invariants derived from quantum groups, Commun. Math. Phys. 127, 1-26 (1990)

[Wa] Wallace, A.H.: Modifications and cobounding manifolds, Can. J. Math. 12 (1960), 503-528

[W] Witten, E.: Quantum field theory and the Jones polynomials, Commun. Math. Phys. 121, 351-399 (1989)

INDEX

GPSR Compliance

The European Union's (EU) General Product Safety Regulation (GPSR) is a set of rules that requires consumer products to be safe and our obligations to ensure this.

If you have any concerns about our products, you can contact us on ProductSafety@springernature.com

In case Publisher is established outside the EU, the EU authorized representative is:

Springer Nature Customer Service Center GmbH
Europaplatz 3
69115 Heidelberg, Germany

Batch number: 10755284

Printed by Printforce, the Netherlands